Immobilized Enzymes
in Analytical and
Clinical Chemistry

CHEMICAL ANALYSIS

A SERIES OF MONOGRAPHS ON ANALYTICAL CHEMISTRY AND ITS APPLICATIONS

VOLUME 56

A WILEY-INTERSCIENCE PUBLICATION

JOHN WILEY & SONS
New York / Chichester / Brisbane / Toronto

Immobilized Enzymes in Analytical and Clinical Chemistry

Fundamentals And Applications

PETER W. CARR
University of Minnesota
Department of Chemistry
Minneapolis, Minnesota

LARRY D. BOWERS
University of Minnesota
Department of Laboratory Medicine & Pathology
Minneapolis, Minnesota

A WILEY-INTERSCIENCE PUBLICATION

JOHN WILEY & SONS
New York / Chichester / Brisbane / Toronto

Library of Congress Cataloging in Publication Data:

Carr, Peter W
  Immobilized enzymes in analytical and clinical
chemistry.

  (Chemical analysis; v. 56)
  "A Wiley-Interscience publication."
  Includes bibliographies and index.
  1. Immobilized enzymes. 2. Biological
chemistry—Technique. 3. Chemistry, Clinical—
Technique. 4. Chemistry, Analytic—Technique.
I. Bowers, Larry D., joint author. II. Title.
III. Series. [DNLM: 1. Enzymes, Immobilized.
QU135 C321i]

QP601.C33   574.19'285   80-13694
ISBN 0-471-04919-0

Printed in the United States of America

10 9 8 7 6 5 4 3 2 1

**To our Parents
and Teachers**

PREFACE

The intent of this monograph is to serve as a guide and introduction to the area of immobilized enzymes for scientists involved in analysis. Immobilized-enzyme technology has been expanding at a rapid rate for nearly a decade. Although there are many excellent books and reviews on the subject, these are predominantly devoted to nonanalytical topics. Many of these are uncorrelated symposia proceedings which emphasize recent advances in the four major areas of immobilized enzyme research: (1) basic biochemical research where immobilized enzymes are used as a model of intracellular membrane-bound enzymes, (2) *in vivo* pharmaceutical systems, (3) chemical and biochemical engineering, and (4) fermentation technology and food science.

There are many advantages to the use of immobilized enzyme systems in place of soluble enzymes in an analytical system. A major emphasis has been on the increased stability of these preparations which allows ease of reuse and greater economy. We feel that the potential of economical and rapid equilibrium analysis is a major advantage of immobilized-enzyme technology. Another subtle but significant advantage is that the enzyme becomes a semipermanent part of the entire analytical system; thus the enzyme component can be calibrated along with the system. One can calibrate the enzyme component in terms of system sensitivity or specificity. Indeed it may be reasonable to purify the enzyme since analytical specificity is often a question of purity rather than nature of the enzyme.

A very intriguing potential advantage of immobilized enzymes is the fact that an inhibitor must interact with the protein in order to alter its activity. In a two-phase system, this is equivalent to partitioning. For a flowing stream such as would be used in a tubular or packed bed reactor, the enzyme is somewhat immune to inhibitors due to the continued partitioning of the inhibitor between the enzyme and the mobile liquid phase. Thus although the inhibitor may eventually cause the demise of the reactor, the analytical reaction can occur in another part of the reactor not affected by the inhibitor. In essence a separation will occur in the reactor for any inhibitor which is relatively more strongly bound to the enzyme. Whether or not sufficient binding capacity exists in the immobilized enzyme reactor to realize this potential advantage remains to be shown. A second advantageous consequence of using a two-phase system is that diffusive and convective mass transfer often become the rate-limiting processes. These processes are always first order with respect to concentration and are at worst moderately dependent on such variables as temperature, buffer ion, and pH. As a result, the linear dynamic range can be extended significantly beyond the K_M of the enzyme—a feature *never* observed in solution-phase enzyme analysis.

In view of all of these potential advantages, why have all soluble enzyme assays not disappeared? We believe that this rests on some of the basic, pragmatic tenets of analytical chemistry. First, a new method which performs the same function as an old one will be adopted only if it is less expensive or more convenient; in effect, more economical. Second, the new method will be accepted only if it resolves without undue additional expense, a problem such as the presence of interferences or a high limit of detection that plagues the old method. The direction of immobilized-enzyme technology in analysis is therefore set.

Although a number of analytical advantages accrue from immobilizing an enzyme, these advantages are obtained at the price of increased complexity. A simple homogeneous enzyme system is transformed into a two-phase system with concomitant problems of mass transfer and diffusion. In addition, the characteristics and effectiveness of the detection system as well as the contribution of the immobilized-enzyme reactor to analytical usefulness should be considered. This manuscript is the result of our efforts to assess systematically the strengths and weaknesses of various immobilized-enzyme techniques and to establish a theoretical framework from which to procede in the development of analytical immobilized-enzyme systems.

The focus and organization of this monograph includes the wide range of material necessary for a detailed understanding of analytical uses of immobilized enzymes. There are several topics of biochemical interest, including basic enzymology and enzyme kinetics, which would not have been included had the intended readership been restricted to biological chemists. These topics were included primarily to make this a single introductory source for analytical chemists. Similarly, a number of topics which would be considered basic analytical chemistry have been discussed in the chapter on analytical principles to provide background for those not familiar with enzyme-based analysis techniques. In view of the two-phase nature of these systems and their use in flow reactors, we felt that treatment of these topics, which is primarily the venue of chemical engineers, could not be avoided. Thus, we have included a discussion of some of the complexities which occur when reactions are carried out in open systems, and of the process of dispersion in flowing streams and packed columns in both single-phase and air-segmented (two-phase) flow.

This material, along with the biochemical and analytical sections, is quite basic and not restricted to immobilized-enzyme systems. Although these topics are essential to a detailed understanding of analytical immobilized enzymology, we have treated the specific immobilized-enzyme topics in such a way that they should be intelligible without a detailed understanding of the foregoing material.

The coverage of immobilized enzymes per se is broken into three distinct topics: immobilization chemistry, enzyme electrodes, and immobilized-enzyme reactors. In the chapter on immobilization schemes, we have attempted to survey those methods which are relatively simple and of benefit to the analyst. In addition, we have compared selected enzyme-immobilization schemes with respect to those characteristics which are of interest analytically. The effect of immobilization on enzyme stability and kinetics has also been discussed. Enzyme electrodes, where an enzyme layer is fixed on the surface of an electrode or other sensor, have been extensively investigated for analytical use. We have devoted a complete chapter to their theory and applications. The remainder of the applications have been grouped into a chapter on immobilized-enzyme reactors and membranes. Special emphasis has been placed on packed columns and open tubular reactors because of their applications to automated analysis. We anticipate that this organization will allow the novice to ascertain if and how his particular application has been approached and to prepare immobilized enzymes in the most beneficial form. As more difficult applications are attempted, the material in the basic chapters can be used to optimize the enzyme system.

In particular we want to express our gratitude to Dr. Richard S. Schifreen for many valuable discussions and insights, to Mrs. Becky McRorie at the University of Georgia and Mrs. Christa Elguther at the University of Minnesota for preparing typed manuscript from one notoriously bad penman. We owe a great debt of gratitude and appreciation to our wives Leah Carr and Janet Bowers as well as to our children Sean, Erin and Kelly Carr and Geoff Bowers for their help, patience and understanding but mainly for putting up with us during the final stages of the writing.

PETER CARR
LARRY BOWERS

Minneapolis, Minnesota
May 1980

CONTENTS

Immobilized Enzymes
in Analytical and
Clinical Chemistry

1

BASIC ENZYMOLOGY

Enzymes are high molecular weight biochemicals which catalyze numerous reactions which are the basis of life itself. Without exception, enzymes are members of a class of compounds called proteins, but are distinguished from others of the class by the existence of a geometric area in the structure which facilitates the catalysis. One of the properties of enzyme-mediated catalysis is the specificity with which the enzyme acts. Although the specificity is somewhat dependent on the type of reaction catalyzed, the fact that over 2000 enzymes have been isolated and characterized allows the choice of a specific reaction for species ranging from phosphate and nitrate ions to macromolecules (1, 2).

The analytical use of enzymes has been well documented (3, 4). The drawbacks which have limited the use of these biochemicals have been their scarcity, expense, instability, and the general lack of familiarity with their behavior. In the past decade, advances in the isolation and purification of proteins have dramatically increased the availability of enzymes while at the same time decreasing their cost. The confinement or immobilization onto a water-insoluble matrix conveys several additional analytical advantages frequently including a decrease in the fragility of the enzyme. At the very least, immobilized enzymes facilitate transfer into and out of a reaction mixture. Although our major objective here will be to provide an understanding of the immobilization processes and the analytical uses of the matrix-bound enzymes, it is necessary to understand some basic enzymology before delving into the manipulation of enzymes.

1.1 ENZYME STRUCTURE

As mentioned previously, enzymes constitute one type of protein. In fact, the study of enzymes has in large measure contributed to the present understanding of proteins in general. The essential structure of proteins is the result of the condensation of their fundamental constituents, the amino acids.

1.1.1 Amino Acids

The α-amino acids, which predominate in plants and animals, have the basic structure:

1

$$H_3\overset{+}{N}-\underset{\underset{H}{|}}{\overset{\overset{R}{|}}{C}}-COO^-$$

at neutral pH values. The zwitterion structure is, of course, amphoteric, and at acid or alkaline pH values exhibits a net positive or net negative charge (carboxyl pK = 4, amino pK = 9). In the protein structure, the charges due to these carboxyl and amino groups are eliminated due to the involvement of these groups in the formation of the peptide bond. The groups at the end of the protein chain are free, but contribute relatively little to the overall charge on the protein. The portion of the structure which is of interest in regard to protein structure and enzyme function is the R sidechain which differentiates the various amino acids. The most common amino acids are listed in Table 1.1 along with the pK values for their respective R substituents. These sidechains are conveniently grouped into four categories: nonpolar, uncharged polar, and positively and negatively charged polar. As will be shown, this polarity greatly influences the structure and function of the protein.

The basis of the long, unbranched polymeric chain which is the peptide backbone is the peptide bond, which unites the carboxyl group of one amino acid with the amino group of the next. Thus the peptide backbone can be represented as:

$$\cdots N-\underset{\underset{H}{|}}{\overset{\overset{R}{|}}{C}}-\overset{\overset{O}{\|}}{C}-N-\underset{\underset{H}{|}}{\overset{\overset{R'}{|}}{C}}-\overset{\overset{O}{\|}}{C}-N-\underset{\underset{H}{|}}{\overset{\overset{R''}{|}}{C}}-\overset{\overset{O}{\|}}{C}\cdots$$

The number of amino acids and their specific position in the sequence determine in great measure the function, size, and shape of the protein. Thus, a great deal of effort has been expended in sequencing proteins, and in particular enzymes, to elucidate the molecular basis of their function (5, 6).

1.1.2 Primary, Secondary, and Tertiary Structure

Although the amino acid sequence, or primary structure, of the protein is of tremendous importance, and indeed, gives rise to the three-dimensional structure of the protein, the spatial relationships of the constituents are the key to peptide function. There have been three classes of spatial relationships recognized: secondary structure, which gives the relation-

Table 1.1 Common Amino Acids

Amino acid	R Group	pK

Glycine

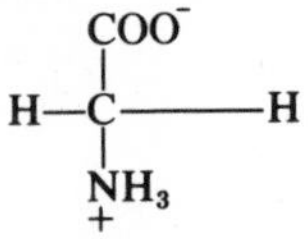

$$H-\underset{\overset{|}{\underset{+}{NH_3}}}{\overset{\overset{COO^-}{|}}{C}}-H$$

—

Alanine

$$H-\underset{\overset{|}{\underset{+}{NH_3}}}{\overset{\overset{COO^-}{|}}{C}}-CH_3$$

—

Valine

$$H-\underset{\overset{|}{\underset{+}{NH_3}}}{\overset{\overset{COO^-}{|}}{C}}-CH\underset{CH_3}{\overset{CH_3}{}}$$

—

Leucine

$$H-\underset{\overset{|}{\underset{+}{NH_3}}}{\overset{\overset{COO^-}{|}}{C}}-CH_2-\underset{\overset{|}{}}{\overset{CH_3}{CH_2}}-CH_3$$

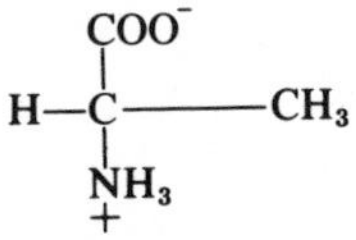

—

Isoleucine

$$H-\underset{\overset{|}{\underset{+}{NH_3}}}{\overset{\overset{COO^-}{|}}{C}}-\underset{\overset{|}{}}{\overset{CH_3}{CH}}-CH_2-CH_3$$

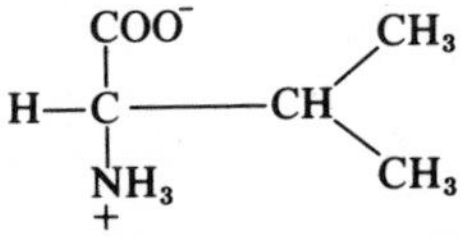

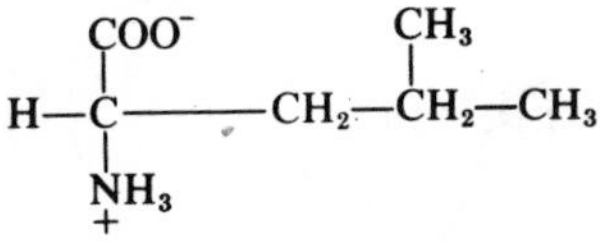

—

Proline

$$H-\overset{\overset{COO^-}{|}}{C}\overset{CH_2-CH_2}{\underset{\overset{+}{NH_2}-CH_2}{}}$$

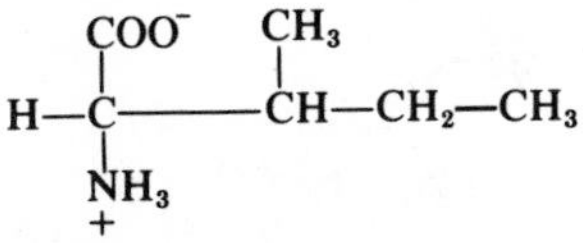

—

Hydroxyproline

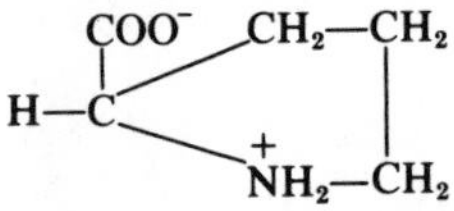

$$H-\overset{\overset{COO^-}{|}}{C}\overset{CH_2-CHOH}{\underset{\underset{+}{NH_2}-CH_2}{}}$$

—

Phenylalanine

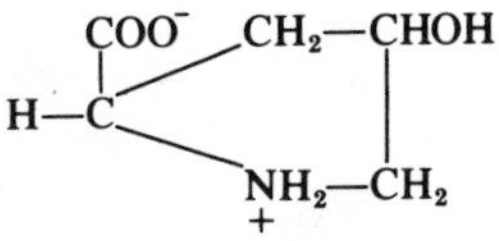

$$H-\underset{\overset{|}{\underset{+}{NH_3}}}{\overset{\overset{COO^-}{|}}{C}}-CH_2-C_6H_5$$

—

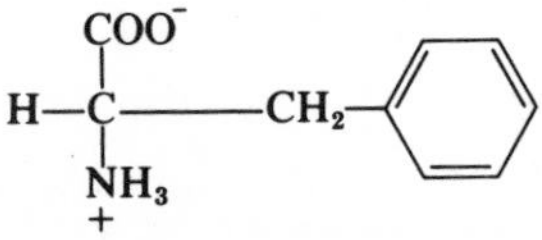

Tryptophan

$$H-\underset{\overset{|}{\underset{+}{NH_3}}}{\overset{\overset{COO^-}{|}}{C}}-CH_2-C\text{(indole ring)}$$

—

3

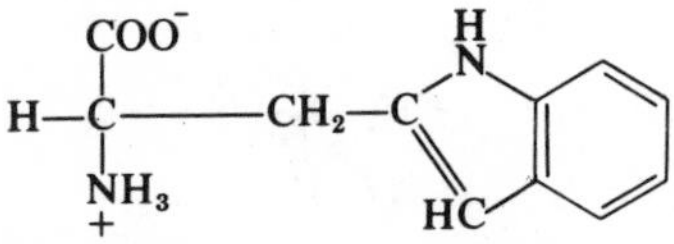

Table 1.1 (*Continued*)

Amino acid	R Group	pK
Methionine	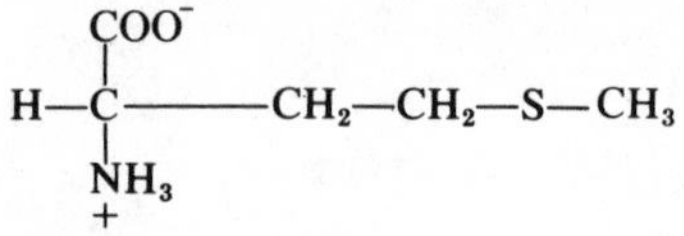 $H-\overset{\displaystyle COO^-}{\underset{\displaystyle \overset{+}{N}H_3}{C}}-CH_2-CH_2-S-CH_3$	—
Serine	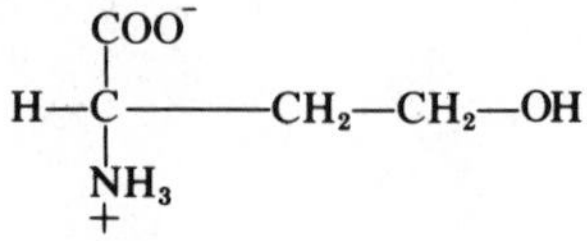 $H-\overset{\displaystyle COO^-}{\underset{\displaystyle \overset{+}{N}H_3}{C}}-CH_2-CH_2-OH$	—
Threonine	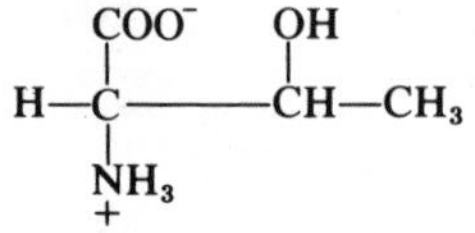 $H-\overset{\displaystyle COO^-}{\underset{\displaystyle \overset{+}{N}H_3}{C}}-\overset{\displaystyle OH}{CH}-CH_3$	—
Cysteine	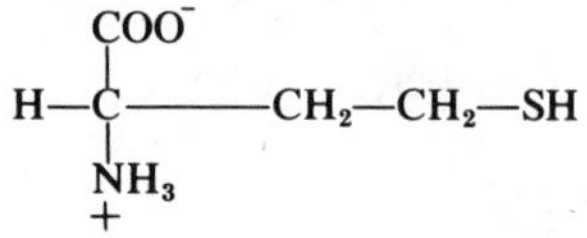 $H-\overset{\displaystyle COO^-}{\underset{\displaystyle \overset{+}{N}H_3}{C}}-CH_2-CH_2-SH$	$pK_a = 8.2$
Tyrosine	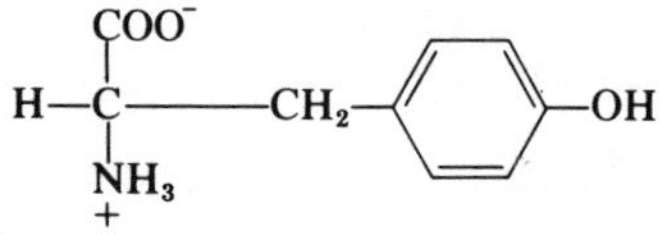 $H-\overset{\displaystyle COO^-}{\underset{\displaystyle \overset{+}{N}H_3}{C}}-CH_2-\bigcirc-OH$	$pK_a = 10.1$
Asparagine	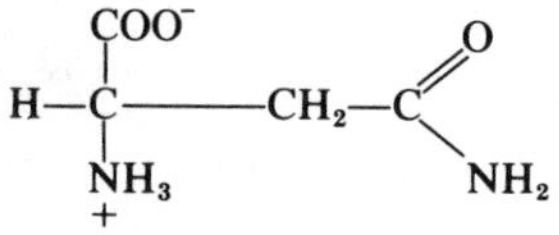 $H-\overset{\displaystyle COO^-}{\underset{\displaystyle \overset{+}{N}H_3}{C}}-CH_2-\overset{\displaystyle O}{\underset{\displaystyle NH_2}{C}}$	—
Glutamine	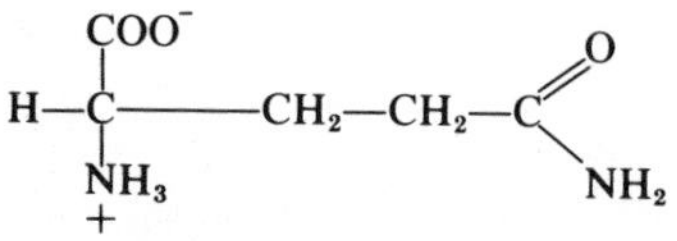 $H-\overset{\displaystyle COO^-}{\underset{\displaystyle \overset{+}{N}H_3}{C}}-CH_2-CH_2-\overset{\displaystyle O}{\underset{\displaystyle NH_2}{C}}$	—
Lysine	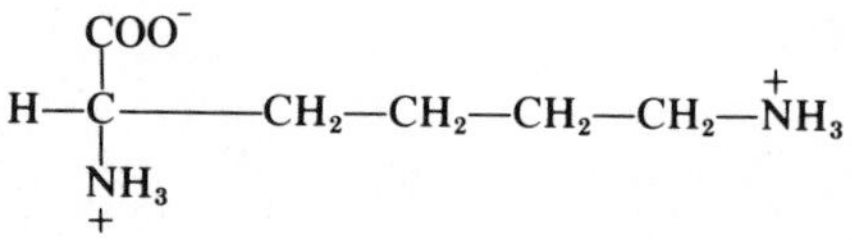 $H-\overset{\displaystyle COO^-}{\underset{\displaystyle \overset{+}{N}H_3}{C}}-CH_2-CH_2-CH_2-CH_2-\overset{+}{N}H_3$	$pK_a = 10.5$
Arginine	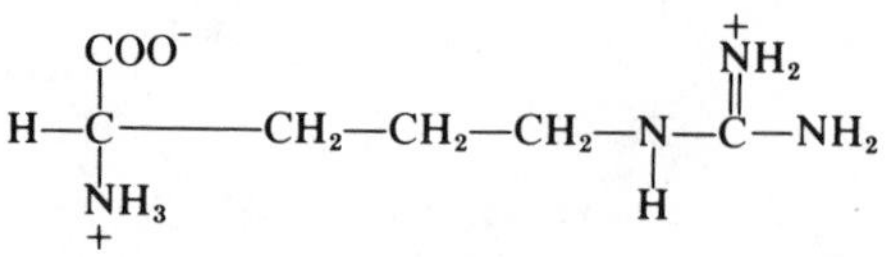 $H-\overset{\displaystyle COO^-}{\underset{\displaystyle \overset{+}{N}H_3}{C}}-CH_2-CH_2-CH_2-\underset{\displaystyle H}{N}-\overset{\displaystyle \overset{+}{N}H_2}{C}-NH_2$	$pK_a = 12.5$

4

Table 1.1 *(Continued)*

Amino acid	R Group	pK
Histidine	H—C(COO⁻)(NH₃⁺)——CH₂—C=C(HN—C(H)=NH⁺)	$pK_a = 6.0$
Aspartic Acid	H—C(COO⁻)(NH₃⁺)——CH₂—C(=O)(O⁻)	$pK_a = 3.6$
Glutamic Acid	H—C(COO⁻)(NH₃⁺)——CH₂—CH₂—C(=O)(O⁻)	$pK_a = 4.2$

ship of amino acids which were close to one another in the primary structure; tertiary structure, which is the relation of constituents which are far from each other in the primary structure; and quaternary structure, which yields the spatial relationships of the peptide chains in a multichain protein.

There have been two types of secondary structures suggested on the basis of bond angles and the maximum probability of hydrogen bonding. The first of these, the α-helix, was proposed by Pauling and Corey (7) in the early 1950's. In this arrangement, each amino group of the peptide backbone is hydrogen bonded to the carbonyl group of the amino acid four residues beyond it. The structure of the right-handed helix is shown in Figure 1.1. Both right-handed and left-handed helices have been postulated, and both have been observed in X-ray crystallographic studies. The right-handed α-helix appears to be more stable (8) and has been observed more frequently in protein structure. It was originally thought that the α-helix comprized the major portion of the structure, but in fact, the amount of helical content is quite variable.

The second type of secondary structure is called the β-pleated sheet. In this case, the hydrogen bonding is between parallel sections of peptide chain rather than between the close neighbors as in the α-helix. The axial distance between adjacent amino acids is fully extended rather than tightly coiled. The strands of the sheet may be either parallel, in which case the

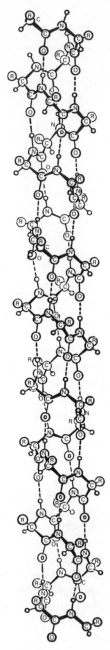

Fig. 1.1 Structure of a right-hand alpha helix. [Reprinted from reference (7) with permission.]

6

carboxyl terminal ends of the chain would be beside each other, or antiparallel, in which case the segments are in opposition. The latter is more frequently observed since the hydrogen bonding angle is the ideal of 180°. This structure is observed in many peptides when the backbone makes a 180° turn and doubles back on itself. This points up a difficulty with the definition of secondary, tertiary, and quaternary structure. In many biochemical texts, the β-pleated sheet structure is shown for interchain hydrogen bonding, which, by definition, is quaternary structure. It is also feasible that relatively far removed sections of a peptide chain could exhibit β-pleated sheet structure and be in a tertiary relationship. The main point, then, is that the differentiation between secondary and tertiary structure is rather arbitrary and that care must be taken when using the nomenclature.

The importance of the position of individual amino acids in the primary structure in determining the three-dimensional structure was alluded to above. The most elementary example of this effect is the termination of the α-helix with the presence of the amino acid proline. Proline is the only common amino acid in which the amino group is a secondary amine. As a result, the angle between the amino and carboxyl functions is fixed in such a way that an α-helical structure is impossible. Steric hindrance also acts to change the secondary structure. Most amino acids have as the first portion of their R substituent a methylene ($-CH_2-$) group. Although no interference is observed for these amino acids, the substitution of a carbon atom for one of the hydrogen atoms (valine, isoleucine, threonine) does cause steric repulsion and results in destabilization of the α-helix. A similar effect is observed when D-amino acids are substituted for L-amino acids. In contradistinction, more flexibility is added to the helix when glycine is present, and in many cases glycine forms a segment of a β-pleated sheet structure. Another source of modification of the secondary structure occurs as a result of the formation of hydrogen bonds which are outside of the α-helix. Such hydrogen bonding has been observed with the sidechain of serine.

If only the interactions described above were present, proteins would exist as essentially tubular or planar structures. Although a few fibrous proteins exist, the majority of these polymers exhibit a spherical or globular shape. The forces that maintain the tertiary structure are basically the same forces that generate the secondary structure of proteins (9, 10):

1. The hydrogen bond has been discussed previously as the major force in the formation of the α-helix and the β-pleated sheet. The existence of hydrogen bonding in the interior of the globular structure has been postulated from X-ray crystallographic data. The great importance of

these bonds in the interior of the protein is due to the relative nonpolarity of the interior where no hydrogen bonding competition exists with the water molecules.

2. Coulombic forces also play an important role in the three-dimensional structure. A majority of the charged sidechains in the protein structure exist at the protein–water interface. The forces of attraction and repulsion at the surface are decreased by the dielectric constant of the water, the complexation of the groups with anions and cations in the solution, and the strong hydration layer which surrounds the protein. In the interior, however, no such interactions exist. X-ray crystallographic studies of protein structure have indicated the presence of salt bridges from the pairing of amino acids such as aspartate and isoleucine residues in the enzyme chymotrypsin (11).

3. The strongest force involved in the maintenance of the tertiary structure is the covalent-bond formation between cysteine residues to yield a disulfide bridge. The disulfide bond has been observed in many proteins and is often critical to proper biological function. The disulfide bond is also important in maintaining the quaternary structure, an example being the five disulfide bridges holding together the three peptide chains of chymotrypsin.

4. The existence of hydrophobic bonding in proteins has been well documented (12–14). Basically, the term hydrophobic bonding is used to describe the arrangement of nonpolar groups, which maximizes the hydrogen bonding of the water molecules. The formation of the hydrophobic bond is driven by the positive entropy of the resulting system. With respect to protein structure, the effect of hydrophobic bonding is realized by the tendency of the nonpolar constituents to remain in the interior of the structure.

The actual three-dimensional shape of the protein is the result of the competition and cooperation of these various forces. The resultant molecule is a fragile balance of thermodynamics with the polar groups on the outside forming hydrogen bonds with the water molecules, the carboxyl and amino groups of the peptide backbone forming interchain hydrogen bonds, and the nonpolar residues inside the structure. With the known effects of the dielectric constant, ionic strength, pH, and temperature on the London and coulombic forces and on hydrogen bonding, the difficulties encountered in handling proteins and enzymes should be understandable.

1.1.3 Active Site Considerations

The dependence of the catalytic properties of enzymes on the three-dimensional structure of the protein was first suggested by Kunitz and

Northrop (15), who related the loss of enzyme activity to the degree of denaturation of the protein. Subsequent experiments linked the enzyme activity to specific amino acid residues and a specific region of the protein called the active site. The characteristics of the active site can be summarized as follows: (1) the active site constitutes a small portion of the overall protein structure; (2) the active site is a three-dimensional niche in the protein from which, in many cases, water is excluded; (3) the specificity of the enzyme depends on the arrangement of the atoms in the active site; (4) the substrate–enzyme binding process involves a relatively small amount of energy (K_{eq} = 3–12 kcal/mole).

The mechanism by which the active site takes part in the reaction was first postulated by Emil Fischer in 1860. The specificity of the reaction, according to Fischer, was the result of the "lock and key" fit of the enzyme and substrate, that is, stereochemical complementarity. An excellent example of the actions involved in enzyme catalysis can be ascertained from the proposed mechanism of carboxypeptidase A. This enzyme hydrolyzes the carboxy terminal amino acid from a peptide chain provided that the terminal amino acid has an aromatic or bulky aliphatic side chain. The pertinent information about the enzyme is as follows: (1) it is a single peptide chain of 307 residues; (2) there is a zinc atom involved at the active site which is coordinated with two histidine side chains, a glutamate side chain, and a water molecule; and (3) there is a large nonpolar pocket near the zinc atom.

One highlight of the carboxypeptidase A mechanism is the structural rearrangement which occurs upon binding of the substrate. The first interaction appears to be the electrostatic attraction of the negatively charged carboxylate ion of the substrate and the sidechain of arginine residue 145. Subsequently, several other interactions take place: the aromatic or aliphatic sidechain of the substrate is bound to the nonpolar pocket of the enzyme; the amine hydrogen of the peptide bond to be cleaved is hydrogen bonded to the hydroxyl group of tyrosine 248; the carboxyl group of the attacked peptide bond is complexed to the zinc atom; and the amino group of the adjacent peptide bond is hydrogen bonded to glutamate 270. This proposed enzyme–substrate complex is shown in Figure 1.2. The overall changes in the protein structure are shown in Figure 1.3. This "induced fit" model of active site action was first proposed by Koshland in 1962. The largest movement of an amino acid residue is that experienced by tyrosine 248 which is displaced by 12 Å, a distance which is a significant portion of the molecular dimensions. This molecular rearrangement results in the formation of a hydrophobic region which is the active site.

The mechanism of the enzyme reaction has been proposed by Lipscomb from X-ray crystallographic data (16). The carbonyl group of the attacked

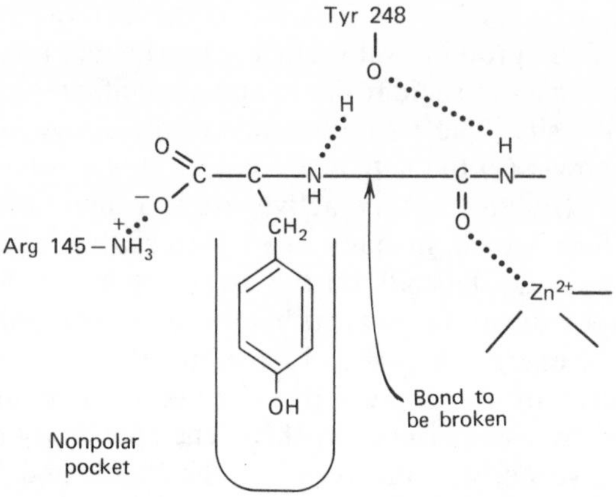

Fig. 1.2 Schematic of enzyme–substrate complex for carboxypeptidase A.

peptide bond is directed towards the zinc atom which serves to polarize the bond. The hydrophobic nature of the active site serves to accentuate the polarization. The hydroxyl group of tyrosine 248 donates a proton to the amino group of the peptide bond to be cleaved. The carbonyl group of this peptide bond is then involved in a nucleophilic attack by the carboxylate group of glutamate 270. The resulting anhydride is hydrolyzed by the addition of water, after which the product molecule is released. There are two major points worthy of discussion: (1) the induced fit in this case is a dynamic recognition process, that is, the molecular conformation change assists in specificity, and (2) the enzyme induces an electronic strain in its substrate to facilitate enhanced reaction rates. The latter point was made succinctly by Pauling in a 1948 lecture (17):

"I think that enzymes are molecules that are complementary in structure to the activated complexes of the reactions that they catalyze, that is, to the molecular configuration that is intermediate between the reacting substances and the products of reaction for these catalyzed processes. The attraction of the enzyme molecule for the activated complex would thus lead to a decrease in its energy and hence to a decrease in the energy of activation of the reaction and to an increase in the rate of reaction."

Numerous other amino acids have been implicated in enzyme catalysis. In chymotrypsin, for example, the proposed mechanism involves the formation of a covalent bond between the substrate and a serine residue of the enzyme with the assistance of two histidine residues (18). Other active site groups include cysteine, methionine, and lysine, to mention a few. The important point is that the catalysis involves the chemical reactivity

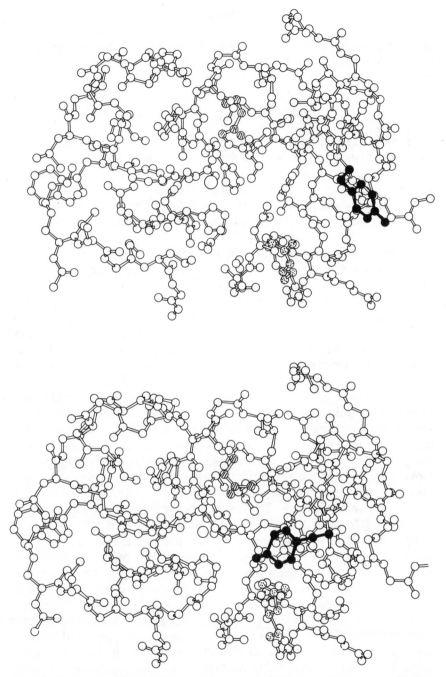

Fig. 1.3 Changes in the three-dimensional structure of carboxypeptidase A upon binding of the substrate. [Reprinted from reference (16) with permission.]

of these groups and the movement of the groups into the proper orientation. These factors have serious implications for the immobilization of enzymes.

1.2 ENZYME CLASSIFICATION

Enzymes are usually named in terms of the reaction which is catalyzed. This is true because although the amount of protein present may be miniscule, the chemical amplification available from the catalysis process makes the enzyme readily measurable. A common practice is to name the enzyme by adding the suffix "-ase" to the name of the stoichiometrically converted reactant or substrate. Thus, the enzyme catalyzing the hydrolysis of urea is urease.

The large number of enzymes alluded to in the previous section makes the systematic arrangement of enzyme nomenclature a necessity. In the early days of enzymology, no such order was apparent. In 1956, the International Commission on Enzymes adopted a scheme based on the single distinctive property of an enzyme, the chemical reaction catalyzed. Two names are provided: a systematic name which describes the action of the enzyme, which has associated with it a specific four-number numerical code, and a practical name which is more amenable to everyday use.

The systematic name of the enzyme consists of two parts: the first part describes the substrate or substrates and the second portion defines the type of reaction catalyzed. All enzymes catalyze one of six types of reactions: (1) oxidation–reduction; (2) functional-group transfers; (3) hydrolysis; (4) lysis; (5) isomerization; and (6) ligation. Lysis reactions remove groups from their substrates, other than by hydrolysis, leaving double bonds. Ligation reactions join two species with the breakdown of the pyrophosphate in ATP. The first number of the Enzyme Commission number is defined by the class of reaction above. The next two numbers indicate the subclass and sub-subclass to which the enzyme belongs, respectively. For example, these may differentiate amino transferase reactions from those involving phosphate transfers, and alcohol group receptors from carboxyl receptors. The final number is simply to distinguish enzymes in the same class, subclass, and sub-subclass. Some of the classifications are given in Table 1.2 along with some of the more commonly used enzymes. It is important to note that the second and third number convey information about what kind of analytical techniques are applicable. For example, any oxidoreductases with a sub-subclass of 3 involve O_2 as an acceptor and thus can be quantified with O_2-measuring techniques.

Table 1.2 List of Commonly Used Enzymes and their Nomenclature

Enzyme commission number	Systematic name	Common name
1.1.1.27	L-lactate:NAD oxidoreductase	Lactate dehydrogenase
1.1.3.4	β-D-glucose:oxygen oxidoreductase	Glucose oxidase
1.4.3.2	L-amino acid:oxygen oxidoreductase (deaminating)	L-Amino acid oxidase
1.6.6.1	Reduced-NAD(P):nitrate oxidoreductase	Nitrate reductase (NAD(P))
1.9.6.1	Ferrocytochrome:nitrate oxidoreductase	Nitrate reductase (cyto)
2.7.1.1	ATP:D-hexose 6-phosphotransferase	Hexokinase
2.7.3.2	ATP:creatine phosphotransferase	Creatine kinase
3.1.1.3	Glycerol ester hydrolase	Lipase
3.2.1.1	α-1,4-glucan-4-glucano-hydrolase	α-Amylase
3.4.2.1	Peptidyl-L-amino acid hydrolase	Carboxypeptidase A
3.5.1.5	Urea amidohydrolase	Urease

The classification of enzymes also gives valuable information about another analytically important feature—specificity. Some enzymes exhibit the ability to catalyze a single unique reaction. Pyruvate kinase, for example, enhances the transfer of phosphate between phosphopyruvate and ADP and catalyzes no other phosphate transfer. The analytical implications of such a reaction are obvious. More frequently, however, enzymes exhibit a group specificity, catalyzing a number of reactions involving a common functional group. The phosphatases are exemplary of this concept, catalyzing the hydrolysis of a large number of organic phosphate esters. It should be pointed out that the rate of reaction of each of the substrates may be different.

A third group of enzymes exhibit a wide range of catalytic activity. Esterases and proteases are relatively nonspecific with respect to the overall substrate structure, being specific for the type of bond cleaved. Some enzymes, notably trypsin, exhibit both esterolytic and proteolytic activity. A final consideration in the use of enzymes is the relatively commonly observed stereospecificity. All of the enzymes involved in the metabolism of glucose are specific for only the D-stereoisomer of glucose and its derivatives. In some cases, not only is the stereoisomerism of the compound important, but also the spatial relationship of the functional groups at or near the site of enzyme action. In the reaction of glucose with oxygen in the presence of glucose oxidase, for example, only the α-form of the D-glucose isomer is oxidized.

1.3   ENZYME NOMENCLATURE

Some enzymes possess groups which are not amino acids. These conjugated enzymes, called holoenzymes, are comprised of the protein fraction, called the apoenzyme, and the organic portion called the prosthetic group. An example of this arrangement is the enzyme glucose oxidase which consists of a bound electron transfer moiety, FAD^+, and a protein. In many cases, as in the carboxypeptidase A example above, a metal ion functions as part of the active site. It may either be tightly bound to the enzyme, be part of a larger prosthetic group such as protoporphyrin, or be complexed with a substrate to form the true reactive substrate. Prosthetic groups are part of a larger group of compounds known as cofactors or coenzymes. A coenzyme is a substrate which is cycled, usually in a metabolic pathway, and thus is not stoichiometrically consumed in vivo as is the substrate. In analytical applications, the use of this definition is somewhat ambiguous since the coenzyme will also be stoichiometrically consumed. Prosthetic groups differ from other members of this classification in that they are tightly bound to the protein and they are generally regenerated as part of the enzyme mechanism. Many different definitions of these terms have been given, and in relation to a single enzyme reaction most of the differences are purely semantic. In general, however, coenzymes function by accepting or donating groups involved in the catalysis. The type of coenzyme used in an enzyme reaction helps to classify the enzyme as shown in Table 1.2. Some of these cofactors are shown in Table 1.3 and discussed below.

The coenzyme adenosine-5'-triphosphate, ATP, is well known and is involved in phosphoryl transfer reactions. The biochemistry of ATP is well detailed in all biochemistry texts. The use of ATP as a reagent requires some discussion, however, since pH, temperature, concentration, and the presence of various ions influence its stability. The effect of hydrogen ions on the hydrolysis of ATP is well known (19, 20). In addition, it has been well documented that more concentrated solutions of ATP are more stable than lower concentrations over a wide range of temperatures (21). The presence of ions such as citrate, glycine, and ethylenediamine tetraacetic acid (EDTA) seem to affect the stability of ATP in solution as well; citrate by accelerating its hydrolysis, and glycine and EDTA by their metal-complexing ability. Small amounts of metals deleteriously affect the stability of ATP. Mg(II), Cu(II), Zn(II), and Mn(II) have been implicated in the decomposition of ATP, but the mechanism of these agents is not well understood. Mn(II) has the additional disadvantage of precipitating the ATP if the metal concentration exceeds 10^{-3} M.

The nucleotides nicotinamide adenine dinucleotide (NAD^+) and nicotinamide adenine dinucleotide phosphate ($NADP^+$) are probably the best known coenzymes. These compounds function as electron acceptors. They can be reduced by enzyme dehydrogenases or dithionite, requiring two equivalents of hydrogen ions per mole of nucleotide. The great utility of NAD^+ and $NADP^+$ is the fact that in the reduced form only they absorb light at 340 nm due to the quinonoid structure of the reduced nicotinamide ring. The mechanism of the reduction is the addition of two hydrogen atoms to the *para* carbon and nitrogen atoms followed by the dissociation of the hydrogen atom from the nitrogen. The stability of the oxidized and reduced forms of NAD is quite different, with NAD^+ more stable at neutral and acidic pH's (22). The reduced form shows the opposite characteristics—stability in alkaline media and instability in acids. An interesting sidelight is the fact that on freezing and thawing, a potent inhibitor is formed from nucleotide solutions (23). This has been observed in particular for the enzyme lactate dehydrogenase in serum where freezing the sample causes a significant decrease in the enzyme activity.

Flavin groups are coenzymes which are usually covalently bound to the enzyme and which take part in electron-transfer processes. The structures of the two flavin compounds, flavin mononucleotide (FMN) and flavin adenine dinucleotide (FAD^+), are shown in Table 1.3, the former not containing the adenosine-5-phosphate group. Of the enzymes which contain flavin moieties, FMN appears most frequently, the best known example of which is glucose oxidase. The function of the flavin compounds is mediated by the addition of two hydrogen atoms to the dimethylisoalloxazine ring structure. The flavin is then reoxidized by the reduction of O_2 to H_2O_2 or reduction of another species. The enzyme is then ready for another cycle. It should be pointed out that many, but not all, flavin enzymes have a metal ion associated with the flavin structure.

Coenzyme A acts as an acyl group carrier in acyl-transfer reactions. The functional part of the molecule is the sulfhydryl group. This compound is very important in metabolism as reflected in the fact that over 60 enzymes utilize it as a coenzyme. The transfer of acyl groups may be to other sulfhydryls, or from a nitrogen, oxygen, carbon, or phosphate group. In addition, coenzyme A is involved in elimination reactions and in addition reactions involving carbonyl and ethylene groups. The latter two functions are not observed in enzyme-catalyzed reactions.

Pyridoxyl phosphate takes part in amino group transfers and the decarboxylation of amino acids. It has been found that the pyridoxyl group is bound to the protein through a link between the aldehyde function and the ε-amino group of lysine. This in effect activates the pyridoxyl phosphate, which then forms a Schiff's base with the substrate amino acid. In the transaminases, the hydrolysis yields pyridoxamine phosphate and

Table 1.3 Coenzymes and Their Structures[a]

Coenzyme	Function	Structure
Adenosine phosphate (ATP, ADP, AMP)	High-energy compound; transfer of phosphate groups	
Nicotinamide-adenine dinucleotide (NAD, NADP)	Transfer of hydrogen	

Flavin nucleotides (FMN, FAD)   Transfer of hydrogen

Coenzyme A   Transfer of acyl groups

17

Table 1.3 (*Continued*)

Coenzyme	Function	Structure
Pyridoxyl phosphate	Transfer of amino groups; decarboxylation of amino acids	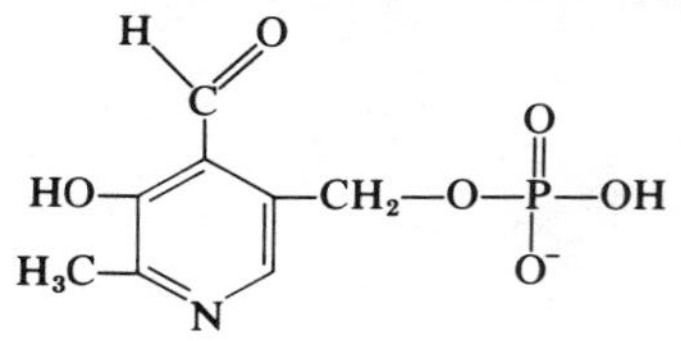
Thiamine pyrophosphate	Decarboxylation of oxo acids	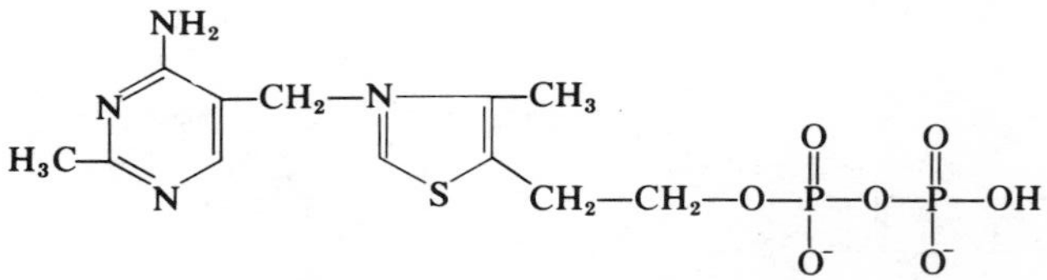

[a] Incomplete bands indicate $-OH$ groups, R signifies either OH or OPO_3H, P signifies $-OPO_3H$

18

a ketoacid. By reversing the reactions, the amino function is transferred. In the decarboxylases, CO_2 is eliminated and the amine hydrolyzed. An extensive discussion can be found in such texts as Dixon and Webb (22a) or Bernhard (10). Thiamine pyrophosphate contains an active thiazolium ring. Its function is thought to be a nucleophilic attack of the thiamine on the substrate. It is interesting to note that both pyridoxyl phosphate and thiamine pyrophosphate will catalyze these reactions in the absence of the protein.

Other coenzymes exist but it is beyond the scope of this monograph to discuss all of the coenzymes. One important group of coenzymes not listed in the table that deserves mention are the heme and metal-atom coenzymes. The presence of a metal ion in the center of a tetrapyrrole ring is a relatively common coenzyme. The best known example is catalase which contains a ferric tetrapyrrole which catalyzes the decomposition of hydrogen peroxide. It is interesting to note that here again, the coenzyme is capable of catalyzing the reaction alone, but the presence of the protein moiety increases the rate of the reaction. Other heme groups include other metal atoms such as zinc and cobalt. As in the example of carboxypeptidase A above, in some cases the metal atom is bound directly to the protein and functions presumably as an electrophilic agent.

One other structural component of proteins must be mentioned. Many enzymes have carbohydrate moieties covalently attached to their amino acid sidechains. In the case of peroxidase, oxidation of the carbohydrate fraction with periodate removes the catalytic capacity of the enzyme, leading one to postulate that the carbohydrate groups are involved in the active site. In other enzymes, the number and type of carbohydrate residues can be used to differentiate isoenzymes. Isoenzymes are enzymes which catalyze the same chemical reaction but differ in some structural property which allows them to be separated, for example, by ion-exchange chromatography. The best example of the carbohydrate differences in isoenzymes is alkaline phosphatase, in which the number of sialic acid residues can be related to the tissue from which the isoenzyme originated. A detailed discussion of the carbohydrate groups is beyond the scope of this book, but it should be noted that these functionalities can be used advantageously in immobilizing enzymes.

1.4   ISOLATION AND PURIFICATION

The tremendous number of enzymes which have been discovered, and the reactions associated with them, makes a wide variety of enzyme-catalyzed analysis procedures available. Unfortunately, only a very small fraction of the known enzymes are commercially available. This makes the isolation and purification of enzymes desirable. Indeed, this would expedite the evolution of analytical enzymology considerably through new and better enzyme preparations.

The first step in the isolation process is to locate a rich source of the enzyme. This serves two purposes: (1) the required separation may be simplified if the desired protein is the predominant one in the matrix, and (2) the losses which occur during the separation will be offset and a reasonable amount of the enzyme obtained. In general, the biochemical literature is an excellent source of such information. Two compilations of enzyme data, Barman's *Enzyme Handbook* (1) and *Enzyme Nomenclature* (2), provide extensive lists and references for over 2000 enzymes. There are several additional factors to be considered in the isolation process. It would be advisable, for example, to use a microbial source rather than to use a specific tissue source of a large animal if large amounts of the enzyme are desired, simply from the point of view of raising and maintaining the source organism. In addition, using growth media containing the desired analyte may act to enrich the source. An example of this is the preparation of a gentimicin-adenylating enzyme by culturing *E. coli* resistant to gentimicin in a medium containing this antibiotic. Another consideration is the specificity of the enzyme obtained from various sources. Lipase from the microorganism *Rhizopus delemar,* which is used in the enzymatic determination of triglycerides, was found to cleave only the α and α' fatty acid esters. This required the addition of a second enzyme to remove the β fatty acid in order to quantify the glycerol. Subsequently, it has been found that lipase from *Rhizopus arrhizus* removes all of the fatty acids in a single step. Such differences may considerably simplify an analytical scheme.

The next step is to remove the enzyme from its environment. Enzymes may exist in various cellular compartments, such as the mitochondria, or may be bound to cell membranes. The location of the enzyme may have an effect on the ease with which it is extracted. In general, simple disruption of the cell membrane is sufficient. For animal tissues, simple homogenization in a Waring blender using a buffer solution as the enzyme recipient is sufficient. In the case of micro-organisms, a more vigorous method may be required to lyse the cell walls. Such techniques include grinding the microbe with sand or finely ground glass beads; ultrasonic disruption; pressure as in a French press; rapid freezing and thawing; or enzymatic lysis of the membrane. In general, the latter two are undesirable if an alternative method exists. An extensive and detailed discussion of these techniques is found in various references.

The separation of the enzyme from the other proteinaceous cell constituents can be achieved by numerous means. The initial step after obtaining an aqueous solution of the enzyme is to eliminate the cell fragments and other particulate matter. This is generally accomplished with relatively high speed centrifugation. In the case of animal tissue, incubating the homogenate at pH 5.5 before centrifugation may help to remove some of the nuclear matter. The cleared homogenate may then be processed using salt fractionation, organic solvents, heat, dialysis, or ion-exchange, partition, gel-permeation, or affinity chromatography or any combination of the above.

Salt fractionation is frequently used for preliminary separation of the protein fractions. The basis of the technique is the fact that globular proteins precipitate in the presence of about 50% saturated solutions of ammonium sulfate. Other salts have been used however. Individual proteins appear to precipitate at slightly different salt concentrations. The technique requires the addition of a weighed portion of salt to the enzyme solution, a 15 min incubation for dissolution, and high-speed centrifugation due to the marginal solubility of the protein. The supernatant is removed, the precipitate redissolved, and the resulting solution assayed for enzyme activity. It is important to note that the supernatant should not be assayed since the high ionic strength will effect the reaction. If the protein precipitate contains the activity, the purification may proceed by other techniques. If the enzyme is not present, incremental additions of salt may be made to isolate the enzyme. A number of factors are important in reproducing enzyme purification by salt fractionation: the pH, temperature, nature of the salt, and the concentration of the enzyme (24, 25).

The use of organic solvents to fractionate proteins has been used since the purification of urease by Sumner. Although ethanol has been used, acetone is more commonly employed. Important parameters include the use of low temperatures to prevent denaturation by the solvent and low electrolyte concentration to improve the resolution of the protein fractions. The latter point may require prior dialysis. Another less commonly used technique is heat fractionation. Although it can be used to remove a fairly large amount of protein in one step, problems with temperature homogeneity and limitations on the volume of sample that can be purified have combined to reduce the utility of the method. Dialysis is, of course, a prime method of protein purification. In the case of enzymes, the dialysis can be used to remove reversible inhibitors as well as other small ionic species.

The newest methods of enzyme purification involve column chromatographic techniques. The use of ion-exchange resins such as DEAE-Sephadex and others have been used extensively in the separation of proteins. Other protein techniques have included gel permeation or gel filtration (26) on Sephadex, Biogel, or Sepharose, and adsorption chro-

Table 1.4   Isolation of the MB Fraction of Creatine Kinase (CK) from Human
Heart Tissue

Sample	Procedure	Activity (μmoles/min·l)	Recovery (%)
18 g fat-trimmed heart	Homogenize in 250 mmol/l sucrose, 10 mmol/l KCl, 1 mmol/l EDTA, 1 mmol/l β-mercaptoethanol, pH 8.0 Filter, centrifuge		
Homogenate	Ammonium sulfate fractionation, collect 50–85% saturation at 0°C.	100,520	100
Dialysate	DEAE-sephadex ion-exchange chromatography, step gradient elution with 25 mmol/l Tris	49,700	49
Eluate		4,400	4.8

matography with calcium phosphate and alumina (27). The basic principles of these techniques can be reviewed in any separation science text. Affinity chromatography is a relatively new technique which is based on the reversible binding of macromolecules to a specific substance (28, 29). For enzymes, for example, inhibitors, substrates and substrate analogs, and coenzymes have been used to separate the desired molecule specifically.

It should be emphasized that purification of the enzyme requires the development of a relatively simple, rapid, and reliable method of determining enzyme activity. Two representative examples of enzyme purifications are given in Tables 1.4 and 1.5. The specific activity is a commonly used term in enzyme analysis and is the amount of enzyme activity per gram of protein. The main features of merit in the separations are the decrease in the total amount of protein, the loss of absolute activity, and the improvement in the specific activity.

Table 1.5   Purification of the Major Isoenzyme of Phosphoglucomutase

Purification technique	Total activity (μmol/min)	Specific activity (units/mg)	Recovery (%)	Purification (-fold)
Autolysis	35,238	0.081	100.0	1.0
Heat treatment	35,206	0.11	99.9	1.4
Ammonium sulfate fractionation	31,784	0.60	90.0	7.4
CM-cellulose chromatography	21,788	31.40	62.0	388.0
DEAE-cellulose chromatography	12,000	138.00	34.0	1704.0
Sephadex gel permeation	7,785	205.00	22.0	2531.0

1.5   ENZYME DENATURATION

One final topic requires discussion to complete a general survey of enzymology—denaturation. Denaturation is defined as the breakdown of the numerous interactions which maintain the biologically active conformation. Due to the cooperative nature of the forces which sustain the ordered structure, denaturation generally results in an essentially random conformation. Several common techniques are used to intentionally denature a protein, including contact with guanidine hydrochloride, urea, organic solvents, detergents, and temperature, ionic strength, and pH changes. Each technique requires some comment.

The use of guanidine hydrochloride and urea as denaturation agents comes from their strong hydrogen bonding character in concentrated solutions. In the presence of 3 mol/l guanidine hydrochloride or 8 mol/l urea, the intramolecular hydrogen bonds of the protein are broken and subsequently reformed with the denaturant. In the case of urea, a second proposed mechanism is the production of a relatively nonpolar environment in which the hydrophobic interactions of the polypeptide are disrupted. As such, urea retains a function somewhere between guanidine hydrochloride and an organic chloride. The end result of these processes is the randomization of the polypeptide chain.

Another approach to opening the ordered structure is the addition of a detergent. The most popular detergent is sodium dodecylsulfate (SDS). The mechanism is thought to be the result of the interaction of the nonpolar portion of the detergent molecule with the hydrophobic constituents of the protein, after which the polar region of the detergent makes the entire structure hydrophilic. Again, the result of the removal of the forces which stabilize the natural conformation is randomization of the structure.

One troublesome source of denaturation is due to changes in the ionic strength of the protein solution. In the native conformation, most of the charged amino acids will be on the surface of the protein exposed to the solvent. If, for example, a large number of lysine residues are in close proximity, the presence of an anion may serve to alleviate some of the ionic repulsion. Thus, as the anion concentration, and the ionic strength, is reduced, the repulsive forces would increase possibly causing destabilization of the entire structure. Cations may also have specific effects due to complex formation with particular sidechains, but this will be discussed in more detail later.

The effects of pH and temperature, although mechanistically different, are closely related, that is, the temperature-dependent denaturation of a protein which is stable at pH 7.0 at room temperature may denature at pH 4.0. By varying both the pH and temperature, this same protein may be stable at pH 4.0. The effect of pH and ionic strength seems to be on the rate of the process rather than on the equilibrium. The mechanism by which temperature-change denaturation occurs is simply to supply the energy necessary for the breakdown of the numerous hydrogen bonds. The high free energy of activation (50,000 kcal/mole) appears to be due to a very high positive entropy of activation. One interesting feature of temperature denaturation is the fact that, at least for trypsin, the enthalpy of the overall process is higher than the enthalpy of activation; a rather unique feature of protein denaturation. In many cases, the heat denaturation is reversible upon returning the solution to room temperature. This has led to a number of postulates about the thermodynamics of the peptide-chain folding, which is beyond the scope of this monograph.

The denaturation of a protein due to pH changes is most likely the result of increased coulombic repulsions between the charged amino acids on the protein–water interface. In the effects of pH and temperature on the protein structure, care must be exercised so that effects on functional parameters, such as enzyme activity, are distinguished from true structural changes. The former are discussed in Chapter 2.

It is interesting that in some cases a substrate, coenzyme, or reversible inhibitor may increase the stability of an enzyme in the presence of denaturing compounds. For example, trypsin and chymotrypsin are protected by their substrates; dehydrogenases requiring NAD and pyridoxyl phosphate-dependent enzymes are stabilized by their respective cofactors; and urease is maintained by the reversible inhibitor p-chloromercuribenzoate. In other enzymes, particularly those under allosteric control, the presence of the control moiety may actually facilitate denaturation. Stabilization may result from the interaction between the active-site amino acids and the substrate, eliminating the possibility of denaturation originating in that region of the molecule. In contradistinction, the binding of an allosteric moiety is proposed to cause a destabilization of the active site, which makes that region susceptible to denaturation. As mentioned before, once denaturation begins, it generally becomes complete due to the cooperative nature of the intramolecular forces.

References

1. T. E. Barman, *Enzyme Handbook,* Vol. 1–2, Suppl. I (Springer Verlag, New York, 1974).
2. "Enzyme Nomenclature *Recommendations (1978) of the Commission on Biochemical Nomenclature and Classification of Enzymes together with their Units and the Symbols of Enzyme Kinetics"* (Elsevier, Amsterdam, 1978).

3. G. G. Guilbault, *Anal. Chem.* **42**, 334R (1970).

4. H. U. Bergmeyer, Ed., *Methods of Enzymatic Analysis,* Vol. 1–4, (Academic, New York, 1975).

5. W. H. Stein and S. Moore, *Scientific American,* February, 1961.

6. C. H. W. Hirs and S. N. Timasheff, Eds., *Methods in Enzymology,* Vol. 25, (Academic, New York, 1972).

7. L. Pauling, R. B. Corey, and H. R. Branson, *Proc. Natl. Acad. Sci.* **37**, 205 (1951).

8. C. Ramakrishnan and G. N. Ramachandran, *Biophys. J.* **5**, 909 (1965).

9. F. M. Richards, *Ann. Rev. Biochem.* **32**, 269 (1963).

10. S. A. Bernhard, *The Structure and Function of Enzymes,* (W. A. Benjamin, Menlo Park, CA., 1968).

11. D. M. Blow, J. J. Birktoft, and B. S. Hartley, *Nature* **221**, 337 (1969).

12. W. J. Kauzman, *Adv. Protein Chem.* **14**, 1 (1959).

13. C. B. Anfinsen, *Science* **181**, 223 (1973).

14. I. M. Klotz, *Brookhaven Symp. Biol.* 25 (1960).

15. M. Kunitz and J. H. Northrop, *J. Gen. Physiol.* **19**, 991 (1936).

16. W. N. Lipscomb, *Proc. Robert A. Welch Found. Conf. Chem. Res.* **15**, 140 (1971).

17. L. Struyer, *Biochemistry,* (Freeman, San Francisco, 1975).

18. M. L. Bender, *J. Am. Chem. Soc.* **84**, 2580 (1962).

19. R. A. Alberty, R. M. Smith, and R. M. Bock, *J. Biol. Chem.* **193**, 425 (1951).

20. H. Mahler and E. H. Cordes, *Biological Chemistry,* (Harper & Row, New York, 1966).

21. S. Lin and H. P. Cohen, *Anal. Biochem.* **24**, 531 (1968).

22. O. H. Lowry, J. V. Passonneau, and M. K. Rick, *J. Biol. Chem.* **236**, 2756 (1961).

22a. M. Dixon and E. C. Webb, *The Enzymes* (Academic, New York, 1964).

23. K. Danziel, *Biochem. J.* **84**, 240 (1962).

24. P. R. Foster, P. Dunhill, and M. D. Lilly, *Biotechnol. Bioeng.* **18**, 545 (1976).

25. T. P. King, *Biochemistry* **11**, 367 (1972).

26. O. Levin, *Methods in Enzymology,* S. P. Colowick and N. O. Kaplan, Eds., Vol. V (Academic, New York, 1962), p. 27.

27. P. Andrews, *Biochem. J.* **96**, 595 (1965).

28. H. H. Weetall, *Anal. Chem.* **48**, 44R (1976).

29. W. B. Jakoby and M. Wilchek, Eds., *Methods in Enzymology,* Vol. 34 (Academic, New York, 1974).

CHAPTER

2

ENZYME KINETICS

2.1 SIMPLE MICHAELIS–MENTEN KINETICS

Most of the interest in enzymes is the result of their unique catalytic function. Indeed, the apparent purpose of these proteins is to increase the velocity of certain reactions which are important to the organism containing them. Since enzymes function as catalysts, the determination of the enzyme concentration is most easily accomplished by measuring the effect of its presence on the rate of a chemical reaction. Thus, we must be concerned with enzyme kinetics. It should also be apparent that to use the enzyme as a reagent, one must also possess a knowledge of the kinetic factors involved. It will be assumed that the reader has a fundamental knowledge of general kinetics.

2.1.1 Single-Substrate Reaction

The observation by Fisher that the enzyme and its substrate were stereochemically complementary led Michaelis and Menten (1) to postulate the formation of an enzyme–substrate complex to account for the rectangular hyperbolic relationship between the substrate concentration and the reaction velocity. The mechanistic scheme most frequently written involves a single substrate and a single intermediate given by:

$$E + S \underset{k_{-1}}{\overset{k_{+1}}{\rightleftharpoons}} ES \xrightarrow{k_{+2}} E + P \qquad (2.1)$$

where E is the free enzyme, S is the substrate, ES is the enzyme–substrate complex, and P is the reaction product. Thus, Michaelis and Menten derived a rate equation based on the assumption that the rate of breakdown of the enzyme–substrate complex to product was much slower than the dissociation of ES into enzyme and substrate, that is $k_{+2} \ll k_{-1}$.

A more valid derivation was proposed by Briggs and Haldane (2) based on the initial rate of the reaction. Almost immediately after the reaction begins, the rate of change of concentration of the enzyme–substrate complex with time is essentially zero, that is, the enzyme–substrate complex achieves a steady-state condition. The rate equation for the intermediate,

$$\frac{d[ES]}{dt} = k_{+1}[E][S] - k_{-1}[ES] - k_{+2}[ES] = 0 \qquad (2.2)$$

26

can be solved for the concentration of ES. Since $[E]_0$, the total enzyme concentration, can be given at any time as the sum of $[E]$ and $[ES]$,

$$\frac{d[ES]}{dt} = k_{+1}[E]_0[S] - (k_{+1}[S] + k_{-1} + k_{+2})[ES] = 0 \qquad (2.3)$$

Dividing through equation (2.3) by k_{+1} and solving for $[ES]$,

$$[ES] = \frac{[E]_0[S]}{(k_{-1} + k_{+2})/k_{+1} + [S]} \qquad (2.4)$$

and the rate equation for the formation of product is

$$v = k_2[ES] = \frac{k_2[E]_0[S]}{(k_{-1} + k_{+2})/k_{+1} + [S]} = \frac{V[S]}{K_M + [S]} \qquad (2.5)$$

There are a number of interesting features about this equation. First, K_M, the Michaelis constant, is the equilibrium constant for the dissociation of the enzyme–substrate complex and is inversely related to the "affinity" of the enzyme for the substrate. Second, it should be noted that the maximum velocity will be attained when all of the enzyme is in the form of enzyme–substrate complex ($E_0 = ES$). Third, it should be apparent that the $[S]$ term in equation (2.5) is the total concentration of substrate. If we write a mass balance equation for S, we can see that:

$$[S]_{total} = [S]_{free} + [ES]$$

if $[S]_0 \gg [E]_0$, then the concentration of S in equation (2.5) can be assumed to be equal to the initial substrate concentration at the beginning of the reaction. From the above considerations, $[S]_0$ should be much greater than $[E]_0$ for the Michaelis–Menten kinetic scheme to apply. The validity of this assumption has been questioned, particularly for immobilized enzyme kinetics.

Finally, the steady-state assumption requires that $d[ES]/dt = 0$. From equation (2.3), however, it can be seen that if

$$\frac{d[ES]}{dt} \ll (k_{-1} + k_{+2} + k_{+1}[S])[ES]$$

the subsequent equations are valid. This inequality will always be maintained if the substrate concentration is kept sufficiently high. It has been shown that the steady-state assumption is satisfactory as long as $[S]/[E]_0 > 10^3$.

The initial rate of reaction as a function of the substrate concentration at constant enzyme concentration is shown in Figure 2.1. As predicted by equation (2.5), at low substrate concentrations, the rate is first order with respect to substrate concentration. If $[S] = 0.1\ K_M$, the deviation of the reaction rate from first-order conditions is 9%. At high substrate

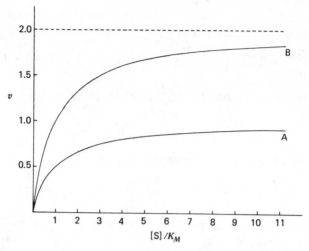

Fig. 2.1 Graph of a rectangular hyperbola of the type observed in an enzyme reaction. The velocity on the abscissa is the initial velocity of the reaction. Curve B has twice the enzyme concentration of curve A.

concentrations, the rate is predictably zero order with respect to the substrate concentration. Again it should be noted that if $[S] = 10\ K_M$, the deviation from zero-order kinetics is 9%. These deviations must be borne in mind when kinetic assays are performed. Finally, it should be pointed out that the definition of the Michaelis constant is given directly in equation (5) since when $v = \frac{1}{2}\ V$, $[S] = K_M$.

To evaluate these parameters in an enzyme preparation, a linear relation would be more useful. The most frequently used transposition is that due to Lineweaver and Burke (3) where equation (2.5) is written in the form

$$\frac{1}{v} = \frac{K_M}{V} \cdot \frac{1}{[S]} + \frac{1}{V} \tag{2.6}$$

A plot of $1/v$ as a function of $1/[S]$ will give a straight line of slope K_M/V and a y intercept of $1/V$ (Figure 2.2). It is then straightforward to calculate K_M and V. The results obtained from the Lineweaver–Burke plot must be carefully analyzed since the reciprocal function tends to put all of the data in one section of the plot. Statistical analysis of the intercept (4) and slope should be used to determine the appropriate concentrations for quantitation of the parameters of interest.

Two other types of plots are used on occasion. One is obtained by multiplying equation (2.6) by $v \cdot V$ and is due to Hofstee (5) and Eadie (6).

The equation,

$$v = -K_M \frac{v}{[S]} + V \tag{2.7}$$

is plotted as v versus $v/[S]$ with the resulting slope and intercept being $-K_M$ and V respectively. The major advantage of this method is the more even spacing of the experimental results. The second alternative, proposed by Hanes (7), is a plot of the equation

$$\frac{[S]}{v} = \frac{1}{V} [S] + \frac{K_M}{V} \tag{2.8}$$

where the variables would be $[S]/v$ and $[S]$. Each of these last two methods is of use in enzymology and should be recognized.

It should be pointed out that although the Michaelis–Menten type mechanism applies for many enzyme reactions, a significant number of enzymes do not follow this simple model. In the following sections, some of the exceptions will be discussed. Another simple mechanism which finds frequent use is that involving two intermediates, that is,

$$E + S \underset{k_{-1}}{\overset{k_{+1}}{\rightleftharpoons}} ES \underset{k_{-2}}{\overset{k_{+2}}{\rightleftharpoons}} EP \underset{k_{-3}}{\overset{k_{+3}}{\rightleftharpoons}} E + P \tag{2.9}$$

A rate equation can be derived in a manner analogous to that used for the single intermediate mechanism using the steady-state assumption. The

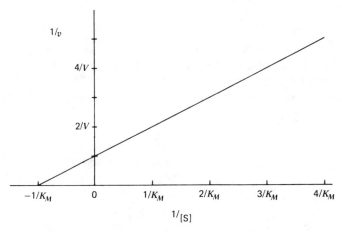

Fig. 2.2. Graph of a Lineweaver–Burke plot for an enzyme.

resulting equation is given below.

$$\frac{k_{+2}k_{+3}/(k_{+2} + k_{-2} + k_{+3})([E]_0[S]_0)}{(k_{-1}k_{-2} + k_{-1}k_{-3} + k_{+2}k_{+3})/(k_{+1}(k_{+2} + k_{-2} + k_{+3}) + [S]}$$

$$= \frac{V[S]}{K_S + [S]} \qquad (2.10)$$

It should be noted that the two intermediate mechanisms can yield the rate constants of the simple system. For example, if $k_2 \ll k_{+3}$ then K_S, a more general form for K_M, is given by $(k_{-1} + k_{+2})/k_1$ independent of the relative values of k_{+2} and k_{-1}. This should be recognized as the K_M of the simple single intermediate rate equation. If $k_{+2} \gg k_{-3}$, K_S is given by $[(k_{-1} + k_{+2})/k_{+1} \cdot k_{+3}/k_{+2}$, again independent of the relative values of k_{+2} and k_{-1}. We will make use of the two intermediate models in subsequent discussions.

2.1.2   Two-Substrate Reaction

Many enzyme reactions require two or more substrates. The most common examples of this type of system are enzymes which use coenzymes as a second substrate. Although the mechanism of these reactions may be quite complex, pseudo-first-order conditions as expressed in equation (2.5) are generally followed if the level of all substrates, except the one of interest, is held constant.

The derivation of rate equations for two-substrate systems is somewhat more complicated due to the number of mechanisms that can be postulated. For example, if an enzyme requires two substrates, A and B, there may be binary or ternary complexes formed. If the substrates add in an ordered sequence such as

$$E + A \rightleftharpoons EA + B \rightleftharpoons (EAB \rightleftharpoons EPQ) \rightleftharpoons EP + Q \rightleftharpoons E + P + Q \qquad (2.11)$$

the mechanism is called ordered. The formation of a number of ternary complexes (EAB, EPQ) further complicates the kinetics of this system. Another mechanism, known as random, allows the formation of a ternary complex from either of the binary complexes as illustrated by:

$$(2.12)$$

In other circumstances, there has been no evidence for the formation of a ternary complex at all. One of the basic mechanisms involves the release of a product molecule during the substrate addition sequence, for example,

$$E + A \rightleftharpoons EA \rightleftharpoons FP \rightleftharpoons F + P + B \rightleftharpoons FB + P \rightleftharpoons FQ \rightleftharpoons E + Q \quad (2.13)$$

where F is another form of the enzyme. This mechanism has been designated the ping-pong mechanism. The nomenclature above was introduced by Cleland (8) and is commonly used to describe the various mechanisms. A further means of defining the mechanism involves the stipulation of the number of important substrates as they appear on each side of the equation. For example, a random reaction with a single reactant and two products would be called a random uni bi reaction.

As the mechanisms of these reactions become more complex, a great deal of effort is required to obtain the final rate expression. A simple and reliable method was introduced by King and Altman (9). The method will not be described here, but the interested reader is referred to the original reference. All of the various rate equations can be derived via the King and Altman method and the rate equations for two-substrate enzyme systems are summarized in Table 2.1. As can be seen from the table, with the exception of the ping-pong mechanism, the kinetic equation can be written in the form:

$$v = \frac{V[A][B]}{K_A K_B + K_B[A] + K_A[B] + [A][B]} \quad (2.14)$$

2.1.3   Integrated Michaelis–Menten Equation

The relationship between the change in the substrate concentration and time is an important aspect of enzymology, particularly with immobilized enzymes. Up to this point, we have used equations dealing only with initial rates because of the difficulties involved in accurately detailing the course of the reaction in the presence of product inhibition or reversible reactions. On the other hand, in order to develop equilibrium methodologies using enzyme reactions, one must obtain a progress curve for the reaction, bearing in mind that is is a true representation only when the decrease in the reaction rate is due to a decrease in the substrate concentration.

Using the single intermdiate model, integration yields the equation

$$([S]_0 - [S]) + K_M \ln \left(\frac{[S]_0}{[S]} \right) = Vt \quad (2.15)$$

where $[S]_0$ is the initial substrate concentration and $[S]$ is the concentration

Table 2.1 Summary of Bimolecular Rate Equations

Name	Rate Equation
Random bi bi	$$v = \dfrac{k_{+3}[E]_0[A][B]}{K_A K_B + K_b[A] + K_A[B] + [A][B]}$$ $$\left(\dfrac{k_{+3}k_{+4}}{k_{+3}+k_{+4}}\right)[E]_0[A][B]$$
Ordered bi bi	$$v = \dfrac{k_{+1}k_{+2}k_{+3}k_{+4}[E]_0[A][B]}{\dfrac{k_{+4}(k_{-1}k_{-2}+k_{-1}k_{+3})}{k_{+1}k_{+2}(k_{+3}+k_{+4})} + \dfrac{k_{+4}(k_{-2}+k_{+3})}{k_{+2}(k_{-3}+k_{+4})}[A] + \dfrac{k_{+3}k_{+4}}{k_{+1}(k_{+3}+k_{+4})}[B] + [A][B]}$$
Ping pong	$$v = \dfrac{k_{+1}k_{+2}k_{+3}k_{+4}[E]_0[A][B]}{k_{+1}k_{+2}k_{+4}(k_{-1}+k_{+2})[B] + k_{+1}k_{+3}k_{+4}[A][B] + k_{+1}k_{+2}k_{+3}k_{+4}[A][B]}$$

32

at time t. It can be seen that under conditions of high substrate concentration, zero-order kinetics are followed and $[S]$ is proportional to Vt. When $[S]_0$ is smaller than K_M, the first-order relation is followed and $[S]$ is proportional to $(V/K_M)t$. Thus, the Michaelis constant and the maximal velocity can be determined from the measurement of the substrate concentration at various times during the reaction as well as from the initial rate data. For a more detailed discussion of the integrated equations, the reader is referred to Chapter 3.

2.2 NON-MICHAELIS–MENTEN KINETICS

2.2.1 Transient-Phase Kinetics

Under some conditions, the equation presented by Michaelis and Menten does not apply. The initial stages of an enzyme reaction, before the steady-state condition is attained, have also influenced our knowledge of the kinetics of these reactions. With the availability of both stop-flow and perturbation techniques such as T-jump, the rate constants for the early part of the reaction are measurable. Although the simple Michaelis–Menten single-intermediate model is not universally applicable, consideration of it will provide relevant general principles.

The mechanism for the reaction is, again,

$$E + S \underset{k_{-1}}{\overset{k_{+1}}{\rightleftharpoons}} ES \xrightarrow{k_{+2}} E + P \tag{2.16}$$

A rate equation for the change in the concentration of the complex and product with time can be written as:

$$\frac{d[ES]}{dt} = k_{+1}([E]_0 - [ES])([S]_0 - [ES] - [P]) - (k_{-1} + k_{+2})[ES]$$

$$\frac{d[P]}{dt} = k_{+2}[ES] \tag{2.17}$$

and

$$\frac{d^2[P]}{dt^2} = k_{+1}k_{+2}[E]_0([S]_0 - [ES] - [P])$$

$$- (k_{-1} + k_{+2} + k_{+1}([S]_0 - [ES] - [P])) \tag{2.18}$$

If we assume that the concentration of substrate is not changed from the initial concentration, that is, the reaction has not proceeded very far,

$$\frac{d^2[P]}{dt^2} + (k_{-1} + k_{+2} + k_{+1}[S]_0)\frac{d[P]}{dt} - k_{+1} + k_{+2}[E]_0[S]_0 = 0 \tag{2.19}$$

Solving for [P],

$$[P] = \frac{k_{+2}[E]_0[S]_0}{(k_{-1} + k_{+2})/k_{+1} + [S]_0} t$$

$$- \frac{k_{+1}k_{+2}[E]_0[S]_0}{(k_{-1} + k_{+2} + k_{+1}[S]_0)^2} [\exp(-(k_{-1} + k_{+2} + k_{+1}[S]_0)t - 1)] \quad (2.20)$$

Writing the exponential term in the form of a series and allowing t to become very small, the initial rate is given by

$$[P] = \tfrac{1}{2} k_{+1}k_{+2}[E]_0[S]_0 t^2 \quad (2.21)$$

These relations were derived by Roughton (10) and illustrate the utility of transient-state determinations. At very early times, the value of $k_{+1}k_{+2}$ may be determined. From steady-state measurements at high substrate concentrations, the value of k_{+2} can be determined and thus the value of k_{+1} calculated. From the Michaelis constant, k_{-1} can be calculated and thus all of the rate constants determined.

Another feature of this model is the induction period observed in the early stages of the reaction. From equation (2.20) it can be seen that at small values of t, an exponential component influences the buildup of product. The extrapolation of the linear portion of the curve yields:

$$\tau = \frac{1}{k_{-1} + k_{+2} + k_{+1}[S]_0} \quad (2.22)$$

Determinations of the induction period, τ, at various substrate concentrations will allow confirmatory measurement of the rate constants to be made.

2.2.2   Reversible Reactions

We have tacitly assumed in all previous equations that the decomposition of the enzyme–substrate complex was an irreversible process. A rate equation can be derived based on the model

$$E + S \underset{k_{-1}}{\overset{k_{+1}}{\rightleftharpoons}} ES \underset{k_{-2}}{\overset{k_{+2}}{\rightleftharpoons}} EP \underset{k_{-3}}{\overset{k_{+3}}{\rightleftharpoons}} E + P \quad (2.23)$$

Applying the steady-state assumption to both of the intermediate complexes,

$$\frac{d[ES]}{dt} = k_{+1}[E][S] + k_{-2}[EP] - (k_{-1} + k_{+2})[ES] = 0 \quad (2.24)$$

$$\frac{d[EP]}{dt} = k_{+2}[ES] + k_{-3}[E][P] - (k_{+3} + k_{-2})[EP] = 0$$

If we again substitute for the total enzyme concentration and assume that the rate-limiting step is the decomposition of ES to EP, then

$$v = k_{+2}[ES] - k_{-2}[EP] \tag{2.26}$$

$$v = \frac{(k_{+1}k_{+2}k_{+3}[S] - k_{-1}k_{-2}k_{-3}[P])[E]_0}{k_{-1}k_{-2} + k_{-1}k_{+3} + k_{+2}k_{+3} + k_{+1}(k_{+2} + k_{-2} + k_{+3})[S] + k_{-3}(k_{-1} + k_{+2} + k_{-2})[P]} \tag{2.27}$$

This equation can be written in a more familiar form

$$v = \frac{V_S[S] - V_P[P]}{K_S K_P + K_P[S] + K_S[P]} \tag{2.28}$$

The evaluation of the constants is left to the reader. At the initial rate with no product present (or vice versa), the reaction is rectangular hyperbolic but the values of the constants are different from those of the single-intermediate model. If k_{+3} is very large, the equation reduces to that of the two-intermediate case. It should be pointed out that a similar equation can be derived for the single-intermediate case. The most important point about this equation is that the effects of not obtaining initial-rate data before the product builds up can be quantitatively determined if the constants are known.

2.2.3 Allosteric Effects

A significant number of enzymes have been found that display a sigmoidal relationship between the rate of reaction and the substrate concentration. One explanation of this type of behavior is allostery, that is, there are effectors which bind to a site separate from the active site and influence the reaction rate. The first treatment of such a system was presented by Monod, Wyman, and Changeux (11) in which the enzyme was envisioned as an interacting subunit dimer. In their model, each subunit had an active site and could exist in one of two conformations, R or T. The basic postulate of the model is that the transition of the dimer is a concerted one, that is, the molecule must exist as either RR or TT, so that symmetry is maintained. An equilibrium constant can then be defined,

$$K_L = \frac{[TT]}{[RR]} = \frac{[E_T]}{[E_R]} \tag{2.29}$$

where $[E_T]$ and $[E_R]$ are the concentrations of free enzyme in the T or R dimer form. We can also write equilibrium constants for the combination of substrate with the two enzyme conformations

$$K_T = \frac{[E_T][S]}{[E_T \cdot S]} \tag{2.30}$$

$$K_R = \frac{[E_R][S]}{[E_R \cdot S]} = \frac{[E_R \cdot S][S]}{[E_R \cdot S_2]} \tag{2.31}$$

If $K_R \gg K_T$, the rate equation can be derived such that

$$v = \frac{([S]/K_R)(1 + [S]/K_R)}{K_L + (1 + [S]/K_R)^2} V_{max} \tag{2.32}$$

The variation of v with [S] is sigmoidal, indicating the cooperativity of the substrate binding.

In order to explain true allosteric effects, the binding site for the effector must be separate from the substrate-binding site, which is not a prerequisite of the above derivation. It is, however, rather straightforward to extrapolate to a true allosteric situation. In this case, the addition of an allosteric inhibitor would favor the formation of the T dimer and thus increase K_L in equation (2.32). The opposite relation would hold in the presence of an allosteric activator.

Three important facts are of interest at this point. First, the substrate cooperativity suggested by Monod's model has been observed and is known as a "homotropic effect" as opposed to the "heterotropic effect" noted as allosteric behavior. Another popular model involving sequential transformation of the subunits has been used to describe the latter effect (12). Second, it should be apparent that although allostery is an explanation for sigmoidal kinetics under the appropriate conditions, sigmoidal kinetics are not a necessary and sufficient condition for the presence of allosteric effects. Other reasons which may lead to a sigmoidal relation between the rate of the enzyme reaction and the substrate concentration include the presence of a substrate-binding impurity in the enzyme preparation (13), the existence of two forms of the enzyme which have different activities (14), the binding of two molecules of substrate (15), the action of the substrate as an inhibitor at a separate site (16), and alternate pathways for the mechanism of the reaction which proceed at different rates (17).

2.3 EFFECTS OF ACTIVATION AND INHIBITION

It has been observed that substances which are not involved in the catalyzed reaction influence the rate of enzyme-catalyzed reactions. When the rate is decreased, there is said to be inhibition and the substance causing the effect is called an inhibitor. Conversely, an effector which causes an increase in the rate is known as an activator. In general, activators and inhibitors are referred to as reaction modulators. Analytically, a calibration curve is used to measure the amount of inhibitor by plotting the degree of inhibition as a function of the inhibitor concentration. The degree of inhibition is given by the relation

$$i = \frac{v_0 - v}{v_0} \tag{2.33}$$

where v_0 is the rate of the reaction under the same conditions with no inhibitor present and v is the rate when a modulator is present. Activation can be measured in an analogous manner using the relation

$$a = \frac{v - v_0}{v_0} \tag{2.34}$$

2.3.1 Inhibition

In dealing with inhibition, there are a number of important distinctions which must be recognized. First, two types of inhibition are observed— reversible and irreversible. Reversible inhibition is defined as inhibition which can be removed rapidly through dialysis, that is, the inhibitor is at equilibrium with the enzyme. Irreversible inhibition causes an irreversible change in the enzyme which eliminates the catalytic capacity of the enzyme. Irreversible inhibition is progressive, eventually removing the catalysis completely if the inhibitor concentration is greater than the enzyme concentration. The effectiveness of irreversible inhibition is expressed in terms of velocity instead of an equilibrium constant. The best known irreversible inhibition is the effect of nerve gas on cholinesterase.
In reversible inhibition, three situations can be distinguished:

1. The degree of inhibition is not a function of the substrate concentration in which case the inhibition is known as noncompetitive;
2. The degree of inhibition is a function of substrate concentration, where the degree of inhibition is reduced as substrate concentration increases. The case is known as competitive inhibition.
3. The degree of inhibition is a function of substrate concentration where inhibition is increased as substrate concentration is increased in which case the inhibition is known as uncompetitive.

It is important to distinguish between empirical observations and the mechanisms which are proposed to explain them. Competitive inhibition, for example, was thought to be the result of inhibitor and substrate molecules competing for the active site of the enzyme. This approach is valid only for the simple single-intermediate model presented earlier. Although mechanistically incorrect, the classical mechanisms do yield empirically useful results. The model envisaged for competitive inhibition involves the binding of either an inhibitor or substrate molecule at the active site and can be presented as

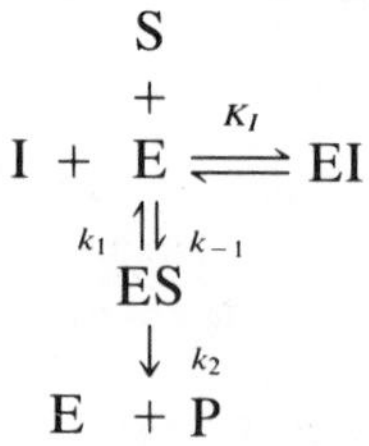

$$
\begin{array}{c}
\text{S} \\
+ \quad {\scriptstyle K_I} \\
\text{I} + \text{E} \rightleftharpoons \text{EI} \\
{\scriptstyle k_1} \updownarrow {\scriptstyle k_{-1}} \\
\text{ES} \\
\downarrow {\scriptstyle k_2} \\
\text{E} + \text{P}
\end{array}
$$

where K_I is the equilibrium constant for the dissociation of EI. Using the steady-state approximation, the relation

$$
v = \frac{k_{+2}[\text{E}][\text{S}]}{(K_M + [\text{S}])(1 + [\text{I}]/K_I)} \tag{2.35}
$$

can be derived for pure noncompetitive inhibition where [I] is the concentration of inhibitor and K_I is the inhibition constant obtained in a manner analogous to the Michaelis constant for substrates. Since the degree of inhibition is given by equation (2.33), the degree of inhibition can be obtained as a function of the relative inhibitor concentration:

$$
i = 1 - \frac{v}{v_0} \tag{2.36}
$$

$$
i = \frac{\gamma}{1 + \gamma} \tag{2.37}
$$

where γ is the relative inhibitor concentration given by $[\text{I}]/K_I$. As can be seen, the degree of inhibition is independent of the substrate concentration and follows the general shape of a rectangular hyperbola.

Competitive inhibition can be represented by equation (2.38):

$$
v = \frac{k_2[\text{E}][\text{S}]}{K_M(1 + [\text{I}]/K_I) + [\text{S}]} \tag{2.38}
$$

Solving again for the degree of inhibition,

$$
i = \frac{\gamma}{(\gamma + 1) + \beta} \tag{2.39}
$$

where $\beta = [S]/K_M$. As can be seen, the relation between the degree of inhibition and the inhibitor concentration is linear only at low substrate and inhibitor concentrations. It has been reported that irreversible inhibition can be observed as a pseudo-competitive inhibition as a result of the protective interaction of the substrate with the active site. The major difference is that the irreversible case eventually eliminates all of the enzyme activity since the protection of the active site is an equilibrium process.

The third type of behavior given above, uncompetitive, can be approximated by the relation

$$v = \frac{k_2[E][S]}{K_M + (1 + [I]/K_I)[S]} \tag{2.40}$$

and the degree of inhibition will be given by

$$i = \frac{\beta\gamma}{1 + (1 + \gamma)\beta} \tag{2.41}$$

In this case, low substrate concentrations will yield a linear relation between i and the inhibitor concentration, as will the combination of high substrate concentration and a low inhibitor concentration. Under the appropriate conditions, all of the types of inhibition yield a rectangular hyperbolic relationship between the degree of inhibition and the inhibitor concentration.

It should be noted that the degree of inhibition determinations require the measurement of an uninhibited-reaction rate, the proper conditions for which one must know the type of inhibition being exhibited. Fortuitously, the type of inhibition and the inhibition constant, K_I, can be simultaneously determined for simple systems. This is particularly helpful since the same enzyme can present different types of inhibition with the same inhibitor and different substrates as illustrated by Krupka and Laidler for acetylcholinesterase (18). If the initial reaction rate is plotted as a function of substrate concentration by the method of Lineweaver and Burke, a straight line for each inhibitor concentration is obtained. The intersection of the lines with the ordinate, abscissa, and each other give information about the type of inhibition and the constants involved, as shown in Table 2.2 and Figure 2.3.

Up to this point, we have considered only pure systems. It can be shown that if k_2 makes a contribution to K_M, some type of mixed inhibition must occur. The general equation which applies is

$$v = \frac{k_2[E][S]}{K_M(1 + [I]/K_I) + (1 + [I]/rK_I)[S]} \tag{2.42}$$

Table 2.2 **Inhibition Information Available From Lineweaver–Burke Plots**

Type of Inhibition	Abscissa Intercept	Ordinate Intercept	Intersection of Inhibition Lines
Competitive	$-\dfrac{1}{K_M(1 + [I]/K_I)}$	$\dfrac{1}{V}$	Ordinate
Noncompetitive	$-\dfrac{1}{K_M}$	$\dfrac{(1 + [I]/K_I)}{V}$	Abscissa
Uncompetitive	$-\dfrac{1 + [I]/K_I}{K_M}$	$\dfrac{(1 + [I]/K_I)}{V}$	No intersection; lines parallel
Mixed	$\dfrac{(1 + [I]/rK_I)}{K_M(1 + [I]/K_I)}$	$\dfrac{(1 + [I]/rK_I)}{V}$	Quadrant II

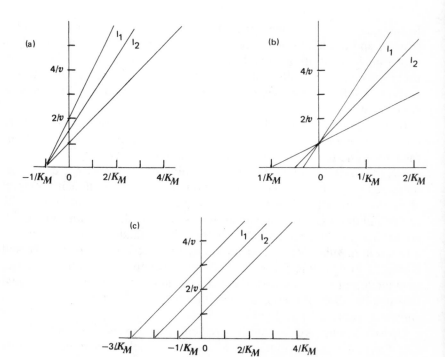

Fig. 2.3. Lineweaver–Burke plots for the three common types of inhibition: (a) noncompetitive, (b) competitive, and (c) uncompetitive. For all graphs, $[I_1] > [I_2]$ The unlabeled curve represents the uninhibited case.

where r is the ratio of the inhibition constants obtained from the slope and intercept of a Lineweaver–Burke plot. The degree of inhibition can be shown to be

$$v = \frac{\gamma(1 + \beta/r)}{(1 + \gamma) + (1 + \gamma/r)\beta} \tag{2.43}$$

It can be seen that under the conditions where r approaches 1, 0, and ∞, the inhibition approaches noncompetitive, uncompetitive, and competitive, respectively. Kaplan and Laidler (19) have presented a general treatment of inhibition involving the individual rate constants. Unfortunately, it is rather complex and not of general utility.

The inhibition of two-substrate reactions has been used, as has the inhibition of simple reactions, to differentiate between the various mechanisms discussed earlier. The equations observed for these mechanisms are given in Table 2.3. It should be noted that under the appropriate conditions, multi-substrate systems can approximate the simple relations derived above.

One further important consideration is the action of high-affinity inhibitors. All of the above equations were derived assuming that the total concentration of inhibitor was approximately equal to the free inhibitor concentration. In the case where the inhibitor concentration is about the same as the enzyme concentration, for example, 10^{-8} mole/l, the above equations do not apply. This would be the case for many trace pesticide determinations using enzyme-inhibition methods. It can be shown that the degree of inhibition for such a noncompetitive system is given by the

Table 2.3   Rate Equations for the Inhibition of Two-Substrate Reactions

Name	Rate equation
Random ternary	$v = \dfrac{k_{+3}[A][B]\dfrac{[E]_0}{K_A'K_B}}{K_A'K_B\left(1 + \dfrac{[I]}{K_I}\right) + K_B[A]\left(1 + \dfrac{[I]}{K_I^A}\right) + K_A[B]\left(1 + \dfrac{[I]}{K_I^B}\right) + [A][B]}$
Ordered ternary	$v = \dfrac{k_{+3}[E]_0[A][B]}{(K_A'K_B + K_A[B])\left(1 + \dfrac{[I]}{K_I}\right) + K_B[A]\left(1 + \dfrac{[I]}{K_I^A}\right) + [A][B]}$
Ping pong	$v = \dfrac{k_{+1}k_{+2}k_{+3}k_{+4}[E]_0[A][B]}{K_B[A]\left(1 + \dfrac{[I]}{K_Q^A}\right) + K_A[B]\left(1 + \dfrac{[I]}{K_I}\right) + \dfrac{[A][B]}{k_{+2} + k_{+4}}\left\{k_{+4}\left(1 + \dfrac{[I]}{K_Q^A}\right) + k_{+2}\right\}}$

relation

$$i = \frac{\alpha i_t}{K_I + \alpha[E]} \qquad (2.44)$$

where α is the fraction of enzyme not inhibited. It can be seen that if the amount of enzyme is small compared to K_I, the relation will be linear. On the other hand, high enzyme concentrations will yield very nonlinear results even at low inhibitor concentrations.

2.3.2 Activation

In the general sense, the same type of rate expressions derived for inhibition should apply to substances which enhance the reaction rate. Specifically, activators are cofactors which are not themselves altered in the enzyme reaction. Most of the work that has been done in this area involves simple electrolytes. The work almost exclusively involves the investigation of enzymes which are completely inactive in the absence of the activator. Thus, data in this area is rather limited and the reactions are quite complex, some even exhibiting both activation and inhibition at different levels of the same modulator.

Of the electrolyte effects studied, anion activation has been investigated under the most varied conditions. Anions have a fairly nonspecific effect, the best known example being the activation of α-amylase by chloride ions. The effect of the anion has been postulated by Alberty to involve the ionization equilibrium of the enzyme, and thus is a function of pH, and the rate of breakdown of the enzyme–substrate complex (20). The equations derived for the maximum rate and substrate affinity are of the form

$$V = \frac{k_2[E] + k_2'[E]\delta r}{f_{ES} + \delta[E]f_{EAS}} \qquad (2.45)$$

and

$$K_S' = K_S \left(\frac{f_E + \delta f_{EA}}{f_{ES} + \delta \, r f_{EAs}} \right) \qquad (2.46)$$

where δ is the relative concentration of the anion, that is, $[A]/K_A$, r is the ratio of the formation constants for the formation of the enzyme–substrate complex in the absence and presence of the anion, k_2' is the rate of complex breakdown with anion present, K_S is the equilibrium constant, and f represents the Michaelis pH function for the subscript species (see below). The overall effect is to increase the observed activity and/or increase or decrease the breadth of the pH–activity curve. It should be noted that the buffer anions may have such an effect, thereby making a specific buffer "optimal."

Of much greater interest has been the activation of enzymes by metal cations. All of the metal activators have been found to have an atomic number between 11 and 55, with the large majority between 19 and 30. This is probably a result of atomic size. It is of interest to note that chemical similarity is not sufficient cause for activation since, for example, pyruvate kinase is activated by K^+ and inhibited by Na^+. Another consideration is that the metal activation can be affected by the substrate concentration, pH, and even the purity of the enzyme. For example, partially purified preparations of arginase are activated by cobalt, manganese, and nickel while the purified enzyme is only activated by manganese.

A general equation for metal-ion activation can be derived from the following equilibria:

$$E + M \rightleftharpoons EM \qquad K_m \qquad\qquad (2.47)$$

$$E + MS \rightleftharpoons EMS \qquad K_{mS} \qquad\qquad (2.48)$$

$$M + S \rightleftharpoons MS \qquad K_0 \qquad\qquad (2.49)$$

$$EM + S \rightleftharpoons EMS \qquad K_{S'} \qquad\qquad (2.50)$$

$$ES + M \rightleftharpoons EMS \qquad K'_m \qquad\qquad (2.51)$$

$$E + S \rightleftharpoons ES \qquad K_S \qquad\qquad (2.52)$$

From microscopic reversibility,

$$K_m K_S' = K_S K'_m = K_0 K_{mS}$$

$$K_m K_{S'} = K_S K_{m'} = K_0 K_{mS'}$$

The velocity of the reaction can then be written

$$v = \frac{k_2[E]_0}{1 + (K_0 K_{mS}/[M][S]) \left(1 + \dfrac{[M]}{K_m} + \dfrac{[S]}{K_S}\right)} \qquad\qquad (2.53)$$

where [M] is the concentration of metal ion. Various special cases can be derived from the substitution of the appropriate conditions. One such case is the apparent rectangular hyperbolic function, that is $v = (1/(1 + K_m/[M])$, observed with many cations. An example is shown in Figure

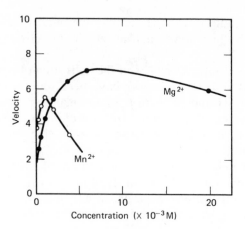

Fig. 2.4. The rectangular hyperbolic relationship between the rate of the enzyme reaction and the concentration of metal ion for isocitrate dehydrogenase. [Reproduced with permission from M. Dixon and D. C. Webb, *The Enzymes* (Academic Press, New York, 1964). Copyright by Academic Press, Inc.]

2.4 from the activation at constant substrate concentration of isocitrate dehydrogenase by Mg^{2+} and Mn^{2+}.

The assumption in the above relationships is again that the free metal-ion concentration is equal to the total metal-ion concentration. Obviously, for trace-metal determinations, this assumption does not hold. An excellent example is the trace-metal determination of Mg^{2+} and Mn^{2+} using isocitrate dehydrogenase (21). The mechanism of the enzyme reaction requires the formation of a metal–NAD complex, which is the true substrate. Using equations (2.48), (2.49), and (2.53) to represent the mechanism, and expressing the total concentrations of the metal ion and substrate by the relations

$$[M] = [M]_{free} + [MS] \qquad (2.54)$$

$$[S] = [S]_{free} + [MS] \qquad (2.55)$$

one can derive the equation

$$v = \frac{k_2[E]_0}{1 + K_{ms}\{(K_0 + [M] + [S]) - \sqrt{(K_0 + [M] + [S])^2 - 4[M][S]}\}^{-1}} \qquad (2.56)$$

It can be readily seen that the relation between the velocity and the metal-ion concentration does not follow the rectangular hyperbolic function. Thus, one would not expect to observe a linear relation between the metal-ion concentration and the rate of reaction. Indeed, this has been found to be the case for isocitrate dehydrogenase (22). Other reaction mechanisms, do, of course, occur and appropriate rate equations can be derived (23).

It should be noted that other factors may affect the functionality of the rate–metal-ion relation. Obviously, a change in the mechanism would cause a nonlinear response. We have assumed a simple single-intermediate model for these derivations, which is not always applicable. The point of this discussion is that for trace-metal determinations involving enzymes, there is no a priori reason to expect a linear calibration curve.

2.3.3 Substrate Inhibition

One further case of activation and inhibition requires a brief discussion. In some single-substrate reactions, the rate passes through a maximum instead of exhibiting hyperbolic behavior. One example of this type of behavior is the enzyme urease which catalyzes the hydrolysis of urea (24). The simplest mechanism for this phenomenon was proposed by Haldane (25) and by Murray (26). It involves the addition of a second substrate molecule to the enzyme–substrate complex, that is,

$$E + S \underset{k_{-1}}{\overset{k_{+1}}{\rightleftharpoons}} ES \xrightarrow{k_{+2}} E + P$$

$$+$$

$$S$$

$$k_{+3} \Big\updownarrow k_{-3}$$

$$ES_2 \xrightarrow{k_{+4}} ES + P$$

Thus at high substrate concentration, the rate-limiting step is the decomposition of the ES_2 complex. Steady-state analysis of the mechanism leads to the relation

$$v = k_{+2}[ES] + k_{+4}[ES_2]$$

$$= \left(k_{+2} + \frac{k_{+3}[S]}{(k_{-3} + k_{+4})} \right)[ES] \tag{2.57}$$

where [ES] is the steady-state concentration of the ES complex. The above equation can be reduced to the form

$$v = \frac{(k_{+2} + k_{+4}K_{S'}S)[E]_0[S]}{K_{M'} + [S] + K_3[S]^2} \tag{2.58}$$

where $K_{S'} = (k_{+3}/k_{-3} + k_4)$, $K_{M'}$ is a modified Michaelis constant, and K_3 is the equilibrium constant for the formation of ES_2 from ES and S.

Three special cases of equation (2.58) can be observed in the behavior of enzyme systems at high substrate concentration. When the rate of breakdown of the ES_2 is slow relative to that of ES, a curve with a maxima is observed. If the rates are roughly equal, classical rectangular hyperbolic functionality will be observed. When the rate of the ES_2 complex breakdown is high, the activity observed increases beyond the point where the assymptote of the hyperbolic function would be expected. In this case, the substrate is acting as an activator. (27)

2.4 EFFECT OF TEMPERATURE

The fact that the rate of an enzyme reaction passes through a maximum as the temperature is increased has been known for many years. The temperature optimum is a function of a number of parameters such as pH and enzyme concentration. It is not, however, a physicochemical property of the enzyme. It is generally accepted that the form of the activity–temperature curve is the result of two distinct processes: enzyme inactivation and the reaction rate. (See Figure 2.5)

Even if the effects of enzyme inactivation (denaturation) are eliminated, temperature moderation of the reaction may be complex. The observed

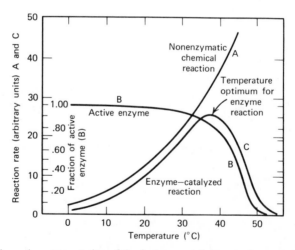

Fig. 2.5. Schematic representation of the factors contributing to the temperature dependence of the rate of an enzyme-catalyzed reaction. [Reproduced with permission from N. Teitz, *Fundamentals of Clinical Chemistry*, 2nd ed. (W. B. Saunders Co., Philadelphia, 1977). Copyright by W. B. Saunders Company.]

change may be due to a change in the velocity of the breakdown of the enzyme–substrate complex, a change in the affinity of the enzyme for the substrate whether due to a change in k_{+1} and k_{-1} or a change in the ionization of the substrate, a change in the affinity for activators or inhibitors, a change in the solubility or stability of the substrate, or a change in the pH of the buffer due to its temperature coefficient. Many of the above deviations are due to changes in the activation enthalpy of the process as a function of temperature.

The effect of temperature on the rate of a reaction is given by the Arrhenius relation

$$k = Ae^{-E_a/RT} \qquad (2.59)$$

where the terms have their customary meaning. Strictly speaking, the rate of each of the reactions in the simple single-intermediate mechanism should follow the Arrhenius relationship. It has been found empirically that in many cases the overall enzyme reaction obeys this law. A commonly used term in enzymology is the temperature coefficient, Q_{10}, or the ratio of rates of the reaction observed over a 10 degree temperature interval, that is,

$$\ln Q_{10} = \ln \left(\frac{k_{T_2}}{k_{T_1}} \right) = -\left(\frac{E_a}{R} \right) \left(\frac{1}{T_2} - \frac{1}{T_1} \right) \qquad (2.60)$$

Since in most cases the activation energy for an enzyme reaction is about 10 kcal/mole, Q_{10} is generally found to be about 2.0 with a range of between 1.7 and 2.5.

2.5   EFFECT OF pH

The rate of an enzyme reaction, in most cases, passes through a maximum as a function of pH as shown in Figure 2.6. The pH corresponding to the maximum rate is known as the optimal pH, but again, this optimum is a function of other parameters such as temperature and the substrate concentration. At the extremes of pH, the enzyme may undergo an irreversible denaturation where the activity may not be restored even after readjustment of the pH to the optimal value. Of greater interest is the reversible behavior exhibited in the general area of the pH optimum.

The initial explanation for the shape of the pH curve was proposed by Michaelis and Davidsohn (28). It was based on the amphoteric nature of the amino acids which make up the enzyme, specifically those affecting the binding of the substrate, and the effect of protonization on the rates of formation of the enzyme–substrate complex. Later, Michaelis and

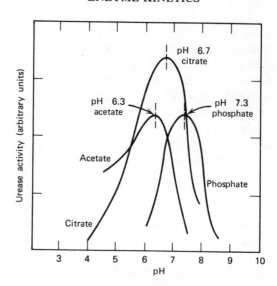

Fig. 2.6. pH activity curves for the enzyme urease illustrating both the optimum and the effect of buffer on the pH optimum. [Reproduced with permission from N. Teitz, *Fundamentals of Clinical Chemistry*, 2nd ed. (W. B. Saunders Co., Philadelphia, 1977). Copyright by W. B. Saunders Company.]

Rothstein (29) extended the concept to include the decomposition of the active complex. Finally, a general equation encompassing both of these concepts was proposed (30), based on the mechanism

$$EH_2 \underset{H^+}{\overset{K_b}{\rightleftharpoons}} EH^- \underset{H^+}{\overset{K_a}{\rightleftharpoons}} E^{2-}$$
$$+$$
$$S$$
$$\updownarrow$$
$$EH_2S \underset{H^+}{\overset{K_{b'}}{\rightleftharpoons}} EHS \underset{k_{+2}}{\overset{K_{a'}}{\rightleftharpoons}} ES$$
$$\downarrow$$
$$EH + P$$

The rate equation for the reaction then becomes

$$v = k_2[E]_0[S] \left/ \left\{ K_M \left(1 + \frac{K_a}{[H^+]} + \frac{[H^+]}{K_b} \right) \right.\right.$$
$$\left. + \left(1 + \frac{K_{a'}}{[H^+]} + \frac{[H^+]}{K_{b'}} \right)[S] \right\} \quad (2.61)$$

Several interesting conclusions can be obtained from this equation. First, the equation does give rise to a curve with a pH optimum. Second, depending on the ratio of the equilibrium constants of importance, that is, K_b and $K_{b'}$ at acidic pH's, the shape of the pH curve may be a function of the substrate concentration. Laidler has introduced a more general approach to this problem allowing the reaction of both the deprotonated and protonated forms of the enzyme to determine the appropriateness of the above model. Under the conditions where $k_{+2} \ll k_{-1}$ or $K_{a'} = 0$ and $K_{b'} = \infty$, or the reactions in the above mechanism are faster than those for the reaction of substrate with the other forms of protonization, the above equation was reasonable (31).

One of the important outgrowths of the above theory was the exploration of the active site. A definitive discussion of this topic is beyond the scope of this monograph and the interested reader is referred to the literature (32–35). There is, however, one parameter in the area with which the reader should be familiar, the so-called Michaelis pH function. Referring to the mechanism on the previous page, the concentration of the species EH_2, EH^-, and E^{2-} can be calculated from the equilibrium constants. Since the total enzyme concentration is given by the relation

$$[E]_0 = [E^{2-}] + [EH^-] + [EH_2]$$

then

$$[E]_0 = [E^{2-}] + \frac{[E^{2-}][H^+]}{K_a} + \frac{[EH^-][H^+]}{K_b}$$

$$= E^{2-}(1 + [H^+]/K_a + [H^+]^2/K_aK_b)$$

$$= f_E[E^{2-}] = f_{EH}[EH^-] = f_{EH_2}[EH_2]$$

where f_{EH_2}, f_{EH}, and f_E are the Michaelis pH functions. The Michaelis pH function is inversely related to the fraction of its species which is present. It is interesting that the K_a and K_b represent not the group ionization constants, but rather the stage of ionization of the protein. Thus, the ionization of groups involved in the active site can be studied.

2.6 EFFECT OF ENZYME CONCENTRATION

In most cases, any two molecules of enzyme function independently and thus the velocity of the reaction is directly proportional to the enzyme concentration. There are several conditions under which this proportionality may not occur. First, if large amounts of enzyme are used, any coenzyme that is added may be bound to the enzyme and not be available for reaction. Second, if there is an inhibitor, either reversible or irreversible, in one of the assay solutions, any of a variety of relations may be observed. Thus it is imperative in any assay system to be sure that there are no inhibitors present and that the reaction is linear with respect to enzyme concentration.

A special effect is noted with proteases and is sometimes referred to as "Schutz law" (36). In this case, the initial rate is found to be proportional to the square root of the enzyme concentration, that is,

$$v = k[E]^{1/2}$$

Of the various explanations proposed, the most reasonable involves the rate of diffusion of the macromolecular substrate (37). Thus, the enzyme reaction is not the rate-limiting step. This proposal is supported by the fact that the effect is observed only with large substrates such as the reaction of lipase with micellular triglycerides (38) and pepsin with casein (39), that it occurs at relatively high enzyme concentration, and that the use of smaller substrates such as benzoyl ethyl ester give linear plots with the same enzyme. In any case, the great majority of solution-phase enzyme reactions show a linear relation. This phenomenon has not been studied in any detail for immobilized enzymes, but the fact that mass transfer can limit the enzyme catalysis would be consistent with a nonlinear relationship.

A final comment should be made with regard to all of the above derivations. All of these equations assume soluble substrate species. A notable exception is lipase. It is now generally accepted that lipase acts only on micellular collections of triglycerides. With increasing interest in triglyceride analysis, this is an important point. Some publications have recently appeared which discuss "soluble substrate activity" which is probably due to impurity in the enzyme preparation. In any case, a detailed discussion of this topic is beyond the scope of this monograph and the interested reader is referred to a series of papers by Desnuelle and his coworkers (40–42).

2.7   EFFECT OF BUFFER

It should be appreciated that the nature of the buffer system may influence the enzyme reaction either thermodynamically or kinetically. One example of the effect of buffer ions was discussed earlier in the discussion of anionic activation. It can be seen from Figure 2.6 that the enzyme–pH curve does depend on the buffer species. In addition, the nature of the buffer may play an important role in enzyme stabilization. Phosphate or Tris buffers of the same pH do not have the same surface tension, charge, or many other properties which may affect the conformation of the enzyme and thus its activity and stability. It is commonly observed that carboxylic and sulfonic acid buffers stabilize enzymes. Conversely, in our hands Tris buffer has caused a significant number of problems. A list of common enzyme buffer systems is given in Table 2.4. Ionic strength may also have a strong effect on enzyme stability and activity. Addition of sorbitol, polyvinyl alcohol, and other hydrophilic polymers has also been shown to influence enzyme activity. A commercial firm has begun marketing liquid enzyme preparations stabilized in a polymer matrix which are stable for 1 year. Polyelectrolytes such as polylysine have been shown to prevent subunit dissociation in lactate dehydrogenase and β-amylase and thus increase their stability and activity (43). Tripolyphosphate has significant effects on the activity of desimidases (44). It should be apparent that the choice of the overall enzyme matrix including the buffer species is a rather arbitrary but very important parameter in enzyme kinetics. Optimization of an enzyme-based assay may require careful evaluation of the buffer system.

Perhaps even more important than the effects of the buffer on the activity and stability of the enzyme is their influence on the thermodynamics of the catalyzed reaction. The hydrolysis of urea by urease has been studied in detail by Jespersen (45). The usual reaction scheme is represented as

$$H_2N-\overset{\overset{\textstyle O}{\|}}{C}-NH_2 + H_2O + H^+ \xrightarrow{\text{Urease}} HCO_3^- + 2NH_4^+$$

It was found that in either phosphate or maleate buffer solution, the classical reaction scheme was followed. In citrate or Tris buffer systems, however, an almost quantitative yield of ammonium carbonate was observed. By changing the relative mole fractions of the two types of buffers in a mixed buffer solution, Jespersen was able to change the amount of ammonium ion produced. This data would indicate that the true product of the urease reaction is ammonium carbamate, which is decomposed by a subsequent buffer-mediated proton transfer. A thermodynamic effect such as this may have a substantial influence on analyses—particularly for those with thermochemical detection systems.

Another example of buffer system effects can be obtained from study of the alkaline phosphatase reaction. Alkaline phosphatase is in actuality a group of nonspecific organic phosphate hydrolases which have high

Table 2.4 Common Enzyme Buffer Systems[a]

Buffer Common Name	IUPAC Name	pK_a
Tartrate	—	4.05 (2.70)
Acetate	—	4.76
Citrate	—	4.76 (3.08,6.40)
Succinate	—	5.38 (4.13)
Malonate	—	5.66 (2.82)
MES	2-(N-morpholino)-ethane sulfonic acid	6.15
EDTA	ethylenediamine tetraacetic acid	6.2 (2.2,2.7,10.0)
Carbonate	—	6.46
Bis–Tris	bis-(2-hydroxyethyl)imino-tris-hydroxymethyl methane	6.5
ADA	N-(2-acetamido)-2-iminodiacetic acid	6.62
Pyrophosphate	—	6.63 (9.29)
PIPES	Piperazine-N,N'-bis(2-ethane sulfonic acid)	6.80
Bis–Tris Propane	1,3-bis[tris(hydroxymethyl)methylamino] propane	6.8 (9.0)
ACES	2-[(2-amino-2-oxoethyl)-amino]ethane sulfonic acid	6.88
Imidazole	—	7.09
BES	N,N-bis(2-hydroxyethyl)-2-aminoethane sulfonic acid	7.15
MOPS	Morpholino propane sulfonic acid	7.20
Phosphate	—	7.21 (12.36)
TES	N-tris(hydroxymethyl)methyl-2-aminoethane sulfonic acid	7.50
HEPES	N-(2-hydroxyethyl)-1-piperazine-N'-ethane sulfonic acid	7.55
Ethylglycinate	—	7.57
Glycinamide	—	7.73
HEPPS	4-(2-hydroxyethyl)-1-piperazine propane sulfonic acid	8.00
TRICINE	N-tris(hydroxymethyl) methyl glycine	8.15
Glycylglycine	—	8.21
Tris	tris(hydroxymethyl)amino methane	8.30
BICINE	N,N-bis(2-hydroxyethyl)glycine	8.35
TAPS	tris(hydroxymethyl)methylamino propane sulfonic acid	8.40
AMP	2-amino-2-methyl-1-propanol	9.30
Borate	—	9.78 (2.35)
CHES	2-(N-cyclohexylamino)-ethane sulfonic acid	9.55
Glycine	—	9.78 (2.35)
N,N-dimethylglycine	—	9.95
CAPS	Cyclohexylamino propane sulfonic acid	10.40

[a] The pK_a values given are those most commonly used in buffer preparation. The values in parentheses are other pK_a values for multiprotic acids or bases.

activity at alkaline pH. A reaction scheme can be written

$$R_1O\ PO_3^- + EH \rightleftharpoons E\ PO_3^- + R_1OH$$

$$E\ PO_3^- + H_2O \rightleftharpoons EH + H_2\ PO_4^-$$

$$E\ PO_3^- + R_2OH \rightleftharpoons EH + R_2\ OPO_3$$

where R_1 and R_2 are organic constituents of an alcohol or phosphate ester and E represents the alkaline phosphatase. It can be seen that alkaline phosphatase functions as not only a hydrolase but also as a transphosphorylase, and indeed the latter reaction is often favored. The relative rate of the transphosphorylation depends on the acceptor concentration (R_2OH in the scheme) and the type of receptor (46, 47). This effect is illustrated by Figure 2.7 where carbonate serves as an inert buffer and the three amino alcohol buffers show the increased rate of transphosphorylation. It is also of interest that isoenzymes from different human organs are affected to different degrees by the various phosphate acceptors. It is important to note that an enzyme may catalyze more than one type of reaction and the buffer may play an important role in the determination of the end products. Several other enzymes have had similar effects reported but none of these enzymes has been studied in great detail.

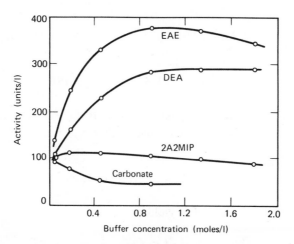

Fig. 2.7. Relationship of activity of alkaline phosphatase to concentration in four different buffers: ethylaminoethanol (EAE), diethanolamine (DEA), 2-amino-2-methyl-1-propanol (2A2M1P), and sodium carbonate. [Reproduced by courtesy of *Clinical Chemistry* (American Association for Clinical Chemistry, publisher) from R. B. McComb and G. N. Bowers, Jr., *Clin. Chem.* **18**, 97 (1972)].

2.8   KINETICS OF TWO-ENZYME SYSTEMS

For many enzyme reactions, none of the products of the reaction are directly detectable. A second enzyme may be used in tandem with the first enzyme to provide a coupled enzyme assay as shown by the reaction scheme

$$S + A \xrightleftharpoons{E_1} P_1 + B$$

$$P_1 + C \xrightleftharpoons{E_2} P_2 + D$$

where the first reaction is generally called the auxiliary reaction when substrate quantification is desired, the second reaction is the indicator reaction (see Chapter 3), and D is the detectable species. Since this type of reaction scheme is of particular interest in enzyme-based analysis, we will discuss the kinetics in some detail.

2.8.1   Homogeneous System Kinetics

When both enzymes are in solution, that is, a homogeneous system, the rate of change of P_1 for reactions obeying first-order Michaelis–Menten kinetics is given by the relation

$$\frac{d[P_1]}{dt} = k_1'[S] - k_2'[P_1] \tag{2.62}$$

where $k_1' = (k_1[E_1]/K_{M_1})$ and $k_2' = (k_2[E_2]/K_{M_2})$. Assuming that at the initiation of the reaction the concentration of P_1 is zero, then

$$[P_1] = \frac{k_1'}{k_2'} \{1 - \exp(-k_2't)\}[S] \tag{2.63}$$

where t is the time elapsed since initiation. If the indicator reaction consumes equimolar amounts of P_1 and C then the rate of change of D is given by

$$\frac{d[D]}{dt} = k_2'[P_1] = k_1'\{1 - \exp(-k_2't)\}[S] \tag{2.64}$$

Integration of equation (2.64) yields

$$[D] = \frac{k_1'}{k_2'} \{k_2't + \exp(-k_2't) - 1\}[S] \tag{2.65}$$

The net result of the above relationships is a *delay* in the appearance of a steady-state rate of production of the detected species. This is commonly referred to as the *lag phase* of the reaction. It should be noted that at any instant in time, the sum of the concentrations of P_1 and D must be equal to the concentration of P_1 which would be formed in a single-enzyme reaction step.

It should be apparent from the above relations that the rate constants of the reactions, k_1' and k_2' and the ratio of the rate constants have a profound effect on the kinetic behavior of the system. The effect of low values of k_1' and k_2', regardless of the ratio of the two, is a prolonged lag phase (see Figure 2.8). This is extremely important to consider in a kinetic assay since the *nonlinear production of the measured substance could cause significant errors*. The ratio of the rate constants is also an important consideration. The higher the ratio k_2'/k_1', the shorter the time elapsed before a steady-state change in the detectable product is observed. Again, referring to Figure 2.8, the lag phase is shorter for the larger ratio of k_2'/k_1'. The influence of this type of nonlinearity on a kinetic assay should again be apparent. For this reason, the ratio of the rate constants for a kinetic assay is generally greater than 5. For an equilibrium assay, this ratio is not as important. It should be noted that when more than one auxiliary reaction is used, the above considerations of rate constants and their ratios apply to each successive step in the analytical scheme.

2.8.2 Heterogeneous System Kinetics

A topic which deserves special recognition in this discussion of enzyme kinetics is the behavior of co-immobilized enzyme systems in which the product of one of the reactions is a substrate for the other. Goldman and Katchalski have presented a theoretical treatment of a two-enzyme system (48). Assuming that the bulk concentration of the substrate for the first reaction was constant and that both reactions were first order, these authors were able to predict the bulk concentration of the product of the first and second reactions for the heterogeneous (immobilized) system in a manner analogous to equations (2.62) through (2.65). The results of these calculations for both homogeneous and heterogeneous systems are shown in Figure 2.8. For the heterogeneous case, appropriate values for the thickness of the diffusion layer, diffusion coefficients, area of enzyme coverage, and rate constants were assumed. The efficiency with which the second enzyme removes the product of the first reaction can also be calculated. The co-immobilized enzyme system is more efficient during the first few minutes of the reaction because the heterogeneous system exhibits little or no lag phase. The reason for this lack of lag phase is shown to be due to a buildup of the intermediate species (P_1) in the dif-

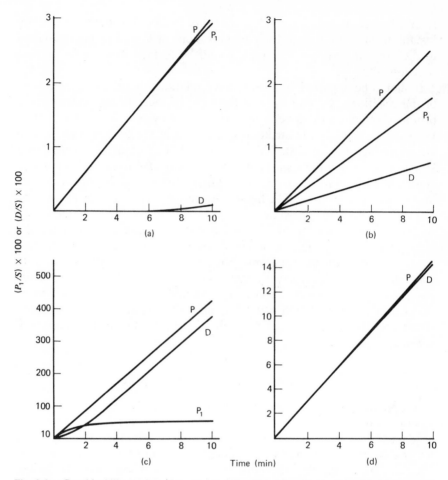

Fig. 2.8. Graphical illustration of the kinetic behavior of homogeneous and heterogeneous two-enzyme systems. P represents the product formed if only the first enzyme is present, P_1 represents the product formed by the first enzyme, and D represents the detectable species. [Redrawn with permission from R. Goldman and E. Katchalski, *J. Theoret. Biol.* **32**, 243 (1971). Copyright by Academic Press, Inc. (London) Ltd.] (a) Homogeneous system: $k_1' = 5 \times 10^{-5}$ sec^{-1}; $k_2'/k_1' = 2$. (b) Heterogeneous system: $k_1 = 5 \times 10^{-4}$ cm/sec; $k_2/k_1 = 2$. (c) Homogeneous system: $k_1' = 7 \times 10^{-3}$ sec^{-1}; $k_2'/k_1' = 2$. (d) Heterogeneous system: $k_1 = 7 \times 10^{-2}$ cm/sec; $k_2/k_1 = 2$.

fusion layer of the membrane or other support which results in an increased rate of the indicator reaction. Goldman and Katchalski showed the effect of increasing and decreasing the thickness of the diffusion layer and, indeed, showed that in the absence of a diffusion layer, no kinetic advantage is observed. The experimental situation approximates the predicted behavior quite closely for a system of hexokinase and glucose-6-phosphate dehydrogenase (49). A three-enzyme system has also been constructed with similar results but with a more pronounced increase of efficiency in the heterogeneous system (50). It has recently been reported that the increased efficiency is maintained throughout the course of the reaction and is not limited to the initial portion of the reaction as implied above (51).

It should be noted that all of the above derivations and experiments have made tacit assumptions about the absence of increased inhibition and pH effects due to the buildup of these species in the film surrounding the immobilized enzyme. Mosbach and his coworkers have studied the effects of enzyme-generated pH changes in the microenvironment (52, 53). As might be anticipated, the buildup of acid or base in the diffusion layer did cause changes in the enzyme activity of the second enzyme. No work has been reported for inhibitors, but it would certainly be reasonable that increased inhibition relative to the homogeneous state would result. The implications of these findings are of paramount importance in the understanding of membrane-bound enzymes in vivo.

2 Definition of Terms

a	degree of activation
[A]	concentration of substrate A (moles/1)
[B]	concentration of substrate B (moles/1)
[E]	concentration of free enzyme (moles/1)
$[E]_0$	total concentration of enzyme (moles/1)
[EP]	concentration of enzyme–product complex (moles/1)
[ES]	concentration of enzyme–substrate complex (moles/1)
F	second form of enzyme in ping-pong mechanism
f_E, f_{EH}, f_{EH_2}	Michaelis pH functions
i	degree of inhibition
[I]	concentration of inhibitor
k_{+1}, k_{-1}, k_{+2}	rate constants of simple Michaelis–Menten mechanism
$k_{-2}, k_{+3}, k_{-3},$ k_{+4}, k_{+5}	rate constants of multistep enzyme reaction schemes
$K_A, K_{B/A}$	Michaelis constants for substrates A, B, and P
K_I	equilibrium constant for the dissociation of enzyme–inhibitor complex
K_L	equilibrium constant for allosteric conformation

K_M	Michaelis constant for one-substrate reaction
K_R	Michaelis constant for R conformation
K_S	general form of the Michaelis constant
K_T	Michaelis constant for T conformation
[M]	concentration of metal ion
[P]	concentration of product (moles/1)
[Q]	concentration of second product in two-substrate reactions (moles/1)
Q_{10}	ratio of reaction rates at 10° temperature interval
r	empirical constant used in mixed-inhibition model
R	conformation of allosteric-effect model
[S]	concentration of substrate
$[S]_0$	initial substrate concentration
T	conformation of allosteric-effect model
v	velocity of the enzyme reaction
v_0	velocity of the reaction in absence of inhibitor or activator
V	maximum velocity of the reaction at substrate saturation
α	fraction of enzyme not irreversibly inhibited
β	substrate concentration relative to K_S
γ	inhibitor concentration relative to K_I
δ	anion concentration relative to K_A
τ	induction period

References

1. L. Michaelis and M. L. Menten, *Biochem. Z.* **49**, 333 (1913).
2. G. E. Briggs and J. B. S. Haldane, *Biochem. J.* **19**, 338 (1925).
3. H. Lineweaver and D. Burke, *J. Am. Chem. Soc.* **56**, 658 (1934).
4. A. Hubaux and G. Vos, *Anal. Chem.* **42**, 849 (1970).
5. G. S. Eadie, *J. Biol. Chem.* **146**, 85 (1942).
6. B. H. J. Hofstee, *Science* **116**, 329 (1952).
7. C. S. Hanes, *Biochem. J.* **26**, 1406 (1932).
8. W. W. Cleland, *Biochim. Biophys. Acta* **67**, 104 (1963).
9. E. L. King and C. Altman, *J. Phys. Chem.* **60**, 1375 (1956).
10. F. J. Roughton, *Trans. Faraday Soc.* **17**, 116 (1954).
11. J. Monod, J. Wyman, and J. P. Changeux, *J. Mol. Biol.* **12**, 88 (1965).
12. D. Koshland, G. Nementhy, and D. Filmer, *Biochemistry* **5**, 365 (1966).
13. J. Westley, *Enzymatic Catalysis* (Harper & Row, New York, 1969.)

14. G. R. Ainsley, J. P. Schill, and K. E. Neet, *J. Biol. Chem.* **297**, 7088 (1972).
15. D. E. Atkinson, J. A. Hathaway, and C. S. Smith, *J. Biol. Chem.* **240**, 2682 (1965).
16. C. Frieden, *J. Biol. Chem.* **239**, 3522 (1964).
17. J. R. Sweeney and J. R. Fischer, *Biochemistry* **7**, 561 (1968).
18. R. M. Krupka and K. J. Laidler, *J. Am. Chem. Soc.* **83**, 1445 (1961).
19. H. Kaplan and K. J. Laidler, *Can. J. Chem.* **45**, 539 (1967).
20. R. A. Alberty, *J. Am. Chem. Soc.* **76**, 2494 (1954).
21. W. J. Blaedel and G. P. Hicks, *Advances in Analytical Chemistry and Instrumentation,* C. N. Reilly, Ed., Vol. 3 (Interscience, New York, 1965), p. 105.
22. P. Baum and R. Czok, *Biochem. Z.* **332**, 121 (1959).
23. M. Dixon and E. C. Webb, *The Enzymes* (Academic, New York, 1964).
24. K. J. Laidler and J. P. Hoare, *J. Am. Chem. Soc.* **71**, 2699 (1949).
25. J. B. S. Haldane, *Enzymes* (MIT Press, Cambridge, MA, 1965).
26. D. R. P. Murray, *Biochem. J.* **24**, 1890 (1930).
27. S. F. Howell and I. B. Sumner, *J. Biol. Chem.* **104**, 619 (1934).
28. L. Michaelis and H. Davidsohn, *Biochem. Z.* **35**, 386 (1911).
29. L. Michaelis and M. Rothstein, *Biochem. Z.* **110**, 217 (1920).
30. S. G. Waley, *Biochim. Biophys. Acta* **10**, 27 (1953).
31. K. J. Laidler, *Trans. Faraday Soc.* **51**, 528 (1955).
32. E. Shaw, *Physiol. Rev.* **50**, 244 (1970).
33. M. Dixon and D. C. Webb, *The Enzymes* (Academic, New York, 1964) pp. 120–150.
34. K. J. Laidler and P. S. Bunting, *The Chemical Kinetics of Enzyme Action,* 2nd ed. (Oxford Press, New York, 1973), p. 142.
35. F. B. Straub, *Enzymologia* **9**, 143 (1940).
36. J. Schutz, *Z. Physiol. Chem.* **30**, 1 (1900).
37. P. S. Bunting and K. J. Laidler, *Biochemistry* **11**, 4477 (1972).
38. S. Arrhenius, *J. Gen. Physiol.* **13**, 46 (1930).
39. W. Van Dam, *Hoppe-Seyler's Z. Physiol. Chem.* **79**, 247 (1912).
40. G. Benzoana and P. Desnuelle, *Biochim. Biophys. Acta* **164**, 47 (1964).
41. B. Entressanles and P. Desnuelle, *Biochim. Biophys. Acta* **164**, 47 (1964).
42. M. Smeriva, G. Benzonana, and P. Desnuelle, *Biochim. Biophys. Acta* **191**, 598 (1968).
43. A. D. Elbein, *Adv. Enz.* **40**, 29 (1974).
44. J. P. Greenstein and F. M. Lenthardt, *Arch. Biochem.* **17**, 105 (1948).
45. N. D. Jespersen, *J. Am. Chem. Soc.* **97**, 1662 (1975).
46. R. B. McComb and G. N. Bowers, *Clin. Chem.* **18**, 97 (1972).
47. R. B. McComb in "Second International Symposium on Clinical Enzymology," N. W. Teitz A. Weinstock, and D. O. Rodgerson, Eds. (American Association for Clinical Chemistry, Washington, D.C., 1976).
48. R. Goldman and E. Katchalski, *J. Theoret. Biol.* **32**, 243 (1971).
49. K. Mosbach and B. Mattiasson, *Acta Chem. Scand.* **24**, 2093 (1970).
50. B. Mattiasson and K. Mosbach, *Biochim. Biophys. Acta* **235**, 253 (1971)

51. P. A. Srere, B. Mattiasson, and K. Mosbach, *Proc. Natl. Acad. Sci.* **70,** 2534 (1973).
52. S. Gestrelius, B. Mattiasson, and K. Mosbach, *Biochim. Biophys. Acta* **276,** 339 (1972).
53. S. Gestrelius, B. Mattiasson, and K. Mosbach, *Eur. J. Biochem.* **36,** 89 (1973).

PRINCIPLES OF KINETIC AND EQUILIBRIUM METHODS OF ANALYSIS

The purpose of the present chapter is to discuss the analytical and instrumental principles involved in measurements based on the use of enzymes as catalysts. Most of this chapter will be devoted to discussions of substrate determination. For the most part we will assume that any activators and cofactors such as nicotine adenine dinucleotide (NAD) and adenosine triphosphate (ATP) or coreagents (O_2) are present in such large quantities that whatever change in their concentration occurs due to consumption of the substrate will have no significant effect on the reaction rate.

Since an enzyme is a catalyst, it cannot alter the equilibrium position of a chemical reaction unless it is present in such large molar concentrations that it forms a significant number of stable "complexes" with either a reactant or product. The major advantages of using enzymes as reagents must be related to their effect on the reaction kinetics. Two chief advantages are (1) the speeding up of a reaction which is normally so slow that it is analytically useless and (2) an improved specificity of a particular reaction by speeding up the rate of the desired reaction *relative* to one or more undesirable processes. Quite often enzymes allow reactions to occur under very mild circumstances (pH, temperature, solvent). These reactions would only occur very slowly under the same conditions or require drastic measures in the absence of the enzyme. For example triglycerides which exist in the body as very complex mixtures of fatty acid esters (oleic, palmitic, stearic) of glycerol can be determined upon saponification for 15–30 minutes in hot, alcoholic, potassium hydroxide. In contrast the enzyme lipase (from pancreatic juices or micro-organisms) will carry out the equivalent reaction at room temperature in nearly neutral buffers (pH 7.5–8.5). Triglycerides hydrolyze very slowly under these mild conditions.

Enzymes can often markedly improve reaction specificity and simplify the determination of a particular substrate. This can take place in two ways. Suppose that under a given set of reaction conditions (pH, reagents, ionic strength, temperature) a substrate can react by either of two paths,

$$R + S \xrightarrow{k} P \text{ (measurable)} \tag{3.1}$$

$$R + S \xrightarrow{k'} P' \text{ (not measurable)} \tag{3.2}$$

and that the second path leads to P' which is not easily measured or makes the measurement of species P very cumbersome. In this situation one could measure the change in the reagent R provided that it can be measured and is stoichiometrically related to the amount of substrate used. Nonetheless it would greatly simplify the determination of S to turn off reaction (3.2) or make reaction (3.1) so fast for example, by specifically catalyzing it, that no P' is formed. Enzyme reagents are very widely used in the biological sciences because they greatly facilitate determinations in the very complex samples which are derived from living systems. In part this is due to their specificity, that is ability to catalyze only certain reactions.

The relative reaction rate of glucose oxidase action and that of hexokinase on several closely related carbohydrates are shown in Table 3.1 (1). The relevant reactions are:

$$\beta\text{-D-glucose} + O_2 \xrightarrow{\text{glucose oxidase}} \text{gluconolactone} + H_2O_2 \quad (3.3)$$

$$\beta\text{-D-glucose} + \text{MgATP} \xrightarrow{\text{hexokinase}} \text{glucose-6-phosphate}$$
$$+ \text{MgADP} + H^+ \quad (3.4)$$

Glucose oxidase is highly selective for β-D-glucose with the sole significant interferant being 2-deoxy-D-glucose. In contrast, hexokinase is not particularly suitable for the analysis of β-D-glucose since several common sugars react almost as rapidly and fructose reacts even faster. However, hexokinase can be combined with the enzyme glucose-6-phosphate de-

Table 3.1   Specificity of Glucose Oxidase and Hexokinase

Substrate	Reaction rate relative to β-D-glucose	
	Glucose oxidase[a]	Hexokinase[b]
β-D-glucose	100	100
2-deoxy-D-glucose	25	100
D-mannose	1.0	80
D-xylose	1.0	0
α-D-glucose	0.6	—
Trehalose	0.3	0
Maltose	0.2	0
D-altrose	0.2	0.3
D-galactose	0.1	0
Fructose	—	180

[a] Reference 1, p. 210.
[b] Reference 1, p. 215.

hydrogenase to obtain a very selective method for glucose determination (2) (see Section 3.3.2). The structures of five hexoses and a hexose amine [α-D-glucose (A), β-D-glucose (B); glucose amine (C), fructose (D), galactose (E), and mannose (F)] which are found at significant concentrations in serum are shown below:

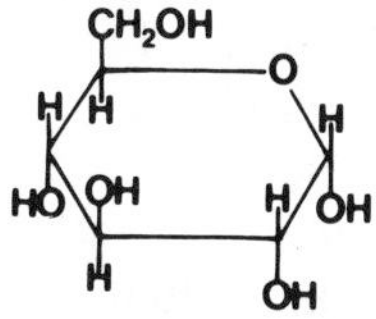

An excellent example of the use of enzymes to achieve specificity is found in the action of urease, for which urea is the only known substrate although many materials will inhibit the reactions. The reaction catalyzed by urease is:

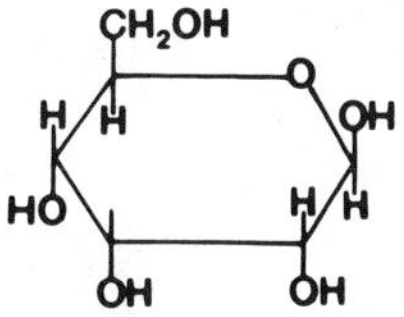

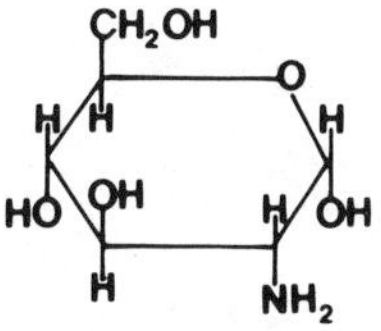

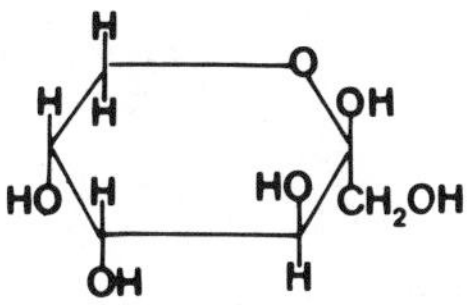

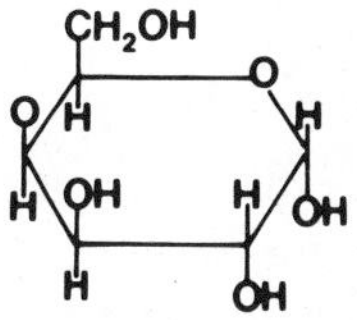

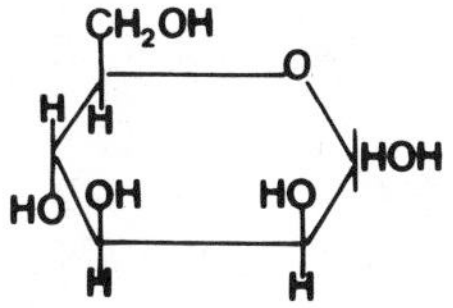

$$NH_2-\underset{\underset{O}{\|}}{C}-NH_2 + H_2O \xrightarrow{\text{urease}} 2NH_3 + CO_2 \qquad (3.5)$$

This is precisely the same reaction which occurs in the well-known Kjeldahl nitrogen assay in which virtually every nitrogenous organic compound will produce some ammonia and carbon dioxide. An unmodified Kjeldahl procedure for urea in serum has no analytical redeeming virtue, yet the urease enzymatic method which is based on the same net chemical reaction is very useful (3). Another reason enzymes are used is that they allow the introduction of a coreactant or coproduct which is often very easy to measure (e.g., NAD, ATP).

Analytical methods which use *enzymes* fall into two distinct categories: *kinetic* methods and *equilibrium* methods. Pardue has recently developed a classification scheme which greatly clarifies the confusion in the literature as to the nature of equilibrium and kinetic methods of analysis (4). We will use the terms kinetic and equilibrium methods in accord with Pardue's definitions, that is, all analytical methods that depend in any way upon data collected during a period when a concentration is time-dependent are kinetic methods, and those methods that depend *only* upon data obtained when all concentrations are time-independent are equilibrium methods.

3.1   EQUILIBRIUM ANALYSIS

3.1.1   Equilibrium Considerations

These methods are analytically the simplest and will be treated before proceeding with a detailed discussion of kinetic based techniques. Let us consider the general overall reaction:

$$r\text{R} + s\text{S} \overset{K_{eq}}{\rightleftharpoons} p\text{P} + q\text{Q} \tag{3.6}$$

R represents a coenzyme or coreactant. As discussed previously R will be in "kinetic excess" when its concentration is about 10 times as great as its Michaelis constant. In order to carry out an equilibrium assay for the substrate, R must be present at a concentration of at least r/s times the initial substrate concentration. If the equilibrium constant of the overall reaction is not large, then the coreactant must be present at considerably higher concentrations in order to force complete consumption of the substrate. The amount of coreactant needed to achieve any desired equilibrium concentration of substrate ($[\text{S}]_{eq}$) can be estimated by assuming that the initial concentration of R ($[\text{R}]_0$) is much greater than that of the substrate ($[\text{S}]_0$):

$$\frac{[\text{S}]_{eq}}{[\text{S}]_0} \cong \frac{1}{[\text{S}]_0} \left(\frac{(p/s)^p (q/s)^q}{K_{eq}} \cdot \frac{[\text{S}]_0^{p+q}}{[\text{R}]_0^r} \right)^{1/s} \tag{3.7}$$

when

$$\frac{[\text{S}]_{eq}}{[\text{S}]_0} \ll 1 \text{ and } \frac{[\text{R}]_0}{[\text{S}]_0} \gg 1.$$

The results for a reaction with equal stoichiometric coefficients for 99, 99.5, and 99.9% conversion are summarized in Table 3.2. As the equilibrium constant increases, the equation underestimates the minimum amount of coreagent and it is necessary to employ a detailed mass balance to estimate the correct ratio.

When a large excess of coreactant is not required, one can utilize the overall change in concentration of any of the species involved in the reaction to effect an analysis for the substrate. A calibration curve can be constructed from a measurement of any desired instrumental measurement (absorbance, pH, conductivity, radiochemical count rate, etc.) which is a function of any species (R, S, P, Q) in the reaction.

A series of theoretical calibration curves showing the effect of equilibrium constant on the linearity of the measured response as a function of the initial amount of sample is shown in Figure 3.1. All of the curves were computed from a detailed mass balance and equilibrium equations with an initial coreactant concentration of 10 mM. A reaction with equal stoichiometry coefficients was assumed. The essential point is that as K_{eq} decreases, the curves become more nonlinear. Even if the measured variable is not *specifically* related to any one of the species or if there is a background signal, the total change in the measured parameter can be used to accurately measure the substrate concentration provided only that reaction (3.6) is the sole process catalyzed by the enzyme. To illustrate this point, suppose the substrate itself absorbs or fluoresces at a particular wavelength but there are other nonreacting species also present which produce a signal. The total signal will be the linear sum of that from the substrate and the background. After the reaction is complete the signal will be due only to the background. The difference in recorded signal will be proportional to the initial substrate concentration if the reaction is complete. A linear relationship between the signal change will also exist even when more than one of the species involved in the enzyme reaction

Table 3.2 Minimum Quantity of Co-reactant Required for Complete Conversion[a]

	[R]$_0$/[S]$_0$		
K_{eq}	99.0%	99.5%	99.9%
1	100	200	1000
5	20	40	200
10	10	20	100
50	2	4	20

[a] Computed from equation (3.7).

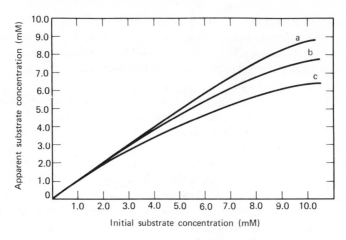

Fig. 3.1. Effect of equilibrium constant on calibration curves for equilibrium assays. Plot of calculated total change in substrate concentration versus initial substrate. Coreactant concentration is 10 mM in all cases. Curve a, $K_{eq} = 50$; Curve b, $K_{eq} = 10$; Curve c, $K_{eq} = 3$.

produce a signal. Suppose that at a particular wavelength, all species contribute to the absorbance, A:

$$A_0 = \epsilon_R[R]_0 + \epsilon_S[S]_0 + \text{background} \qquad (3.8)$$

$$A_{eq} = \epsilon_R([R]_0 - [S]_0) + \epsilon_P[S]_0 + \epsilon_Q[S]_0 + \text{background} \qquad (3.9)$$

It is assumed here that all stoichiometry coefficients are equal, that is, $r = s = p = q$. Note that ϵ_I represents the molar extinction coefficient of the relevant species (I). The change in absorbance $A_{eq} - A_0$ will be proportional to the initial substrate concentration. Further consideration indicates that this will be true for a linear concentration transducer even if some of the products are initially present. The only requirements for an accurate measurement and linear calibration curve in an equilibrium method are:

1. A complete reaction, that is, total consumption of substrate.
2. A linear relationship between the transducer and the concentration of each species to which it is sensitive.
3. Strict additivity of signals from all species.
4. A specific chemical reaction.

The need for a totally specific enzyme reaction is not as great as might be indicated. Suppose that the sample is mixed with the coreactants, buffer, etc., and placed in the measuring cell. An appropriate enzyme is added and a signal–time curve, similar to that shown in Figure 3.2, is recorded. If an impure or nonspecific enzyme were to *slowly* catalyze the reaction of some other constituent of the sample, a slow drift in the recorded signal would occur after the substrate of interest is rapidly consumed. This "linear drift" can often be extrapolated to the time of addition of the enzyme. The difference signal shown in the figure will accurately correct for the extraneous reaction. This is merely a crude form of differential kinetic analysis. Of course, any material which reacts completely, or which produces a nonlinear, signal–time curve, will introduce error. A linear signal–time curve from an interferent can occur only if it is present at concentrations in excess of its K_M, or if the rate of its reaction is so low that its concentration is effectively constant. Obviously the better the specificity of the enzyme, the better will be the overall accuracy of the analysis; there is really no substitute for a pure, specific enzyme. To a very great extent the widespread use of enzymes as analytical reagents is due to the ready availability of purified enzymes from biochemical supply houses.

Despite the very encouraging tone of the preceeding discussion, there are a number of sources of error in equilibrium methods which can be very difficult to detect and handle. For example, when the sample contains endogeneous enzymes and coreactants, the reaction will start before any exogeneous enzyme or cofactor is added. If the analysis is carried out so that the signal is monitored before and after reagents are added this problem will be observed as a drift in baseline *before* addition of the enzyme reagent. This situation could well go unnoticed with some of the "totally

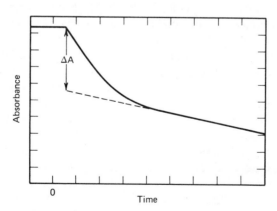

Fig. 3.2. Schematic representation for kinetic correction for a slow-reacting interferant in enzymatic analysis. After completion of fast reaction, extrapolation of the slow drift to zero time corrects for other reactants.

automated" instruments available today. The problem can be cured by removing or destroying the endogeneous enzyme, provided that pains are taken not to alter the substrate except by dilution. That this situation can occur is not very far-fetched because the species most commonly used as analytical "handles" on substrate concentrations are the pyridine nucleotides (NAD, NADP, etc.) which are biochemically connected in an elaborate network of reactions involving a large number of different enzymes. It is the ubiquitous utility of these substances which make them prone to interference from other enzymes and substrates. Other errors are related to the fact that both NADH and NADPH are good reducing agents and are easily oxidized by O_2 in the presence of certain electron transport enzymes (NADH oxidases). This problem can be diminished by adding amytal which will block electron transfer. The excellent monograph by Lowry and Passonneau should be consulted for trouble shooting problems in developing or modifying enzymatic assays (5).

Returning to the question of calibration curves, it is possible to measure specifically or in combination any of the species in the overall reaction. When a large excess of coreagent is used, the relative decrease in its concentration will be small and the situation will generate substantial, random errors. Measurement of a decrease in the substrate concentration is akin to inverse colorimetry, shares the same problems and requires that two measurements on each sample be carried out. Whenever possible it is, of course, generally advisable to measure the increase in concentration of a product or a rise in signal level from a low background level.

Equilibrium methods have many advantages over true kinetic techniques. (1) Reaction time is not an important experimental variable provided that the measured species is stable. (2) If the reaction is at equilibrium and is complete, then even large changes, that is, order-of-magnitude changes, in the equilibrium constant will have almost no influence on the final concentration. It is evident that when either the equilibrium constant or the initial coreactant concentration is large one can tolerate very pronounced variations in these factors which influence K_{eq} without seriously effecting the accuracy. This is not the case in kinetic methods (except those which are controlled by mass transfer), where small changes in pH, ionic strength, and temperature may introduce overwhelming errors. (3) Kinetic methods are sensitive to the presence of adventitious activators and inhibitors. By the definition of catalysts and inhibitors (negative catalysts) equilibrium assays are not so influenced provided that sufficient time is allowed for the reaction to proceed to true equilibrium. Thus equilibrium assays are simpler to implement instrumentally, are generally, but not invariably, more precise (see below), and are less prone to determinate errors than kinetic methods.

Why then should a kinetic analysis be used? First, and most obvious, one cannot measure an enzyme's activity by any approach other than a rate technique (4). For this purpose, equilibrium methods are inherently useless. Similarly enzyme activators and inhibitors can only be measured via their effect on a reaction rate. Second, the quantity of enzyme required to reach equilibrium in an acceptable time limit may be so large that it may be economically prohibitive. Third, with a given quantity of enzyme per sample, a kinetic analysis can be completed in a far shorter time period provided that the instrumentation is sufficiently sensitive to measure a small total change in concentration. Fourth, inhibition by products may be so severe that the reaction becomes so slow that it apparently stops before equilibrium is achieved. For example Desneulles and coworkers (6) have shown that pancreatic lipase is strongly inhibited by the free fatty acid generated by hydrolysis of triglycerides. In this case calcium ions can be added to tie up the product and "inhibit the inhibition." Fifth, smaller, catalytically sufficient quantities of coreactants (coenzymes), which can be more expensive than the enzyme, may be used. Finally, the reaction product may not be sufficiently stable to permit a reaction to go to completion and therefore the reaction can be used only in the early stages.

3.1.2 Minimum Reaction Time

Although there are profound methodological and instrumental differences between kinetic and equilibrium methods of analysis, chemically the distinction is simply a question of time, that is, a reaction ultimately will go to its thermodynamic rest point and an equilibrium assay will be implemented.

The relationship between the quantity of enzyme required to achieve any desired degree of completion and time can be estimated by integrating the Michaelis–Menten rate law:

$$\frac{d[P]}{dt} = -\frac{d[S]}{dt} = \frac{k_2[E][S]}{K_M + [S]}; \quad [S]_{t=0} \equiv [S]_0 \tag{3.10}$$

The final equation, that is, the integrated Michaelis–Menten equation; may be written as follows:

$$[S]_0 - [S] + K_M \ln \frac{[S]_0}{[S]} = +k_2[E]t \tag{3.11}$$

It should be understood that equations (3.10) and (3.11), with all their attendant short comings, including the assumption of irreversibility, single limiting substrate, and the steady-state approximation, are the basis of kinetic substrate analyses. Equation (3.11) has two limiting forms which correspond to the initial substrate concentration greater than K_M (zero-order reaction) and initial substrate well below K_M (first-order reaction:

$$[S]_0 - [S] = k_2[E]t; \quad [S]_0 \gg K_M \tag{3.12}$$

$$K_M \ln \frac{[S]_0}{[S]} = k_2[E]t; \quad [S]_0 \ll K_M \tag{3.13}$$

Equations (3.12) and (3.13) are written in a form so that the time required to bring the reaction to any desired degree of completion is easily computed. Lowry (7) has pointed out that equation (3.11) predicts that the total time needed for an analysis is exactly equal to the sum of a zero-order reaction with rate constant $k_2[E]$ and a first-order reaction with rate constant $k_2[E]/K_M$ to achieve the same desired conversion. The fractional extent of conversion X will be defined as

$$X = \frac{[S]_0 - [S]}{[S]_0} = \frac{[P]}{[S]_0} \tag{3.14}$$

Thus the time required to reach X will be

$$t = \frac{K_M}{k_2[E]} \left\{ \frac{X}{\beta} + \ln \frac{1}{1 - X} \right\} \tag{3.15}$$

where β is a dimensionless Michaelis constant defined as:

$$\beta = K_M/[S]_0 \tag{3.16}$$

which governs whether the reaction is predominantly first order ($\beta \gg 1$) or zero order ($\beta \ll 1$). We should also note that $k_2[E]$ has units of enzyme activity per unit volume, that is, enzyme concentration (in moles/sec·ml) and that the term ($K_M/k_2[E]$) is related to an effective first-order time constant (in seconds).

The dimensionless time required to complete the enzyme–substrate process to any desired degree is given in Table 3.3. As the extent of conversion approaches unity, an excellent approximation can be obtained by setting X equal to 1 in the first term in equation (3.15). A first-order reaction will be 99% complete when $t = 4.6 \, K_M/k_2[E]$; one can show that in general the time needed for 99% conversion ($t_{0.99}$) is approximately:

$$t_{0.99} = \frac{4.6K_M + [S]_0}{k_2[E]} \tag{3.17}$$

The minimum reaction time in minutes can be obtained by multiplying the dimensionless reaction time by K_M in moles/ml and dividing by the

Table 3.3   Minimum Reaction Time for Complete Conversion

X^b	$\infty^{c,d}$	100	10	1	0.10	0.01
			$K_M/[S]_0$			
			Total Reaction Time^a			
0.900	2.30	2.31	2.39	3.20	11.3	92.3
0.950	3.00	3.00	3.09	3.95	12.5	98.0
0.990	4.61	4.62	4.70	5.60	14.5	104
0.995	5.30	5.31	5.40	6.30	15.3	105
0.999	6.90	6.92	7.01	7.91	16.9	107
		Contribution of zero-order reaction to total time				
0.900	0.00	0.00900	0.0900	0.900	9.00	90.0
0.950	0.00	0.00950	0.0950	0.950	9.50	95.0
0.990	0.00	0.00990	0.0990	0.990	9.90	99.0
0.995	0.00	0.00995	0.0995	0.995	9.95	99.5
0.999	0.00	0.00999	0.0999	0.999	9.99	99.9

^{a} Minimum reaction $= k_2[E]t_{min}/K_M$; the required time will be equal to the value in the table times K_M (in moles/ml) divided by enzyme (units/ml). All values are computed from equation (3.15).
^{b} Fractional conversion [see equation 3.14].
^{c} This represents the contribution of the first-order term to the total reaction time under all conditions.
^{d} There is no contribution from the zero-order reaction to the minimum reaction time when the initial substrate concentration is negligible compared to K_M.

number of enzyme units per milliliter (μmoles/ml·min). In the limit of very low substrate concentration, the 99% reaction time can be obtained by dropping $[S]_0$ in equation (3.17). The above calculation is limited to the use of a single rate-limiting enzyme and substrate. The monograph by Lowry and Passonneau and Section 2.8 of this text should be consulted for a discussion of reaction rates when there is more than one rate-limiting species or enzyme.

3.2   FUNDAMENTAL ASPECTS OF KINETIC ANALYSIS

Enzymatic measurements of substrate concentration may be based on the fact that under certain conditions the *velocity* (rate) of an enzymatic reaction is very dependent on substrate concentration. Such methods are therefore kinetic methods of analysis according to Pardue's definition (4).

Equation (3.10), which is the differential form of the Michaelis–Menten rate law, states that the instantaneous reaction rate is a function of the time-dependent substrate concentration. This function is obviously non-linear and is very insensitive to the substrate concentration when it is present in concentrations much greater than the Michaelis–Menten constant (K_M) of the system. To illustrate the basis for kinetic analysis, let us assume that some type of apparatus is available that can directly read out the reaction rate (in moles/ml·sec) very shortly after the reaction is started, that is, before any significant amount of substrate is actually consumed. This is the initial reaction rate and will be denoted as $\mathcal{R}_0$. The initial reaction rate will be controlled by Michaelis–Menten kinetics subject to the assumptions and limitations discussed in previous chapters and will be related to the substrate concentration as follows:

$$\mathcal{R}_0 = -\frac{d[S]}{dt}\bigg|_0 = \frac{d[P]}{dt}\bigg|_0 = \frac{k_2[E][S]_0}{K_M + [S]_0} \qquad (3.18)$$

The above relationship is plotted in Figure 2.1. Suppose that we measured $\mathcal{R}_0$ measured on some unknown sample. We can use the data in this figure as a calibration curve and thereby determine the substrate concentration which corresponds to the measured reaction rate. The reaction rate in kinetic analysis is an exact analog of total absorbance, conductance, and current or potential changes in conventional equilibrium assays. Experimental data similar to those presented in Figure 2.1 are obtained by measuring the reaction rate on a series of standard solutions of the substrate. An unknown concentration is determined by adding the same amount of enzyme under precisely the same experimental conditions (pH, temperature, etc.) as those used to obtain the calibration curve; the unknown substrate may then be read off the calibration curve provided that the rate falls in the linear region. If it does not the sample is diluted and rerun. Clearly the kinetic determination of a substrate is inherently a nonlinear method of analysis in contradistinction to equilibrium assays (with large K_{eq}) whose linearity is dictated solely by the detector. This is a fundamental and inescapable difference between the two approaches.

At this point, it is important to distinguish between two different sets of circumstances which will produce mathematically zero-order, that is, linear, reaction-time curves. An enzymatic reaction will be zero-order when $[S]_0 \gg K_M$; most importantly the rate will be independent of $[S]_0$ and it will not be a useful measure of the substrate concentration. This is a mathematically *zero-order* reaction. In contrast, the very early stages of a first-order process, when there is negligible depletion of the substrate, will appear to be kinetically zero-order, that is, the extent of reaction will increase linearly with time. This is termed a *pseudo-zero-order* process. Most importantly, the reaction rate will vary with sample concentration and will be analytically important.

Although the initial reaction rate was used to illustrate the above point, it is not absolutely necessary to restrict measurements to very short time. There are three factors to consider in deciding whether or not the initial rate can be used. First, certain instrumental approaches to the measurement of reaction rate (principally the variable-time method) will not be accurate unless the kinetics are pseudo-zero-order (the reaction rate is nearly constant), that is, very little substrate is consumed and the product–time curve is linear. Second, since the Michaelis–Menten kinetic scheme is an oversimplification in that all effects of product buildup are disregarded, the rate law may be invalidated if too much product is allowed to accrue. Last, the reaction rate should be not be measured until the initial mixing period is over and any induction period (see Chapter 2) in the enzyme reaction is allowed to dissipate.

A very interesting feature of kinetic methods is that the sensitivity coefficient of the method is proportional to the amount of enzyme added. If a particular sample has so little substrate that the reaction rate is small and therefore cannot be measured due to background drift in either the chemistry or the electronics, more enzyme can be added to improve detectability. There are of course limits to how much one can gain by this approach. A fundamental limit is that the derivation of the Michaelis–Menten kinetics assumes that the total number of moles of enzyme added is quite small compared to the number of moles of substrate.

Automated kinetic analyses require that some instrument be available to calculate the rate of a chemical reaction rapidly and display a signal proportional to the rate. In principle, any signal which is related to any of the reactants or products can be used. We will assume for simplicity that one of the products is measured as a function of time and that its initial concentration is zero. A series of product-concentration–time curves as a function of the amount of substrate present is given in Figure 3.3. The reaction rate could be measured manually by determining the angle between the time axis and the rate curve near zero time or by sampling the data and running a least-squares best straight line through the points, etc. Obviously it is much more desirable to be able to plot the reaction rate directly or read it off a digital panel meter.

A great deal of effort has been devoted over the past decade to the development of both analog and digital devices which will rapidly, precisely, and automatically measure reaction rates (8–28). A large variety of methods of implementing kinetic methods of analysis has arisen due to the intense work in this area. It is not our purpose here to review all methods but merely to present the salient features of some of the main methods. Pardue's (4) classification scheme of kinetic methods of analysis, which provides an excellent overview of the area, is given in Table 3.4.

Table 3.4   Classification of Kinetic Methods of
Analysis[a]

I. Variable sensor-signal methods
  A. Direct response methods
    1. One-point methods
      a. Interrupted reactions
      b. Continuous reactions
    2. Two-point methods
      a. Interrupted reactions
      b. Continuous reactions
        i. Fixed time
        ii. Variable time
    3. Multi-point methods
      a. Delta methods
      b. Regression methods
  B. Derivative methods
  C. Integral methods
  D. Difference methods
II. Fixed sensor-signal methods
  All subgroups of I, except A1 and A2a, are applicable.

[a] Reference 4.

3.2.1   Methods of Kinetic Analysis

With the availability of sophisticated kinetic analysis instrumentation, we can disregard the cruder methods such as the one-point and interrupted-reaction approaches. The concepts of several of the key methods are set forth below:

1. The *derivative method*—in this technique an analog circuit consisting of operational amplifiers is used to automatically differentiate the signal–time curve and plot out a voltage proportional to the instantaneous reaction rate, that is, the output voltage is proportional to the substrate concentration (8, 10).

2. The *fixed-time method* in which the change in a concentration (or a signal) is automatically determined over some pair of preset, fixed times (t_1, t_2) (11–14, 22).

3. The *variable-time method,* which automatically measures the time increment (Δt) for the signal to vary from one predetermined value to a second one. In this approach, the concentration increment is fixed (12, 15–18).

4. The *integration method* involves the measurement of the area under the concentration–time curve over two preset equal time intervals. The

difference in the integrals will be a function of the reaction rate and the initial substrate concentration (23–27).

5. The *kinetic difference method* employs a twin detector system and two reaction chambers. A standard is added to one chamber and sample to the second. Any one of several characteristics of the difference signal–time curve including the initial slope can be used as a measure of reaction rate (28–31).

6. The *signal-stat* or *fixed-sensor method* in which the signal from a transducer is automatically held at a constant value by the addition of a reagent which is consumed by the reaction. Actual rate measurements in this approach may be implemented by any of the techniques outlined above (32–40).

The most recent methodologies (41) are based on the use of laboratory digital computers which fit many points on the reaction curve to an equation which describes the concentration–time relationship. These will be referred to as *multi-point methods*.

The application of rate measurement instruments requires that a transducer be available which will produce a signal that is related to the concentration of one or more reactants or products. In general the simplest situation will result when the sensor measures only the concentration of one product and does so linearly. Some of the above methods (e.g., variable time) have the very distinct advantage that quite nonlinear sensors can be used without any deleterious effect on the accuracy of the method. Other techniques are, under limited circumstances, not particularly sensitive to transducer linearity. The absolute accuracy of the fixed-time and integration methods is directly limited by the transducer's linearity.

The output signal (e) of the sensor may be written as a function of the concentrations of all the species (C_i) to which it responds:

$$e = \sum \alpha_i f(C_i) \tag{3.19}$$

For the sake of simplicity we will assume that the sensor responds in the same way to each species which it detects. The rate of change of the sensor voltage will be:

$$\frac{de}{dt} = \sum \alpha_i \frac{\partial f(C_i)}{\partial C_i} \frac{dC_i}{dt} = \sum \alpha_i \frac{\partial f(C_i)}{\partial C_i} \cdot (\pm \mathscr{R}) \tag{3.20}$$

The correct sign to use depends upon whether the detected species is a reactant or product. The reaction-rate term ($\mathscr{R}$) can be factored outside the summation since it will be the same for all species. Thus,

$$\frac{de}{dt} = \mathcal{R} \cdot \left(\sum \pm \alpha_i \frac{\partial f(C_i)}{\partial C_i} \right) \qquad (3.21)$$

Unless the sensor response is absolutely linear, that is, $\partial f(C_i)/\partial C_i$ is independent of concentration, the proportionality constant between the reaction rate and the rate of signal change will depend upon the total concentration change. Since the variable-time method uses two fixed concentration levels, the nonlinear response does not matter because the extent of curvature is precisely the same in all experiments. In contrast, the fixed-time and integration methods both entail measured concentration changes whose magnitudes vary with the reaction rate and therefore so does the extent of the nonlinearity. The signal-stat method is indifferent to detector nonlinearity because the concentration is held constant, therefore $\partial f(C)/\partial C$ is constant. It is frequently stated that the derivative method is not dependent upon sensor nonlinearity. This will be true only when the derivative is obtained at a point of fixed concentration where $\partial f(C)/\partial C$ is always the same. This will be approximately true near zero time if the transducer is sensitive to the reaction products or a coreactant whose initial concentration is fixed but not if the substrate contributes to the signal.

Sensors such as ion-selection electrodes which are inherently nonlinear can sometimes be used to measure reaction rates if the total extent of reaction is kept quite low. Consider the simplified relationship

$$E = E^0 + \frac{RT}{nF} \ln C_i \qquad (3.22)$$

where R is the gas constant, T is absolute temperature, F is the Faraday and n an integer. For a small change in concentration of species i about some initial value, C_i^0, a Maclaurin expansion indicates that the change in potential will be:

$$\Delta E = \frac{RT}{nF} \left[\frac{\Delta C_i}{C_i^0} - \frac{1}{2} \left(\frac{\Delta C_i}{C_i^0} \right)^2 + \cdots \right] \qquad (3.23)$$

For less than 1% deviation from linearity $\Delta C_i/C_i^0$ must be less than 0.02. In order to obtain a reproducible slope, the initial concentration must be constant. Because the variable-time and signal-stat methods do not require a linear transducer, potentiometric devices can be employed for sensors in these approaches.

The Slope Method

This technique is best illustrated in terms of the product–time curves shown in Figure 3.3 which were calculated with a constant enzyme concentration but different initial amounts of substrate. It is evident that as long as $[S]_0 \ll K_M$, the initial slopes of the curves are proportional to $[S]_0$. For a first-order enzymatic reaction, it is easy to show that the slope of the concentration–time curve is always proportional to the initial substrate concentration

$$[P] = [S]_0 \left[1 - \exp\left(-\frac{k_2[E]t}{K_M} \right) \right] \qquad (3.24)$$

$$\text{slope} = \frac{d[P]}{dt} = |\mathfrak{R}| = \frac{-k_2[E]}{K_M} \left[\exp\left(-\frac{k_2[E]t}{K_M} \right) \right] \cdot [S]_0 \qquad (3.25)$$

Thus the slope should be measured at some fixed time in all experiments to maintain a fixed proportionality constant between the measured instantaneous rate ($\mathfrak{R}$) and the analyte concentration ($[S]_0$). Since the exponential term becomes very small as time progresses, the maximum sensitivity will be obtained when t is kept short and the reaction is conducted under pseudo-zero-order conditions. The product–time curve will be linear to within 1% near zero time when $k_2[E]t/K_M$ is less than 0.01.

Several automatic analog circuits which will differentiate a voltage–time

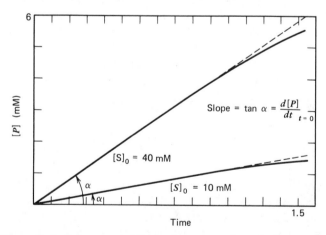

Fig. 3.3. Theoretical product concentration–time curves for a first-order enzymatic reaction illustrating the basis of the slope method of kinetic analysis. ———— = reaction time curve; -------- = initial slope.

curve have been described (8–10). In the complete absence of noise all such devices will work very nicely. Unfortunately, all real detectors will contain noise, generally over a wide frequency domain. A mathematical characteristic of differentiation is that the derivatives of high-frequency signals have a larger amplitude than those of low frequency. Consequently, derivative circuits tend to selectively amplify noise at the expense of the contribution of the reaction rate to the signal. Therefore all practical differentiation circuits contain provision for filtering (damping) high-frequency noise. As the damping time constant increases, the noise decreases. Ultimately the damping can become so excessive that the voltage output of the circuit responds too slowly to the input signal and it no longer truly reflects the rate of change of the input signal. Undoubtedly the chief limitation of automatic slope methods is their extreme sensitivity to noise. As will be seen, the fixed-and variable-time methods and the integration methods are definitely superior to the slope method in their noise-rejection ability but they cannot display the instantaneous reaction rate.

The Fixed-Time (Variable Concentration) Method

This is probably the most familiar method of kinetic analysis in that it is easy to apply to a conventional recorder tracing of a rate curve. In its *simplest* form, a signal is read at zero time (t_1) immediately after mixing and then again after some fixed interval (Δt). A typical set of fixed time measurements are shown in Figure 3.4. The technique is based on the proportional relationship between the change in concentration over the interval Δt and the initial substrate concentration. For a first-order chemical reaction, the change in substrate concentration between t_1 and t_2 which are held constant in all experiments will be

$$\Delta P = -\Delta[S] = [S]_0 \left[\exp\left(-\frac{k_2[E]t_1}{K_M} \right) - \exp\left(-\frac{k_2[E]t_2}{K_M} \right) \right] = [S]_0 \cdot \alpha(t_1, t_2) \quad (3.26)$$

Referring to the above equation the constant of proportionality $[\alpha(t_1, t_2)]$ depends only upon the reaction rate constant and measurement times, which much be rigorously controlled to obtain precise results.

Because the times t_1 and t_2 are fixed, $\Delta[S]$ will be linearly proportional to the initial substrate concentration regardless of the extent of reaction. The slope of a calibration curve, that is, the plot of $\Delta[S]$ versus $[S]_0$ depends very much on the amount of enzyme added and the fixed times.

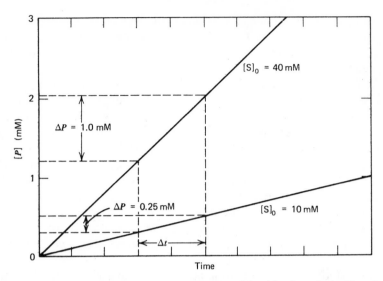

Fig. 3.4. Schematic representation of the basis of fixed-time kinetic analysis. Note that for a zero-order or pseudo-zero-order process, the change in product concentration for the fixed time difference is proportional to the sample concentration.

Optimum sensitivity will be obtained then t_1 is zero and t_2 is very large. Under this condition $\alpha(t_1, t_2) = 1$ and the method is identical to an equilibrium assay. There is a very obvious trade-off between speed of analysis (short $t_2 - t_1$) and sensitivity. It should be noted that the relationship between the experimentally measured variable, the concentration change, and the analyte concentration ($[S]_0$) is always linear regardless of the extent of reaction. The fixed-time method is not restricted to pseudo-zero-order kinetics or small extent of reaction as is the variable-time method (see below). As indicated previously, the transducer must be linear over the entire range of concentration of interest in order that accurate results be obtained. An additional factor which should be considered is that it is not always advisable to set t_1 equal to zero. Some provision for a delay time to allow mixing transients as well as any enzymatic induction period to die out must be made.

Noise is not as serious a problem in the fixed-time method as in the slope method because there is no need to use a circuit which enhances noise. Operational amplifier filters may be used provided that their time constants are short compared to t_1 and $t_2 - t_1$ so that the change in output voltage will be an unbiased estimate of the change in concentration of the measured species. The precision of the fixed-time method is related to both the time interval Δt and the initial substrate concentration. When

[S]$_0$ is very low, the change in [S] over the fixed time may be too small for precise measurement thus the sensitivity coefficient α (t_1, t_2) should be increased by increasing the time interval at the expense of sample throughput.

The fixed-time method is uniquely suited to automation by continuous-flow analysis and by automatic centrifugal analyzers since all samples require the same analysis time. Blaedel (11) has described a continuous-flow analyzer system which uses a differential detector separated by a delay coil. In his system the reagents are mixed and incubated in an up-stream delay coil and measured in an upstream cell (cell 1) after which another delay is encountered before the sample is read in a second detector (cell 2). The actual fixed time is equal to the volume of the intercell delay coil divided by the fixed flow rate. When a given sample occupies the entire volume from cell 1 to cell 2, a simple analog subtraction circuit can be used to measure the concentration change which corresponds to the fixed time interval.

Automated fixed-time instruments for batch analysis require that a detector be readout precisely at t_1, the signal stored, the detector read again at t_2 and the signal subtracted. Signal handling in the analog domain will require two sample and hold amplifiers as well as a subtractor and appropriate timing circuits and switches. This entire process is much more easily implemented with a small computer or microprocesser after the data is converted to digital form. As will be seen the apparatus for the variable-time method is much cheaper and simpler than for the fixed-time method. Additional information describing the fixed-time approach can be found in references 11–14.

The Variable-Time (Fixed-Concentration) Method

The principle of this approach is illustrated in Figure 3.5. This technique and the attendant circuitry are described in references 13–18. As indicated in the figure, the time (Δt) required for the concentration to change from fixed level C_1 to C_2 is determined. When the reaction is fast, very little time is needed whereas when the reaction is slower Δt will be longer, thus Δt is inversely related to the reaction rate. A unique danger in this technique is if C_2 is set too high, there may not be sufficient substrate available for the signal to ever reach this level.

Ingle and Crouch (19) have compared the fixed- and variable-time techniques in terms of their sensitivity to the extent of reaction. Assuming a first-order reaction, the relationship between the calculated initial substrate concentration and the measured time interval Δt can be calculated

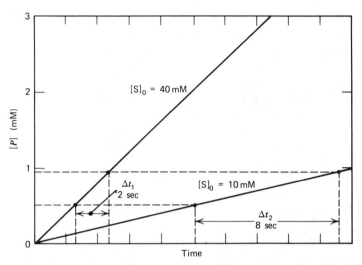

Fig. 3.5. Schematic representation of the basis of variable-time kinetic analysis. Note that the time required for a fixed concentration change is inversely related to the initial substrate concentration.

from equation (3.26) by a slight rearrangement:

$$[S]_0 = -\Delta[S]/\exp\left(-\frac{k_2[E]t_1}{K_M}\right)\left[1 - \exp\left(-\frac{k_2[E]\Delta t}{K_M}\right)\right] \quad (3.27)$$

Only when Δt is small is a simple relationship between $\Delta[S]_0$ and Δt obtained. In the limit of small Δt, a Maclaurin expansion on the exponential term containing Δt may be written as:

$$\exp\left(-\frac{k_2[E]\Delta t}{K_M}\right) \cong 1 - \frac{k_2[E]\Delta t}{K_M} + \frac{1}{2}\left(\frac{k_2[E]\Delta t}{K_M}\right)^2 - \cdots \quad (3.28)$$

After disregarding higher-order terms in Δt

$$[S]_0 = -\left\{\Delta S \Big/ \left[\frac{k_2[E]}{K_M}\exp\left(-\frac{k_2[E]t_1}{K_M}\right)\right]\right\}\frac{1}{\Delta t} \quad (3.29)$$

The initial substrate concentration will be reciprocally related to the measured time increment. In order for this approximation to be valid, the quadratic term in equation (3.28) must be small and the exponential term containing t_1 in equation (3.29) must be very close to unity. These conditions will only be valid when the extent of reaction is so low that the kinetics are essentially zero-order. The relative error involved in the use of a strict proportional relationship between the calculated initial substrate concentration and $1/\Delta t$ will depend upon both the extent of reaction and the initial concentration. Ingle and Crouch have shown that the relative error can exceed 1.2% over a tenfold range in initial concentration even if the extent of reaction is held below 2% for the lowest concentration. Obviously the accuracy of the variable-time method is very sensitive to the extent of reaction, which is not at all the case for the fixed-time method. With the widespread use of small laboratory computers and microprocessors, it should be relatively simple to prepare a program which would sample the data at more than two points, for example, by choosing three concentration levels and measuring the two corresponding time increments a set of two simultaneous equations in which the unknowns are $[S]_0$ and $k_2[E]/K_M$ could be solved. This would have the advantage that the enzyme activity would only have to be controlled during a run, not between runs, and the type of error discussed above would be unimportant.

With existing instrumental approaches, the variable-time method has three distinct advantages over the fixed-time methods:

1. The detector's linearity is not at all important, thus potentiometric measurements and direct light transmittance can be used without preliminary linearization.

2. The variable-time method is easy to implement since all that is required is to start a clock when the signal reaches level one and turn off the clock at level two. It is not necessary to store any voltage levels. When direct concentration readout is required, a voltage must be developed which is proportional to $1/\Delta t$ but this is easy to do in the analog domain with any of a number of so-called "reciprocal time computers."

3. Instrumentally the variable-time technique has a wider dynamic range than the fixed-time method (12). This results because it is much easier, more accurate, and more precise to resolve and measure a time interval corresponding to two fixed signal levels than the converse process. James and Pardue (12) state that over a decade in initial substrate concentration, both approaches can be made to work equally well, but that over a wider range, the variable-time method will be superior. Their analysis of the method assumes that all chemical and instrumental parameters are held constant. Obviously one could change the amount of enzyme or adjust various circuit components to permit measurements over a very wide dynamic range for the fixed-time approach.

The effect of noise on the fixed-time method is illustrated in Figure 3.6; similar effects exist in the variable-time procedure in that a noisy signal

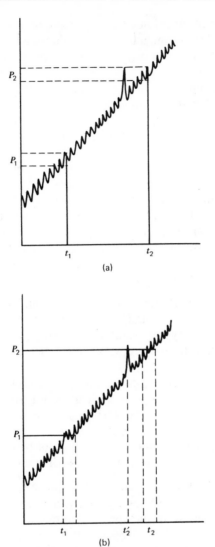

Fig. 3.6. Noise error propagation in fixed-time and variable-time kinetic analysis. [Reprinted from H. V. Malmstadt, C. Delaney, and E. Cordos, *Anal. Chem.* **42** (12), 79A (1972). Copyright by the American Chemical Society.]

may trigger the clock prematurely or after it should have been triggered. The same considerations pertain to stopping the clock at signal level two. When rate-measurement instruments, be they slope, fixed-time, or variable-time, are applied to pure noise-free signals, the results are generally at least an order of magnitude more precise (0.01–0.1%) than data obtained from actual experimental runs with photometric equipment (1–2%). This indicates that random errors in sample preparation and/or noise in the transducers are responsible for the limiting precision. James and Pardue have pointed out that a precision of 5–10% in kinetic determinations with manual data manipulation is often found (12). Similarly, Malmstadt, Delaney, and Cordos (21) point out that a major fraction of the overall poor reproducibility involved in kinetic analysis is due to indeterminate operator bias.

Ingle and Crouch (22) carried out a detailed error propagation for photometric reaction-rate measurements. The variable-time method could not be analyzed because it would require knowledge of the frequency distribution of the noise; therefore their results were restricted to consideration of the fixed-time approach. They point out that major sources of random error in reaction-rate measurements with photometric sensors are very different than the principal error sources in conventional absorption spectroscopy for the following reasons:

1. Since measurements are usually carried out near 100% T, the photocathode currents can be much larger than in conventional work. Spectral bandpass is not very important therefore high light levels can be used by opening the slits. In essence, this means that dark-current shot noise is negligible.

2. In the optimum transmittance range, a readout resolution of 0.5% will suffice for 1% overall precision. This is not true near 100% T.

3. Shot and flicker noise are very important when the total readout variance is small.

4. Long integration or filtering times can not be used in rate measurements because they will distort the signal–time curve.

Their overall results indicate that signal-to-noise ratios better than 100:1 are very difficult to obtain for small changes in transmittance near 100% T and the random noise in the spectrophotometric system will limit the analytical precision to about 1% for a total change in transmittance of 1%. Fast reaction-rate measurements will be limited by photocurrent shot noise. Slower reaction rates are generally limited by source flicker noise.

The above points out a distinct advantage of equilibrium methods. The same considerations as to spectral bandpass apply to these since the chemical specificity of the method is not supplied by the photometer but by the enzyme. Consequently, the much larger transmittance changes which occur in equilibrium methods in combination with the same high photocathode currents will allow the theoretical precision to be much better than 1%.

The Integration Method

As will develop shortly, the integration method as originally introduced was a modification of the fixed-time method. Its major purpose was to reduce noise. In order to provide some smoothing and noise reduction Cordos, Crouch, and Malmstadt introduced the first integration rate meter (23) which is based on the difference in two integrals over different segments of the signal–time curve. More elaborate and accurate versions of their original analog circuit have been described (24–27). The principle of the technique is illustrated in Figure 3.7 which is an exploded view of a zero-order reaction rate curve. When a signal is integrated over two contiguous time intervals $(t_2 - t_1)$ and $(t_3 - t_2)$, it is very easy to show that the difference in areas (ΔA) of the two segments will be

$$\Delta A = \mathcal{R} \cdot (\Delta t)^2 \qquad (3.30)$$

where $\mathcal{R}$ is the reaction rate (mmoles/sec·ml) corresponding to the zero-order kinetic curve. Since Δt is established by the instrument's settings and $\mathcal{R}$ is proportional to the substrate concentration, it is evident that ΔA will be proportional to the substrate concentration. It is interesting to note that the measured signal (ΔA) builds up as the square of the time increment whereas in the fixed-time method with zero-order kinetics, the signal

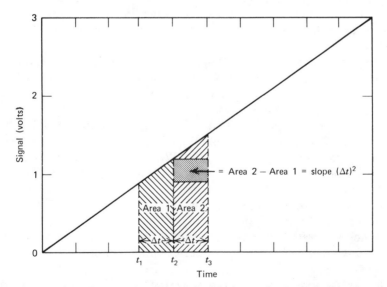

Fig. 3.7. Schematic representation of the basis of the integration method of kinetic analysis. The difference in area under two equal-time increments is proportional to the slope of the concentration–time or signal–time curve and therefore to the initial substrate concentration.

builds up as the first power of the time interval. In addition the integrator will tend to average out noise whereas the fixed-time approach has no such inherent capability.

Equation (3.30) exaggerates the rate of signal buildup. Actually the reaction will be first-order when Δt is sufficiently long. Integration of any of the true concentration–time curves [e.g., see equation 3.24)] indicates that the difference in area under any two adjacent equal time intervals will be:

$$\Delta A = \frac{[S]_0}{k_2[E]/K_M} \left[\exp\left(-\frac{k_2[E]t_1}{K_M}\right) \right]$$

$$\left[1 - 2\exp\left(-\frac{k_2[E]\Delta t}{K_M}\right) + \exp\left(-\frac{2k_2[E]\Delta t}{K_m}\right) \right] \quad (3.31)$$

In the limit of small Δt, this reduces to the result for zero-order kinetics. When all of the experimental factors (t_1, Δt, enzyme activity) are held constant, it is evident that the measured signal will be proportional to the initial substrate concentration. The slope of a calibration curve will be greatest when t_1 is zero and Δt is made quite long relative to the reaction half-life. The accuracy of the method does not depend upon a small value of Δt, but merely a fixed value; thus the technique is more closely related to the fixed-time procedure than the variable-time technique. The signal transducer must be linear over the range of concentrations of the measured species.

The chief advantage of this approach is the fact that random noise, which has as many positive and negative components will tend to average out. Analysis of purely sinusoidal noise sources (23) indicates the maximum error will decrease with the *second* power of Δt for the integration method, but only as the first power of Δt in the fixed-time method. Ingle and Crouch (25) designed a totally digital device based on a voltage-to-frequency converter as an integrator. With a 2 sec integration interval and 60 Hz noise, an increase in noise amplitude from 0.01 to 1 V peak-to-peak increased the relative standard deviation of the measurement of a 100 mV/sec slope from 0.02% to only 0.12%. The latter case represents a noise-to-signal ratio of almost 10:1. With modern digital circuitry, a difference integrator can be built using many inexpensive components such as a voltage-to-frequency converter, an oscillator time base, and an up–down counter and a decimal readout driver system. The integration method is superior to the fixed-time method with respect to both noise rejection and ease of implementation with modern electronics.

The Kinetic Difference Method

This technique, which has been developed by Weisz and his coworkers (28–31), is primarily useful in the measurement of reaction catalysts for example, enzymes. In principle, it could be applied to the determination of substrates, but there is no real advantage in this case. It can certainly be applied to the measurement of enzymes, and their activators and inhibitors. Essentially two identical detectors are placed in two separate but identical solutions. A standard solution of catalyst is added to the first cell and an unknown concentration of catalyst to the second. An electronic system measures the difference signal between the two cells. At zero time, the signal difference is zero and at the cessation of reaction the signal difference will be zero. Remember the only distinction between the two cells is the catalyst concentration. Any of several properties of the recorded signal including the magnitude of the maximum difference, or the area or the slope of the curves could be used to determine the ratio of catalyst concentrations in the two cells. Since the catalyst concentration in the other cell is known, the concentrations in the test solution can be deduced. As yet the method has not been applied to the determination of enzymes.

The Signal-Stat or Fixed Sensor-Signal Method

Conceptually this approach is not analogous to the methods described above. Rather, it is a special measurement approach to which the standard kinetic approaches can be applied. In the signal-stat system, *one* of the reactants or products is monitored and steps are taken to hold its concentration constant by addition or removal of the substance. The amount of this material is plotted as a function of reaction time. Since the signal is maintained at a fixed level, the transducer need not be linear, and thus potentiometric sensors and the current from a phototube rather than absorbance can be used. Weisz and his coworkers have developed the signal-stat technique in recent years and used it in conjunction with potentiometric, absorbance, bi-amperometric and chemiluminescent detectors (32–35). Christian has developed enzymatic analysis based on an amperostat (36) and Malmstadt and Peipmeier demonstrated the use of a pH stat for the kinetic determination of urea and glucose (37). The major application of the method has involved the control of pH by addition of strong acid or base. Obviously a pH stat can only be used to follow those reactions which generate acid or base, or at least a reaction which replaces a species with a second material with a different pK_a value. In principle one could "stat" any convenient reaction variable. It makes more sense to "stat" a reactant than a product so as to maintain the reaction rate constant and it is evidently easier to add a reactant than to remove a product.

Fortunately a very large number of enzymatic processes generate acids or bases. There is no question that the field of biochemistry exploits the pH stat much more than any other area except perhaps industrial chemistry. The output of a stat system will be a plot or recording of the amount (either moles or volume) of reagent required to maintain control. Provided that the stat is "in control," which requires a fast system-response time, the slope of the output curve will be proportional to the reaction rate. One can apply any of the mathematical and instrumental procedures outlined above to obtain the reaction rate. An unfortunate complication of this approach is that the added volume changes the concentrations of the reactants; this mandates that some correction for dilution be introduced. Recently several groups have introduced fast, totally electronic, that is, nonmechanical, coulometric pH stats, one of which detects pH changes via an indicator dye and colorimetry (38) and the other of which uses glass-electrode detection of the pH change (39). A review of the pH stat approach is available (40).

3.2.2   Precision of Kinetic Determination of Substrates in Two-Point Fixed-Time Methods

There are two additional approaches to kinetic analysis which will be taken up shortly. We will refer to these methods as the "optimized fixed-time" method and the "multi-point" kinetic method. In order to understand their importance, it is essential to grasp the influence that chemical variables can have on the precision of kinetic methods of analysis. Due to the inherent complexity of kinetic determinations, quite a number of factors can determine or limit their ultimate precision. These factors include: electronic noise emanating from the transducer or elsewhere in the measurement system, the sample pipeting and dilution system involved in preparing the reagents, imprecision in measuring time or controlling a time interval, and subtle changes in those factors which can change the reaction rate constant or Michaelis constant. Many of these factors will also affect the ultimate precision in an equilibrium assay technique, that is, the electronic noise, and errors in pipeting. In the present discussion we will focus on the *effect of the fluctuations in the rate constant per se* on the random errors in measurement of the substrate concentration.

A detailed treatment of the effect of rate-parameter fluctuations in the case of first-order and Michaelis–Menten order kinetics has been presented (42). The results given below pertain only to a pure first-order reaction.

A single kinetic run consists of a series of measurements of the concentration as a function of time. We will assume from an instrumental point of view that two measurements are made: one at zero time and one at some arbitrarily fixed later time in the reaction curve. In Pardue's classification scheme, this amounts to a two-point, fixed-time variable sensor signal kinetic assay. For the sake of simplicity, we will take the reaction to be complete at infinite time, that is, the equilibrium constant is very large and excess reagent is used. A set of theoretical calibration curves for a pure first-order reaction rate law, at five distinct measurement times, is shown in Figure 3.8. One should note from this figure that as time increases, the slope of the calibration curve becomes progressively *less sensitive* to changes in the rate parameter. The slope of this calibration curve will depend in general on the reaction rate constant, the time of measurement, and the functional dependence of the reaction rate on C_S^0 which is the *true initial concentration*. Generalizing the concepts illustrated in Figure 3.8, we may state that the *measured* concentration of unknown, designated C^*_S, will be given by

$$C^*_S = \alpha \cdot \Delta C_M; \quad \alpha = f(\Delta C_M, k, t_j) \qquad (3.32)$$

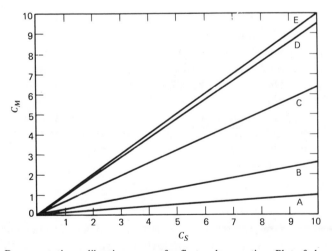

Fig. 3.8. Representative calibration curves for first-order reaction. Plot of change in concentration of the measured variable (ΔC_M) versus true initial sample concentration (C_S^0). Curve A, $kt = 0.1$ ($X = 0.095$); B, $kt = 0.2$ ($X = 0.181$); C, $kt = 1.0$ ($X = 0.632$); D, $kt = 3.0$ ($X = 0.950$); E, $kt = 6.0$ ($X = 0.997$).

where ΔC_M represents the change in concentration of the *monitored* species, k represents the first-order rate parameter, and t_j a set of two measured times.

In general, α will be a function of the initial concentration of the sought-for constituent; however, in the case of *zero-order* Michaelis–Menten kinetics ($C^0_S \gg K_M$) the rate will not depend upon the concentration of substrate C^0_S, thus a plot of ΔC_M (ordinate) versus C^0_S will have a slope of zero. Consequently, $dC^*_S/d\Delta C_M$ will be infinite and the substrate concentration will be *indeterminate*. As defined by equation (3.32), α is the slope of the analytical calibration curve (see Figure 3.8). Intuitively, it must be related to the fractional extent of reaction (X), for the case of a reaction with equal stoichiometry coefficients, it is evident that the change in monitored concentration will be given by:

$$\Delta C_M = X \cdot C^0_S \qquad (3.33)$$

We will assume that the first measurement is made at the time of mixing and then X in the above equation is identical to the fractional extent of reaction, regardless of the nature of the reaction (zero-order, first-order, or Michaelis–Menten kinetics). It is evident that α is the inverse of the fractional extent of reaction (based on the first measurement being made at $X = 0$). When both α and ΔC_M are known with *absolute accuracy*, the estimated value of C^*_S will exactly equal C^0_S. Since α is the inverse of the fractional extent of reaction, the slope of the calibration curve (ΔC_M as ordinate, C^*_S as abscissa) increases monotonically with the extent of reaction. By establishing a calibration curve under a given set of experimental conditions (pH, temperature, enzyme concentration, etc.), we are attempting to either determine or to fix the value for X. Any variation in X *between* experiments will cause a commensurate variation in the value of C^*_S. The uncertainty in C^*_S may be obtained by error-propagation mathematics. The important point is that fluctuation in rate parameters will cause changes in X and therefore act as a source of between-run random error in the determination of C^0_S. In the case of a pure first-order reaction (42) it is easy to show that

$$\frac{\sigma_{C^*_S}}{C^0_S} = \frac{(1 - X) \ln (1 - X)}{X} \frac{\sigma_k}{k} \qquad (3.34)$$

where σ_k is the expected fluctuation in the first-order rate parameter. The dependence of $\sigma_{C^*_S}$ on the fractional extent of reaction embodied in equation (3.34) is shown in Figure 3.9. It is evident that for a given rate constant, the precision improves very rapidly as the extent of reaction increases. As kt becomes large, the reaction goes to completion, that is, we gradually achieve the limit of an equilibrium assay technique which

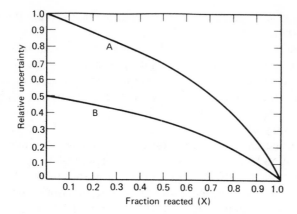

Fig. 3.9. Percent Relative Uncertainty in Estimated Initial Concentration ($C_s{}^*$) vs Extent of Reaction for First-Order Kinetics. A plot of $\sigma_{C_s^{*}}/C_s{}^0$ versus X computed from equation (3.36). A, $\sigma_k/k = 0.01$; B, $\sigma_k/k = 0.005$.

is totally insensitive to any fluctuations in the reaction rate. The sensitivity of the measured substrate concentration to change in k increases as the reaction time decreases. Thus when all other factors are negligible, it is evident that an equilibrium assay will be more precise than a fixed-time two-point kinetic assay.

A very interesting result of the above calculation is the fact that the relative standard deviation does not depend upon the initial substrate concentration! When this is not experimentally observed to be true, this indicates that the precision is being limited by some other factor, for example, electronic noise or readout imprecision, but not the rate constant. The results shown in Figure 3.9 indicate that at low fractional conversion ($X < 0.2$) the precision of substrate measurement is about equal to the relative fluctuation in the rate constant. This means that the precision in the enzyme's rate constants (k_2/K_M) and its concentration must be less than 1% to obtain 1% precision in the measurement of C_S^0. The relative uncertainty in rate constant due to uncertainty in temperature can be obtained by differentiation of the Arrhenius equation and use of error propagation as follows:

$$\frac{\sigma_k}{k} = \frac{|E_a|}{RT^2}\,\sigma_T \tag{3.35}$$

where E_a is the energy of activation, R the gas constant, T the absolute temperature, and σ_T the uncertainty in temperature. This equation indicates an uncertainty of 5.6% per °K per kcal/mol activation energy at

$300°K$. The activation energies of enzymatic reactions are typically 10–12 kcal/mole (corresponding to a Q_{10} of 2).

It should be understood that the temperature effect discussed above pertains only to reaction rate constant per se. The transducer may have a very significant temperature coefficient, for example, amperometry with a dropping mercury electrode has an inherent coefficient of about 2% per °C for a diffusion-controlled current, the potential of ion-selective electrodes will vary several millivolts per °C, and the fluorescence of many compounds decrease with temperature, in particular the pyridine nucleotides (see below) have a temperature coefficient of 1.6% per °C.

Estimation of the effect of pH on reaction-rate parameters is complicated by the fact that, in the case of enzymic reactions, the reaction rate passes through a relative maximum as a function of pH (see Chapter 2). This is often due to ionization of the acidic and basic amino acid sidechains in the protein. At the optimum pH, small fluctuations in pH will have no effect; however, at a pH well removed from this point, one generally sees a dependence on the first power or on the inverse of the hydrogen-ion concentration. When a single enzyme is employed, one can almost always work at the optimum pH; however, this may not be possible with a multienzyme assay. At worst, the relative uncertainty in a rate parameter due to a pH fluctuation will be:

$$\frac{\sigma_k}{k} \cong 2.303 \, \sigma_{\mathrm{pH}} \qquad (3.36)$$

An uncertainty of only 0.01 pH unit will cause a fluctuation of 2.3% in the rate parameter. If one assumes a much weaker dependence of rate parameter on pH, for example, only 10% as effective as that described by equation (3.36), the fluctuation will amount to a few tenths of 1%.

The effect of activators and inhibitors on the rate parameters is quite complicated. These species can be present in low concentration where dependence on their concentration is first-order, or in "saturating" concentrations where the reaction rate is independent of their level. In complex biological fluids, the intrinsic variability in the amount of endogeneous activators, inhibitors, and other rate-controlling species may well establish the net uncertainty in the kinetic parameters.

It is not unduly conservative, in light of all of the above problems, to estimate that rate parameters may fluctuate as much a 1% *between samples or between samples and standards.*

3.2.3   Optimized Fixed-Time Kinetic Analysis

Atwood and Di Cesare considered the question of whether an optimum amount of catalyst exists in a two-point kinetic substrate analysis (43).

For present purposes this means an enzyme level above or below which the measured reaction velocity will be less than at the optimum level. It is evident that, if the first concentration measurement is made at the time of mixing ($t_1 = 0$), then there will be no optimum enzyme level; the measured reaction velocity will merely increase in proportion to the amount of enzyme added. However, Atwood and Di Cesare realized that if the first measurement was not taken at $t_1 = 0$ (this is a very common situation) but at some later time, then a considerable fraction of the substrate may be consumed before the reaction velocity is measured. This can lead to the situation where a peak will occur in a plot of measured reaction rate, at constant initial substrate, versus added catalyst activity.

In the case of a pure first-order reaction ($[S]_0 \ll K_M$) where the product concentration follows equation (3.24), it is easy to show that the reaction velocity decreases exponentially [see equation (3.25)]. The reaction velocity ($\mathcal{R}$) at the time of measurement (t_m) will be

$$\mathcal{R}_{t_m} = \frac{k_2[E]}{K_M} \exp\left(-\frac{k_2[E]t_m}{K_M}\right) \cdot [S]_0 \qquad (3.37)$$

Note that the dependence of $\mathcal{R}_{t_m}$ on the amount of enzyme is quite complex since $[E]$ appears in both the coefficient and the argument of the exponential term. For any value of t_m *other than zero* a peak will occur in a plot of $\mathcal{R}_{t_m}$ versus $[E]$. Thus there is evidently an *optimum amount of enzyme*. It is easily shown by differentiation of equation (3.37) that this optimum enzyme level will be

$$k_2[E]_{opt} = \frac{K_M}{t_m} \qquad (3.38)$$

The analytical implications of the concepts proposed by Atwood and Di Cesare are very significant. First, at the optimum, *small* changes in the amount of enzyme or factors (pH, temperature, etc.) which influence its activity are unimportant since the derivative of $\mathcal{R}$ with respect to $[E]$ at the optimum is zero. Second, the sensitivity of the kinetic analysis, that is, the slope of a plot of $\mathcal{R}$ versus $[S]_0$ can be greatly varied by choice of $[E]$ and t_m.

Hewitt and Davis recently recognized (44) that if there is an optimum level of enzyme for a given measurement time, then, clearly, there will be an *optimum measurement time for any given level of enzyme*. This leads to the concept of optimized two-point fixed-time kinetic analysis. By judicious choice of experimental conditions, particularly of the fraction reacted at the two measurement times, it is possible to achieve kinetic analyses which are *desensitized to fluctuation in rate parameters*. The net result will be much more precise methods of kinetic analysis. At this writing, the technique is in its infancy; we feel that developments in this area will be very important.

3.2.4 Multi-Point Kinetic Analysis

As pointed out above, two point fixed-time kinetic analysis can be much less precise than equilibrium techniques based on the same chemistry. Pardue (45, 46) has recognized the fact that differences in rate constants between samples can be used to *detect errors* in kinetic analysis. Since, in the case of a first-order reaction, one can determine the rate constant without prior knowledge of the initial concentration of rate-limiting species, it should be possible simultaneously to compute both the initial concentration and the rate constant. Clearly, by treating the data in a way that allows the kinetic parameters to vary from sample to sample, one can assign the rate fluctuation due to the rate constant to its rightful source. This should effect a significant improvement in between-run precision. In order to obtain both the rate constant and sample concentration, a multi-point (three points at least) method is called for.

Mieling and Pardue (41) recently introduced a technique based on the above concept. Experimental kinetic curves were fitted over about four to seven half-lives. When some 250 pieces of data obtained in a single run were used, a within-run precision of 0.1% was obtained. More importantly, between-run precision (based on five replicates) was improved dramatically to 0.2%; only slightly poorer than the within-run results. Mieling and Pardue demonstrated that the multi-point curve fitting method was remarkably *insensitive* to pH, temperature, ionic strength, and reagent concentration. We anticipate that far more sophisticated data processing methods, such as the Kalman filter (47), could be used to optimize multi-point kinetic analyses.

3.2.5 Detection Limit for the Kinetic Determination of a Substrate

The detection limit of an analytical method is often defined as the minimum level or amount of material (in moles or moles per unit volume) which is required to produce a signal which exceeds the noise level on a blank with some preassigned degree of confidence. Thus in absolute terms, the detection limit is a rather arbitrary quantity until the confidence limits are assigned. Implicit in the concept is the signal-to-noise level. Since the total signal change in a kinetic assay is often much smaller than that in an equilibrium assay based on the same chemical and physical measurement system, it is safe to say that a kinetic assay will always have a higher detection limit. This is not to imply that kinetic methods are invariably poorer in this regard than equilibrium assays for the same species since, as pointed out above, an equilibrium assay can be afflicted with a number of problems which prohibit its implementation with the same reaction and type of sensor. To a first approximation, we can state that when comparable chemistry and sensors are used, the *ratio of the detection limit of a kinetic assay to an equilibrium assay will be equal to the fractional extent of reaction.* As Ingle and Crouch (22) have pointed out, the bandwidth of an electronic measurement system for reaction-rate methods must be greater than that of an end-point method in order not to distort the shape of the signal–time curve. Thus the blank or background noise may be higher in a kinetic assay thereby increasing the detection limit even further than considerations based on the total change in the signal. An exact comparison of the techniques can only be based on a detailed knowledge of the type of noise encountered, which is evidently related to the sensor and is beyond the scope of the present treatment.

The principle analytical features of the above rate measurement techniques are briefly summarized in Table 3.5. Also included are references to original literature in which the electronic circuitry required to implement the approaches is described in detail. A number of useful reviews of kinetic analysis and enzymatic analysis including the article by Malmstadt (48), the chapters by Pardue (13) and Blaedel (14) and a critical review by Mottola (49), as well as the monographs by Guilbault (50), and Mark and Rechnitz (51), should be consulted for details on instrumentation and applications.

3.3 CHEMICAL TECHNIQUES USED TO VISUALIZE THE EXTENT OF ENZYMATIC REACTIONS

All of the preceeding discussion as to the selection and use of equilibrium and kinetic methods is contingent upon the availability of physical and chemical methods for producing a measurable signal. Just as chemists involved in inorganic analysis naturally favor certain types of chemical reactions and proven instrumentation to obtain measurable signals, biochemists have found that some species involved in biological reactions are more convenient to measure than are other species.

Undoubtedly the most useful system of enzymatic analysis is based on the pyridine nucleotides NAD/NADH (oxidized and reduced nicotinamide adenine dinucleotide) and NADP/NADPH (oxidized and reduced nicotinamide adenine dinucleotide phosphate) which are easily linked to a great number of oxido–reductase enzymes (dehydrogenases) (5, 50).

Table 3.5 Comparison of Instrumental Approaches to the Measurement of Reaction Rates

Method	Measured parameter	Relationship to concentration	Transducer response	Dependence on kinetic law	Sensitivity to noise in signal
A. Derivative	Slope	slope $\propto [S]_0$	Nonlinear	Not important	Very sensitive
B. Fixed time	Change in concentration	Δ conc $\propto [S]_0$	Linear	Not important	Sensitive
C. Variable time	Time increment	$(1/\Delta t) \propto [S]_0$	Nonlinear	Must be close to zero-order	Sensitive
D. Integration	Difference in area	Δ Area $\propto [S]_0$	Linear	Not important	Relatively insensitive
E. Signal-stat	Amount of added material	Any of the above	Nonlinear	Not important	Depends upon implementation
F. Multi-point	Many concentrations versus time	Least-squares fit	Linear	Not very important	Low

Method	Precision	Dynamic range	Amenable to continuous-flow analysis	References
A. Derivative	Poor	Good	No	8–10, 21
B. Fixed time	Good	Good	Yes	11–14, 18, 21, 22
C. Variable time	Good	Best	No	15–18, 21
D. Integration	Very good	Good	No	23–27, 21
E. Signal-stat	Depends upon method	Applicable only to slow reactions	No	32–40
F. Multi-point	Best	Good	No	41, 46, 47

NADH and NADPH have essentially identical absorption spectra in the useful wavelength region (6.22 × 10^6 cm^2/mole for λ_{max} = 340 nm). Even more sensitive analyses can be conducted by use of the intense fluorescence emitted by these materials. By coupling sets of enzyme reactions, it is readily possible to transform a species of interest to a substrate which will consume or generate a pyridine nucleotide. Lowry and Passonneau (5) point out that the pyridine nucleotides are of great general utility because:

1. They are the natural oxidizing and reducing agents which are active in a spectrum of biological reactions. When one or more enzyme systems are used jointly, very few biochemical compounds exist which cannot be linked to the pyridine nucleotides.

2. The reduced nucleotides are fluorescent whereas the oxidized forms are not. Because the fluorescence is easily measured at 10^{-7} M, very small quantities of the nucleotides can be measured.

3. The reduced material can be completely destroyed in acid without changing the concentration of the oxidized form. Conversely the oxidized from can be destroyed in alkali. Thus the excess coenzyme, be it NAD or NADH, can be removed and the product measured by very sensitive methods.

4. Either the oxidized or reduced forms can be converted to a material which is detectable by fluorescence at the 10^{-8} M level.

Examples of other species which are frequently involved in biochemical reactions as either coreactants or products are: adenosine triphosphate (ATP), oxygen, hydrogen peroxide, hydrogen and hydroxide ions, carbon dioxide, ammonia, and orthophosphate. These species are important from an analytical viewpoint because they can be measured very easily with rather low detection limits. For example, 10^{-15} moles of NADH can be measured via bacterial bioluminescence (see below) and 10^{-14} moles of ATP have been determined via the well known "firefly" reaction [equations (3.69)–(3.71)]. Because ADP (adenosine diphosphate) can be converted to ATP by reactions with phosphoenolpyruvate in the presence of pyruvate kinase which is a transphosphorylase, any reaction which generates ADP or uses it up can also be monitored by chemiluminescence.

All of the above easily measurable molecules will be termed *bio-analytical indicators* and the reaction in which their concentrations are measured will be termed *indicator reactions*. Any reaction which converts the substrate of interest to a form which can then be measured will be termed an *auxiliary* reaction. Evidently a series of auxiliary reactions may be needed to lead the sample of interest to the indicator or measurement reaction. Before considering the interaction between indicator and auxiliary reactions, some consideration of the properties of single-step reactions should be given, in particular what can be done to influence the equilibrium position of a thermodynamically unfavorable reaction.

3.3.1 Shifting Equilibrium Constants in One-Step Methods

In all of the kinetic methods discussed thus far, we have neglected the fact that the reaction is ultimately reversible, that is, at some point the substrate will no longer be consumed because the system has achieved its thermodynamic resting place. This is an important consideration in kinetic analysis since the total extent of reaction may be so limited that the total signal change is too small for measurement, or the reaction may stop very quickly, or the reaction may not be pseudo-zero-order over a sufficient time scale to use a particular kinetic method, for example, the variable time approach.

For the purpose of substrate analysis, not determination of enzyme activity, a number of things can be done to improve the measurement thermodynamically:

1. Increase the co-reactant concentration.
2. Change a co-reactant to a thermodynamically more reactive species.
3. Change the pH if H^+ is used or generated.
4. Trap or remove a product.

Each of these techniques can be implemented in several ways. The effect of coreactant concentration of the equilibrium position has already been discussed under end-point assays.

A very elegant approach is to use a different coenzyme. For example the K_{eq} of all reactions which utilize NAD/NADH depend upon the redox potentials of the species involved:

$$H^+ + NAD^+ + 2e^- \rightleftharpoons NADH \quad E_A^{0\prime} = -0.320 \text{ V} \quad (3.39)$$

$$Ox + ne^- \rightleftharpoons Red \quad E_B^{0\prime} \quad (3.40)$$

$$nH^+ + n(NAD^+) + 2Ox \overset{K_{eq}}{\rightleftharpoons} n(NADH) + 2Red$$

The equilibrium constant of the overall reaction (K_{eq}) as written will be:

$$\log_{10} K_{eq} = \frac{2n_B}{0.059} (E_A^{0\prime} - E_B^{0\prime}) \quad \text{at } 25°C \quad (3.41)$$

The cofactor NAD can be changed to acetylpyridine adenine dinucleotide (APAD) which is a stronger oxidizing agent (52–54)

$$H^+ + APAD^+ + 2e^- \rightleftharpoons APADH \quad E_C^{0\prime} = -0.248 \text{ V} \quad (3.42)$$

It is easy to show that all reactions in which n_B is 1e will have their equilibrium constant increased by a factor of 275. In some cases this may be enough to make a promising method into a usable one but will not allow the use of reactions whose K_{eq} with NAD is less than unity. Occasionally with a non-specific enzyme, one can swap coreactants other than the coenzyme. For example in the determination of glyceraldehyde-3-phosphate with glyceraldehyde-3-phosphate dehydrogenase, inorganic phosphate which is used up in the reaction can be replaced with arsenate (55, 56).

Since a great many enzymatic reactions and virtually all NAD/NADH or NADP/NADPH reactions involve the production or usage of protons, the effective equilibrium constant can be shifted by several decades by varying the pH. A definite limitation of this technique is that the enzyme activity may be decreased at a pH where the equilibrium constant is large enough to be useful. An interesting case of a reaction which is rather easily run in either the forward or reverse direction is the lactate–pyruvate transformation. At nearly neutral pH, pyruvate is essentially quantitatively converted to lactate in the presence of lactic acid dehydrogenase (LDH):

$$\text{pyruvate} + \text{NADH} + \text{H}^+ \overset{\text{LDH}}{\rightleftharpoons} \text{lactate} + \text{NAD}^+ \quad K_{eq}$$

$$= 2 \times 10^4 \, (\text{pH} = 7.0) \quad (3.43)$$

If the pH is raised to 9.5 and a trapping agent added to remove pyruvate, for example, hydrazine, the equilibrium constant decreases by about six orders of magnitude and favors formation of pyruvate from lactate. This same approach, that is, addition of hydrazine, can be applied to the measurement of steroids via steroidal dehydrogenases (57).

Trapping agents which tie up or remove a product, thereby forcing the enzymatic reaction to completion, can be carried out in many different ways. As alluded to above, ketones and aldehydes can be trapped by forming either the hydrazine or semicarbazide derivative. It should be noted that such trapping agents can interfere, for example, NAD forms an adduct with semicarbazides and hydrazides which absorb at 340 nm even though NAD does not, thus a blank run must be carried out. It is also possible for the trapping agent to effect the kinetics of the enzyme per se.

Another way to trap reaction products is actually rather general. Either NADH or NADPH which are formed as products can be reacted with oxidizing agents (methylene blue, ferricyanide) thereby reforming more NAD or NADP. When the reduced nucleotide is reacted with a tetrazolium salt, a very intensely absorbing formazan will be generated. This has the net effect of improving the equilibrium and kinetic situation by maintaining the NAD(P) concentration constant and suppressing the NAD(P)H concentration, and improving the reaction sensitivity by forming a species with a higher extinction coefficient. Unfortunately the direct reaction of the reduced nucleotide is kinetically slow and a reaction mediator, that is, electron-transport catalyst, is required. The reaction can be conducted with 5-methyl phenazinium methyl sulfate [reactions (3.44) and 3.45)] or enzymatically with diaphorase:

NADH + (5-methylphenazinium) + $CH_3SO_4^-$ →

NAD$^+$ + (5-methyl-5,10-dihydrophenazine) + $CH_3SO_4^-$ (3.44)

(5-methyl-5,10-dihydrophenazine) + (monotetrazolium salt) →

(5-methylphenazinium) + (monoformazan)   (3.45)

or

$$NADH + H^+ + \text{diaphorase} - FAD \rightleftharpoons NAD$$
$$+ \text{diaphorase} - FADH_2 \qquad (3.46)$$

$$\underset{R_3 \quad N \quad R_1}{\overset{R_2}{\underset{|}{N-N}}} \quad + \text{diaphorase} - FADH_2 \rightleftharpoons$$

$$\text{diaphorase} - FAD + \underset{R_3-C\underset{N}{\nwarrow}N-R_1}{\overset{R_2}{\overset{|}{N}}}H + H^+ \qquad (3.47)$$

In either case the net reaction will be:

$$NADH + \underset{R_3 \quad N \quad R_1}{\overset{R_2}{\underset{|}{N-N}}} \rightarrow NAD + \underset{R_3 \quad N \quad R_1}{\overset{R_2}{\overset{|}{N}}}H \qquad (3.48)$$

It should be noted that either NAD or NADH can be inexpensively consumed by many suitable enzymatic reactions, for example, by ethanol and alcohol dehydrogenase thus shifting an equilibrium system by destruction of a product is not at all uncommon, and the reaction can be carried out with enzymes other than diaphorase. A final trapping method other than the use of reactions which consume nucleotides is based on the reaction of the other reaction products. Any reasonably complete reaction can be employed provided that in the case of a kinetic determination of the substrate of interest, the second reaction is much faster than the first. The kinetic and thermodynamic schemes for both of the above methods of trapping are indicated in Table 3.6, that is, nucleotide recycling and simple consumption. The two forms of the coenzyme (NAD, NADH) are denoted C and C'.

A number of species produced in enzymatic reactions can be removed from solution by gaseous diffusion and other methods, thereby forcing the reaction to completion. For example, in base, ammonia is readily

Table 3.6 Methods for Driving Reactions to Completion

Reaction constraints	Cofactor recycle	Primary product consumption
	$S_1 + C \overset{K_1}{\rightleftharpoons} P_1 + C'$	$S_1 + C \overset{K_1}{\rightleftharpoons} P_1 + C'$
	$C' + S_2 \overset{K_2}{\rightleftharpoons} C + P_2$	$P_1 + S_2 \overset{K_2}{\rightleftharpoons} P_2 + P_3$
Kinetic		$C + S_1 \overset{k_1}{\longrightarrow} P_1 + C'$ $+ S_2$ $\downarrow$ $P_2 + P_3$
Overall	$S_1 + S_2 \rightleftharpoons P_1 + P_2$	$S_1 + C + S_2 \rightleftharpoons P_2 + C' + P_3$
Equilibrium	$K_1 \cdot K_2 \gg 1$	$K_1 \cdot K_2 \gg 1$
Rate constants	$k_1 \ll k_2$	$k_1 \ll k_2$

removed. Carbon dioxide can be removed at low pH, particularly if the solution is purged with an inert gas and the enzyme carbonic anhydrase is present to speed up the reaction:

$$H_2CO_3 \rightleftharpoons CO_2 + H_2O \qquad (3.49)$$

Metal ions can be added to tie up phosphate or CN^-, provided that the metals do not inhibit the enzyme.

3.3.2 Enzymatic Indicator Reactions and Auxiliary Enzymes

In all of the reactions discussed above the direct substrate reaction was sufficient by itself to produce a signal change. Additional reactions were employed only to make the overall reaction equilibrium constant sufficiently large so that the reaction could be measured by either an equilibrium or kinetic technique. Whenever the substrate reaction does not lead directly to a species which is analytically convenient, it is usually (perhaps universally) possible to introduce additional enzymes which will lead the substrate to a species which can be measured.

The auxiliary and indicator reactions can interact in only two ways, that is, the indicator reaction may act on a product of the auxiliary reaction or the indicator reaction may precede the auxiliary reaction (58). In the following discussion, S_1 will always denote the *substrate to be measured*; all other species, whether they are reactants (indicated as R) or products (denoted P) will be numbered in chronological order as the reaction takes place, not in their actual order of addition to a solution.

Succeeding Indicator Reaction

$$S_1 + R_1 \underset{K_{aux}}{\rightleftharpoons} P_1 + P_1' \quad \text{Auxiliary Reaction} \qquad (3.50)$$

$$(P_1 \text{ or } P_1') + R_2 \underset{K_{ind}}{\rightleftharpoons} (P_2 \text{ or } P_2') + P_3 \quad \text{Indicator Reaction} \qquad (3.51)$$

In this case, the decrease in R_2 concentration or the increase in P_2 or P_3 will be measured as a function of time or at the end of the reaction. Note that if the decrease in R_1 or increase in P_2 is measured, this does not truly represent the use of an indicator reaction but only an attempt to shift the overall reaction, which is something the indicator reaction certainly can do. An indicator reaction of this type must have $K_{aux} > 1/K_{ind}$; in fact if K_{ind} is small, the method will still work provided that $K_{aux} \gg 1/K_{ind}$. If a kinetic assay is to be employed, then the net activity in the indicator step ($k_{2,ind} \cdot [E]_{ind}/K_{M,ind}$) must be considerably greater than the net activity in the auxiliary reaction ($k_{2,aux}[E]_{aux}/K_{M,aux}$). The two conditions are summarized in Table 3.7.

When several auxiliary reactions are used, then in order to complete the reaction in a reasonable period of time or to insure the accuracy of a kinetic analysis, the total activity in each successive step should be several times greater than in the first step. A more detailed discussion of this topic can be found in tracts on enzymatic analysis (59).

Specificity in coupled enzyme reactions is a very interesting and surprisingly pleasant topic. The use of multiple enzymes may actually improve the specificity and therefore the accuracy (not reproducibility) when analyzing complex mixtures. Suppose that the *auxiliary* reaction is rather *nonspecific*, for example, the hexokinase-catalyzed phosphorylation of

Table 3.7 Conditions for Accurate Use of Indicator Reaction Systems

System	Equilibrium condition	Kinetic condition
Succeeding indicator reaction	$K_{ind} \cdot K_{aux} \gg 1$	$(k_2[E]/K_M)_{ind} > 5 \cdot (k_2[E]/K_M)_{aux}$
Preceeding indicator reaction	$K_{ind} \ll 1$ and $K_{ind} \cdot K_{aux} \gg 1$	Same as above

glucose or the oxidation of amino acids with a broad-spectrum oxidase, but that a *specific indicator* reaction is used. Since only the substrate of interest will be acted upon to produce the final product, the analysis will be more selective than if only the auxiliary reaction were employed. Since we are speaking of catalytic selectivity, we assume that the measurements are being made in a finite time relative to the complete conversion of all possible reactants. When more than one auxiliary reaction is used to get to the indicator reaction, we actually have a series of "interference filters," each of which has the potential for providing selectivity but, realistically, can also allow interfering species to come into play. As a general rule, the specificity of the overall scheme is set by the most selective enzyme, not the least specific. To some extent the above statement must be modified by a provision for the total enzyme activity. For example, if an enzyme has a very high activity this can to some extent overcome its high specificity. Thus excessively high levels of enzymes should as a rule be avoided on the basis of both economy and avoiding interference. *Obviously any auxiliary enzyme in a reaction stage immediately prior to the indicator reaction which converts a sample contaminant to the measured material can cause problems.*

Preceeding Indicator Reactions

This situation is conceptually a bit puzzling at first glance and experimentally, for reasons which will be evident, not nearly as useful as the application of a succeeding indicator reaction, and therefore should be avoided whenever possible. Use of this technique is contingent entirely upon an *unfavorable* position of the indicator reaction, which is then pulled to completion by a very favorable position of the auxiliary reaction which involves, S, the substrate of interest. The reaction scheme is:

$$R_1 + R_2 \underset{K_{ind}}{\rightleftharpoons} P_1 + P_2 \quad \text{Indicator Reaction} \tag{3.52}$$

$$S + (P_1 \text{ or } P_2) \underset{K_{aux}}{\rightleftharpoons} P_3 + (P_1' \text{ or } P_2') \quad \text{Auxiliary Reaction} \tag{3.53}$$

Imagine that R_1 and R_2 are mixed, but that $K_{ind} \ll 1$; very little of either P_1 or P_2 will be formed. Now some sample is added along with any additional unindicated reagents as needed, and the auxiliary enzyme. If K_{aux} is much greater than K_{ind}, an overall reaction which consumes S as well as R_1 and R_2 and produces P_1 (or P_2), P_3, and P_2' (or P_1') will occur. It is assumed that only R_1, R_2 or one of the products (P_1, P_2) are measurable; otherwise one could simply add P_1 or P_2 in excess without the indicator enzyme and develop a measurable signal. Obviously no side reactions with the auxiliary enzyme of either P_1 or P_2 can be tolerated.

Actual calculations of the analyte concentration are almost always complicated by the fact that K_{ind} is finite and some reaction will occur. Detailed equilibrium calculations show that unless K_{ind} is extremely small, the calibration curve will be nonlinear at low concentration of the substrate (S). The problem cannot be eliminated by a simple blank run because the addition of substrate shifts the position of the unfavorable indicator reaction. In a sense the blank is a function of the sample concentration! Bergmeyer's treatise should be consulted for details on this situation (59). The overall scheme is so complex as to be nearly useless for kinetic analysis since the sample correction factor will be time-dependent. Evidently, the kinetics of the indicator reaction must be very fast compared to the auxiliary reaction to attempt a kinetic assay with a preceeding indicator reaction.

Serial Analysis with the Aid of Auxiliary Enzymes

Example: To illustrate the power of linked auxiliary enzymes to yield highly selective analysis in complex mixtures, let us consider the determination of a mixture of glucose, fructose, lactose, maltose, and sucrose with the enzyme glucose-6-phosphate dehydrogenase. Glucose is easily converted to glucose-6-phosphate (Glu-6-P) by hexokinase(HK)-catalyzed phosphorylation:

$$\text{glucose} + \text{ATP} \xrightarrow{\text{HK}} \text{Glu-6-P} + \text{ATP} \qquad (3.54)$$

The glucose-6-phosphate is readily determined with high specificity via the use of glucose-6-phosphate dehydrogenase (G-6-PDH):

$$\text{Glu-6-P} + \text{NADP} \xrightarrow{\text{G-6-PDH}} \text{6-P-gluconate} + \text{NADH} + \text{H}^+ \qquad (3.55)$$

If fructose were present, it would also be phosphorylated by ATP, but the product fructose-6-P will not react with NADPH. The enzyme fructose-6-phosphate kinase will catalyze its conversion to glucose-6-phosphate. Maltose, lactose, and sucrose can be split into glucose and galactose or fructose by addition of the appropriate enzymes and then analyzed individually by reaction with hexokinase and glucose-6-phosphate dehydrogenase. An absorbance–time curve in which eight separate sugars and sugar phosphates are determined in one series of reactions is shown in Figure 3.10 (59).

3.3.3 Substrate Analysis with Chemical Amplification by Enzyme Cycling

Chemical amplification in analysis (60) refers to the fact that one can often increase measurement sensitivity by reacting the substance of in-

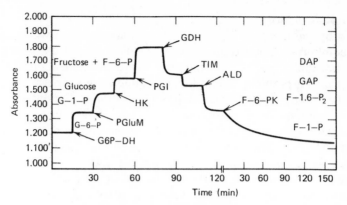

Fig. 3.10. Absorbance–time curve for sequential substrate analysis by serial enzyme additions. The indicated substances can be measured upon addition of the relevant enzyme. [Reproduced with permission from H. U. Bergmeyer, Ed., *Methods of Enzymatic Analysis*, Vol. 1 (Academic, New York, 1974); Copyright by Academic Press]

terest with another species which is consumed in stoichiometric greater quantities. For example in a recent X-ray fluorescence method for inorganic phosphate the phosphate was reacted with molybdate to form a 12-molydophosphoric acid (61). The amount of molybdenum was then determined after extraction and isolation and provided an inherent gain of 12 in sensitivity. Equilibrium chemical amplifications can provide no more chemical gain than the ratio of the stoichiometry coefficients. This is not the case in kinetic amplification which in some cases can provide a gain of 50, and when two-stage amplifiers are used, this can be increased to 50,000 and in some cases to 400,000,000 (5).

The basic principle of enzymatic chemical amplification is the use of enzymes which *regenerate or cycle* the substrate and produce a measurable species. Consider the reaction:

$$\boxed{S_1} + R_1 \xrightarrow{\text{enz 1}} \boxed{P_1} + P_2 \tag{3.56}$$

$$\boxed{P_1} + R_2 \xrightarrow{\text{enz 2}} \boxed{S_1} + P_3 \tag{3.57}$$

The overall reaction will be:

$$R_1 + R_2 \rightarrow P_2 + P_3 \tag{3.58}$$

When the substrate S_1 is rate-limiting then the rate of production of P_2 or P_3 will be limited by S_1, provided that a surplus of the reagents R_1 and R_2 is present, but the total amount of product is not stoichiometrically limited by S_1 because it is regenerated. The amplification will be a function of the turnover number of each enzyme and the reaction time for each cycle. Double cycling can be carried out when one of the products can enter into a cycle. The net gain or amplification will be equal to the product of the gain of each stage. The rate of cycling can be calculated in the steady state very easily since in steady state, the rate of production of P_1 will exactly equal its rate of consumption. Thus at low concentration, where first-order kinetics prevail,

$$\left(\frac{k_2[E]}{K_M}\right)_1 [S] = \left(\frac{k_2[E]}{K_M}\right)_2 [P_1] \tag{3.59}$$

The overall reaction rate is the rate at which the sum of S and P_1 are reacted:

$$k([S] + [P_1]) = \left(\frac{k_2[E]}{K_M}\right)_1 [S] \tag{3.60}$$

Substituting where appropriate to eliminate the concentrations of S and P_1, we find that:

$$k = \frac{(k_2[E]/K_M)_1(k_2[E]/K_M)_2}{(k_2[E]/K_M)_1 + (k_2[E]/K_M)_2} \tag{3.61}$$

This is not unexpected because the two reactions are connected in parallel. This relationship tells us that the actual rate will be limited by that enzyme which is present in smallest amount. The rate constant k is the gain per unit time, that is, the ratio of moles of final product per mole substrate (S_1) per unit time.

Three of the most important enzyme-cycling methods are those which permit measurement of the coenzyme systems (NAD/NADH, NADP/NADPH, and ATP/ADP). Methods for this have been discussed by Lowry and Passonneau (4).

NAD Cycle

$$NAD + lactate \rightarrow NADH + pyruvate + H^+ \tag{3.62}$$

$$NADH + \alpha\text{-ketoglutarate} + NH_4^+ \rightarrow NAD + glutarate \tag{3.63}$$

The enzymes employed here are lactic acid dehydrogenase and glutarate dehydrogenase. A gain of 6000–8000 per hr can be achieved. At the end of the cycle the reaction is quenched by addition of NAD, glutarate dehydrogenase, and peroxide. The peroxide destroys α-ketoglutarate. The pyruvic acid is then measured as per reaction (3.62). Using fluorometry in alkaline solution, one can detect as little as 2×10^{-13} moles of NAD.

NADP Cycle

$$NADP + Glu\text{-}6\text{-}P \rightarrow NADPH + 6\text{-}P\text{-}gluconate + H^+ \qquad (3.64)$$

$$NADPH + \alpha\text{-}ketogluterate + NH_4^+ \rightarrow NAD + glutamate \qquad (3.65)$$

The enzymes used here are glucose-6-phosphate dehydrogenase and glutamate dehydrogenase. A gain of 15,000–20,000 per hr can be achieved. At the end of a cycle (1 hr) the 6-P-gluconate is determined with 6-P-gluconate dehydrogenase and extra NADP. About 1×10^{-13} mole of NADP can be measured using fluorescence.

ADP Cycle

$$ADP + phosphoenol\ pyruvate \rightarrow ATP + pyruvate \qquad (3.66)$$

$$ATP + glucose \rightarrow Glu\text{-}6\text{-}P + ADP \qquad (3.67)$$

The enzymes employed are pyruvate kinase and hexokinase. This is not as efficient as the preceeding cycles since a gain of only 1500 per hr is achieved. The glucose-6-P can be determined as described above. Approximately 3×10^{-13} mole can be determined using fluorescence techniques.

3.3.4   Classification of Enzyme Reactions by Measurement Method

The principle goal of this section is to present an overview of the spectrum of enzymes available for use in biochemical analysis from the vantage point of the species which will actually be measured. As will be seen the International Enzyme Commission's numbering system (for a detailed outline see reference 1) can be used as the basis for an approach for formulating a measurement scheme.

We have already noted that from the *analytical point of view there are really only two kinds of enzyme catalyzed reactions: auxiliary reactions and indicator reactions.* As a sweeping generalization, which has definite limitations, certain enzyme categories will only be useful as auxiliary reactions because they usually do not produce a substance or change which can be easily measured (NADH, ATP, O_2, H^+, etc.) but are limited to interconverting complex molecules (isomerases) or transfer a group from one molecule to another without displacement of a measurable moiety (most transferases). We have already oversimplified real life because thermochemical methods can be used to measure heat and this is certainly an easily measured nearly universal parameter. Similarly, certain transferases which act on phosphorus compounds will liberate orthophosphate and many transferases use up ATP.

Other classes of enzymes invariably lead to an easily measured species. Virtually all oxidoreductases (trivial names: dehydrogenases, oxidases, reductases, peroxidases) involve usage or generation of NAD(P), O_2, or H_2O_2 and many which act on C—N bonds split off NH_3. Such reactions are readily amenable for use as indicator reactions. Obviously they can also be used as auxiliary reactions.

To provide somewhat greater accuracy in classifying enzymes at the price of a more complex system, let us consider enzyme group 3—the hydrolases—in detail. This group is broken down into 9 subclasses, that is, those which act on ester bonds (3.1), glycosyl compounds (3.2), ether bonds (3.3), peptide bonds (3.4), C—N bonds other than peptide bonds (3.5), acid anhydride bonds (3.6), C—C bonds (3.7), halide bonds (3.8) and P—N bonds (3.9).

Enzymes 3.1 act on an ester to hydrolyze it with water to an alcohol plus a carboxylic acid, phosphoric acid, or sulfuric acid. This group contains enzymes such as: the carboxy esterases, lipases, cholinesterases, alkaline phosphatase, and arylsulfatase. Obviously all of them can be measured by pH stat, acid–base titrations, and via the rate of pH change in a lightly buffered solution, measured either electrometrically or colorimetrically. Thus these reactions should be considered indicator reactions.

In contrast, enzymes in subclass 3.2, which act upon glycosyl bonds, decompose carbohydrates and polysaccharides into smaller saccharide units or individual sugars such as: glucose, galactose, mannose, and fructose. These are generally not useful as indicator reactions but are invaluable auxiliary enzymes. It should be noted that here again there are a few exceptions. Sub-subclass 3.2.2, which hydrolyze N-glycosyl compounds, for example, NAD glycohydrolase (3.2.2.5), will hydrolyze NAD to nicotinamide and is therefore useful as an indicator reaction.

Consideration of many of the known and well-studied enzyme groups are presented in Table 3.8. Some subsections of the enzyme-numbering system have not been included because they are rather small and therefore are not particularly important in a broad analytical sense. Other small groups have been collected for the sake of brevity. Those species which can serve as indicators are emphasized in bold print. In other cases a particular product will occur only with some substrates; these will be underlined.

Oxidoreductases

These enzymes are invariably termed dehydrogenases, reductases, oxidases, peroxidases, or oxygenases. By and large, dehydrogenases utilize

Table 3.8 Categories of Enzymes by Reaction Type

1. Oxidoreductases

Subclass: 1.1.1
Trivial names: dehydrogenase, reductase
Oxidant: NAD or NADP
Reductant: CHOH group
Overall reaction type:

$$S + \boxed{NAD} \rightleftharpoons P + \boxed{NADH}$$

Common substrates: alcohols, carbohydrates, carboxylic acids, steroids
All are indicator reactions

Subclass: 1.1.2
Trivial names: dehydrogenase
Oxidant: cytochrome
Overall reaction type:

$$S + cytochrome(ox) \rightleftharpoons P + cytochrome(red)$$

Substrates: L-glycerol-3-phosphate, D-fructose, D-mannitol, lactic acid

---

Subclass: 1.1.3
Trivial name: oxidases
Oxidant: O_2
Overall reaction type:

$$S + \boxed{O_2} \rightleftharpoons P + \boxed{H_2O_2}$$

Substrates: lactic acid, glucose, arylalcohols, galactose
All are indicator reactions

Subclass: 1.2.1
Trivial names: dehydrogenase
Oxidant: NAD or NADP
Reductant: aldehyde or ketogroup
Overall reaction type:

$$S + \boxed{NAD} + \underline{H_2O} \rightleftharpoons P + \boxed{NADH}$$

Substrates: formaldehyde, aldehydes, carboxylates
All are indicator reactions

Subclass: 1.3.1
Trivial names: dehydrogenase, reductase
Oxidant: NAD or NADP
Reactant: CH—CH group
Overall reaction type:

$$S + \boxed{NAD} \rightleftharpoons P + \boxed{NADH}$$

Substrates: uracil derivatives, cortisone derivatives
All are indicator reactions

110

Table 3.8   (*Continued*)

Subclass: 1.4.1
   Trivial names: dehydrogenase, reductase
   Oxidant: NAD or NADP
   Reductant: CH—NH$_2$
   Overall reaction type:

$$S + H_2O + \boxed{NAD} \rightleftharpoons P + \boxed{NH_3} + \boxed{NADH}$$

   Substrates: alanine, glutamate, L-amino acids
   All are indicator reactions via NAD/NADH; some make NH$_3$

Subclass: 1.4.3
   Trivial name: oxidase
   Oxidant: O$_2$
   Reductant: CH—NH$_2$
   Overall reaction type:

$$S + H_2O + \boxed{O_2} \rightleftharpoons P + \boxed{NH_3} + \boxed{H_2O_2}$$

   Substrates: L-amino acids, D-amino acids, diamines, monoamines
   All are indicator reactions; all produce NH$_3$ and H$_2$O$_2$

Subclass: 1.5 (1.5.1–1.5.3)
   Trivial name: dehydrogenase, reductase, oxidase
   Oxidant: NAD, NADP, O$_2$
   Reductant: C—NH
   Overall reaction types:

$$S + \boxed{NAD} \rightleftharpoons P + \boxed{NADH}$$

or

$$S + \boxed{O_2} \rightleftharpoons P + \boxed{H_2O_2}$$

   Substrates: proline, tetrahydrofolate, methylamino acids, amino acids
   All are indicator reactions; some produce H$_2$O$_2$

Subclass: 1.6 (deals with those substances which oxidize NADH)
   Trivial names: reductases, dehydrogenases, diaphorase
   Substrates: NADH or NADPH
   Oxidant: cytochromes, disulfides, quinones, nitrate, nitrite
   All are indicator reactions since NADH is used up

Subclass: 1.7
   Trivial names: oxidases, reductases
   Reductant: nitrogeneous compounds
   Oxidant: O$_2$ or other oxidant
   Substrates: nitroethane, uric acid, ammonia
   Some are indicator reactions; those which use up O$_2$ or NH$_3$

Subclass: 1.8
   Trivial names: dehydrogenases, reductases, oxidases

111

Table 3.8 *(Continued)*

Oxidant: NAD, NADP, O_2, quinones, polyolnitrates, disulfides
Reductant: sulfur groups
Substrates: cysteamine, H_2S, $SO_3^=$, thiols, glutathione
Some are indicator reactions; those which use of NAD or O_2 or produce H_2O_2

Subclass: 1.10
Trivial names: oxidases
Oxidant: O_2
Reductants: diphenols, ascorbate
Overall reaction type:

$$2S + \boxed{O_2} \leftrightharpoons 2P + 2H_2O$$

Substrates: catechol, ascorbate, diphenols
All are indicator reactions (Note: these oxidases do not produce H_2O_2)

Subclass: 1.11
Trivial name: peroxidase, catalase
Oxidant: H_2O_2
Reductant: I^-, cytochromes, NADH, organic reducing agents
Overall reaction type:

$$S + \boxed{H_2O_2} \leftrightharpoons P + 2H_2O$$

All are indicator reactions since H_2O_2 is used up; also organic reductant can be colored

Subclass: 1.99.1
Trivial names: hydroxylases
Oxidant: aromatic amino acids, steroids
Reductant: NADPH
Substrates: phenylalanine, aniline, tryptophan, oestriol, steroids
Many are indicator reactions since NADPH is used up

Subclass 1.99.2
Trivial name: Oxygenases
Oxidant: O_2
Reductant: amino acids, catechol, ...
Overall reaction type:

$$S + \boxed{O_2} \leftrightharpoons P$$

Substrates: catechol, gentisate, homogentisate, inositol, tryptophan
All are indicator reactions; no H_2O_2 is produced

2. Transferases

Subclass: 2.1 (2.1.1–2.1.3)
Trivial names: methyltransferase, formyltransferase, carboxytransferase, hydroxymethyl transferase
Reactions: transfer one methyl, hydroxymethyl, formyl group

112

Table 3.8 (*Continued*)

Substrates: *S*-adenosylmethione + one of the following: catechol, nicotinate, histamine, thiols, homocysteine, acetylserotonin, serine, methylmalonyl-CoA, carboamylphosphate, aspartate, ornithine, etc.
Most are useful as auxilliary reactions only

Subclass: 2.2–2.3
Trivial names: transketolase, transaldolase, acetyl transferase
Reactions: carbonyl transfer, acetyl transfer
Substrates: acetyl-CoA + one of the following: L-aspartate, L-glutamate, imidazole, phosphate sugars, arylamines, H_2S(!), orthophosphate (!), glycine, diglycerides
Most are useful for auxiliary reactions only.

Subclass: 2.4 (2.4.1, 2.4.2)
Trivial names: hexosyltransferase, pentosyltransferase
Reactions: exchange of glucosyl group and addition of orthophosphate. Many involve reaction of nucleosides.

Subclass: 2.6
Trivial name: aminotransferase
Reaction: Transfer of amino group from a 2-oxo-acid (glutarate) to an amino acid
Substrates: aspartate, 2-oxoglutarate, L-alanine, L-cysteine, glycine, L-tyrosine, L-leucine, 2,5-diaminovalerate
Can be used only as auxiliary reactions

Subclass: 2.7
Trivial names: hexokinase, glucokinase, fructokinase, phosphotransferase
Reaction: Transfer of a phosphate group to a sugar or alcohol (2.7.1), to a carbonyl group (2.7.2), to a nitrogeneous compound (2.7.3), to another phospho group (2.7.4), those catalyzing intramolecular transfer (2.7.5), those transferring a pyrophosphate group (2.7.6), those in which a nucleotide in the acceptor (2.7.7) and those which operate on substituted phospho groups (2.7.8). In almost all cases ATP is the donor group and in most ADP is a product.

$\boxed{\text{ATP}}$ + D-hexose $\leftrightharpoons$ ADP + D-hexose-6-phosphate + $\boxed{\text{H}^+}$

$\boxed{\text{ATP}}$ + glycerol $\leftrightharpoons$ ADP + glycerol-3-phosphate + $\boxed{\text{H}^+}$

$\boxed{\text{ATP}}$ + protein $\leftrightharpoons$ ADP + phosphoprotein

$\boxed{\text{ATP}}$ + acetate $\leftrightharpoons$ ADP + acetylphosphate + $\boxed{\text{H}^+}$

$\boxed{\text{ATP}}$ + $\boxed{\text{NH}_3}$ + $\boxed{\text{CO}_2}$ $\leftrightharpoons$ ADP + carbamoylphosphate

$\boxed{\text{ATP}}$ + aspartate $\leftrightharpoons$ ADP + 4-phospho-L-aspartate

$\boxed{\text{ATP}}$ + creatine $\leftrightharpoons$ ADP + phosphocreatine

$\boxed{\text{ATP}}$ + sulfate $\leftrightharpoons$ pyrophosphate + adenylsulfate

In all cases where ATP is a reactant these are indicator reactions; many reactions involve the displacement of H^+

Table 3.8 *(Continued)*

Subclass: 2.8 (2.8.1–2.8.3)
    Trivial names: sulfurtransferases, sulphotransferases, coenzyme A-transferases
    Reactions: transfer of sulfate from 3'-phosphoadenylsulfate to phenols, arylamines, steroids. Also,

$$S_2O_3^{2-} + \boxed{CN^-} \rightleftharpoons SO_3^{2-} + SCN^-$$

All can be used as auxiliary reactions only except reaction of thiosulfate which is catalyzed by rhodanase.

3. Hydrolases

Subclass: 3.1 (3.1.1–3.1.3)
    Trivial names: esterases, hydrolases, peptidase, phosphoesterase, lipases, nucleotidase, phosphodiesterase

Reaction: ester $+$ H_2O $\rightleftharpoons$ $\boxed{acid}$ $+$ alcohol

Substrates: carboxylic esters, glycerol esters, acetylcholine, cholesterol esters, gluconolactone, acetyl CoA, orthophosphoric acid monoesters, mono and diphosphate sugar.

Subclass: 3.2 (3.2.1–3.2.3)
    Trivial names: glycosyl hydrolases, N-glycosyl hydrolases and S-glycosyl hydrolases
    Reaction: hydrolysis of links in oligo-and polysaccharides, converting them to simpler carbohydrates

Subclass 3.4 (3.4.1–3.4.4)
    Trivial names: aminopeptidases, carboxypeptidases, dipeptidases, pepsin, rennin, chymotrypsin, papain, ficin, thrombin, plasmin, etc.
    Reactions: hydrolysis of peptide not to produce smaller peptide.
    Substrates: amino-and carboxypeptidases act only on N and C terminal amino acids; dipeptidases act rather specifically on certain dipeptides, other peptides act on internal bonds of proteins some specifically, some not; many peptidases have a high esterolytic activity
    Virtually all peptidases can be followed via the pH change when the substrate is split and these are therefore indicator reactions

Subclass: 3.5
    Trivial names: asparaginase, glutaminase, amidase, urease, deacyclase
    Reaction: Hydrolysis of C—N bond in known peptides

L-asparagine $+$ H_2O $\rightleftharpoons$ L-aspartate $+$ $\boxed{NH_3}$

monocarboxylic acid amide $+$ H_2O $\rightleftharpoons$ $\boxed{\text{carboxylic acid}}$ $+$ $\boxed{NH_3}$

$$NH_2\!\!-\!\!\underset{\underset{O}{\|}}{C}\!\!-\!\!NH_2 + H_2O \rightleftharpoons \boxed{CO_2} + \boxed{2NH_3}$$

Substrates: L-asparagine, L-glutamine, urea, N-acyl-aspartate, etc.
All reactions yield a carboxylic acid, NH_3, or CO_2; therefore, all are indicator reactions

Table 3.8 *(Continued)*

4. Lyases

Subclass: 4.1.1
Trivial names: carboxy-lyases, decarboxylases, carboxylase
Overall reaction type:

$$S \leftrightharpoons P + \boxed{CO_2}$$

$$S_1 + S_2 \leftrightharpoons P_1 + P_2 + \boxed{CO_2}$$

Substrates: 2-oxo acids, oxalate, acetoacetate, malonyl-CoA aminomalonate, L-aspartate, L-valine, L-glutamate, L-ornithine, L-lysine, L-arginine, L-histidine, *p* or *o* aminobenzoate, L-tyrosine, L-tryptophan, etc.
All are indicator reactions

Subclass: 4.1.2
Trivial names: aldolases, ketolases
Overall reaction type:

$$S \leftrightharpoons P + \text{aldehyde}$$

Substrates: 2-oxo-4-hydroxybutyrate, ribose-5-phosphate, L-threonine, ketose-1-phosphates, some hexose mono and diphosphates
All are classed as auxiliary reactions. Many produce formaldehyde which is used to generate a colorimetric signal. A few generate HCN.

Subclass: 4.1.3
Trivial names: lyases, synthases
Overall reaction type:

$$S \leftrightharpoons P_1 + P_2$$

$$S_1 + S_2 \leftrightharpoons P_1 + P_2$$

Substrates: The ketoacids (L-isocitrate, L-malate, citrate)
These are classed as auxiliary reactions

Subclass: 4.2
Trivial names: dehydratases, hydratases, synthases
Overall reaction types:

$$S \leftrightharpoons P + H_2O$$

or

$$S \leftrightharpoons P + \boxed{NH_3} + H_2O$$

Substrates: H_2CO_3, malate, citrate, L-serine, D-serine, L-homoserine, L-threonine
Most will serve only as auxiliary reactions; some which generate NH_3 can be used as indicator reactions

Subclass: 4.3
Trivial names: lyases, deaminases

115

Table 3.8 (*Continued*)

Overall reaction types:

$$S \rightleftharpoons P + \boxed{\text{NH}_3} + \underline{\text{H}_2\text{S}}$$

Substrates: L-aspartate, L-histidine, L-phenylalanine, β-alanyl-CoA, L-argininosuccinate
All may be classed as indicator reactions

5. Isomerases

Subclass: 5.1.1
Trivial name: amino acid racemases
Overall reaction type:

$$\text{L-amino acid} \rightleftharpoons \text{D-amino acid}$$

Substrates: alanine, methionine, glutamate, proline, lysine, threonine

Subclass: 5.1.3
Trivial names: epimerases, mutarotases
Overall reaction type:

$$\alpha\text{-D-glucose} \rightleftharpoons \beta\text{-D-glucose}$$

Substrates: phosphate sugars, nucleosides, α-D-glucose

Subclasses: 5.2–5.5
Trivial names: isomerases, tautomerase, mutase
Overall reaction types:

$$cis \rightleftharpoons trans$$
$$\text{Aldose} \rightleftharpoons \text{Ketose}$$
$$\text{Keto} \rightleftharpoons \text{Enol}$$
Shift in —C=C—
Intramolecular acryl, phosphoryl transfer

Substrates: maleate, D-glyceraldehyde 3-phosphate, D-erythrose, D-arabinose, D-xylose, D-fructose

6. Ligases

Subclass: 6.2
Trivial names: synthetase
Typical reaction type:

$$\boxed{\text{ATP}} + \text{acid} + \text{CoA} \rightleftharpoons \text{AMP} + \boxed{\text{PO}_4^{3-}} + \text{acyl-CoA}$$

Substrates: acetate, C_4–C_{11} carboxylic acids, succinate, glutarate, cholate
Since almost all use ATP or generate PO_4^{3-}, they may be classed as indicator reactions

Subclass: 6.3
Trivial names: synthetases

Table 3.8   (*Continued*)

Typical reaction:

$$\boxed{\text{ATP}} + \boxed{\text{NH}_3} + \text{acid} \leftrightarrows \text{AMP} + \boxed{\text{PO}_4^{3-}} + \text{amine}$$

Substrates: L-aspartate, L-glutamate, glycine.
Since all use ATP and NH_3 and generate PO_4^{3-}, these may be classed as indicator reactions

---

Subclass: 6.4
Trivial names: carboxylases
Typical reactions:

$$\boxed{\text{ATP}} + \text{S} + \boxed{\text{CO}_2} + \text{H}_2\text{O} \leftrightarrows \text{ADP} + \underline{\text{PO}_4^{3-}} + \text{P}$$

Substrates: pyruvate, acetyl-CoA, propionyl-CoA

---

a pyridine nucleotide, oxidases use up O_2 and generate H_2O_2, peroxidases consume H_2O_2, and oxygenases can be measured by a decrease in O_2 but not by production of H_2O_2. Virtually all enzymes in this class can serve as *indicator reactions*. In some cases (E.C. 1.4.1) the oxidation of a C—NH_2 bond results in production of NH_3. One or two groups are not directly amenable to analysis via the indicator species listed previously. The cytochromes (E.C. 1.1.4) can be lead to consumption of reduced pyridine nucleotides by use of E.C. 1.6, but are also easily measured in the visible region (λ_{max} = 410 nm). As can be seen, this group can be used for the direct measurement of many amines, amino acids, phosphate sugars, alcohols, steroids, carboxylic acids, aldehydes, ketones, diphenols, quinones, and thiols as well as important inorganic species including iodide, peroxide, nitrate, nitrite, sulfide, sulfite, and ammonia. Through the use of auxiliary reactions there are scarcely any biological materials which cannot be determined via the oxidoreductases.

Transferases

With the exception of those enzymes which use up orthophosphate (E.C. 2.3.1.8) and the large group which consumes ATP, the *chief use of transferase will be as auxiliary reactions*, principally for the determination of sugars, some amino acids, and purine nucleosides. Fortunately a huge number of important materials can be phosphorylated with ATP. These reactions generally produce hydrogen ions. Some of the species which can be directly phosphorylated are: glucose, fructose, mannose, galactose, ribose, ribulose, arabinose, xylolose, adenosine, uridine, thymidine, NAD, riboflavin, choline, pyruvate, acetate, ammonia, carbon dioxide, sulfate, aspartate, creatinine, arginine, AMP, nucleoside mono- and diphosphates, and a great number of other materials. In general, we can say that the transferases will be useful in auxiliary reactions.

Hydrolases

As noted above these can be followed easily by acid–base techniques with the exception of the glycoside hydrolases and can be considered *as indicator enzymes.*

Lyases

As a class these are defined as enzymes which remove groups from their substrates by means other than hydrolysis and leave a double bond. Reference to Table 3.8 indicates that in many cases the group which leaves is CO_2 and/or NH_3. By and large the lyases *can be used as indicator reactions.* Many important compounds including a variety of amino acids and dicarboxylic acids are substrates for these enzymes.

Isomerases

By definition all such enzymes can serve *only as auxiliary reactions.* However, one or two isomerases result in the splitting of the compound, for example, aminodeoxyglucose-phosphate isomerase leads to formation of ammonia.

Ligases

This group of enzymes termed synthetases and carboxylases *can be used in indicator reactions* because many of them utilize ATP, NH_3 and CO_2. Most of the enzymes are involved in building nucleotides from simpler phosphate sugars; however analytically important species such as acetate, amino acids, and pyruvate serve as substrates.

The action of the above enzyme groups in terms of species which are easily measured are summarized in Table 3.9.

3.4 COMMON CHEMICAL AND PHYSICAL METHODS FOR MEASUREMENT OF BIOCHEMICAL INDICATOR SPECIES

The preceeding section is based on the generalization that in biological analysis certain (indicator) species are much more commonly measured in routine work than are other species. Techniques have been developed for their determination which have proven to be reasonably reliable and

Table 3.9   Categories of Enzymes by Measured Species

Measured Species	Enzymes
1. NAD/NADP	1.1.1; 1.2.1; 1.3.1; 1.4.1; 1.5.1; 1.5.6; 1.8; 1.99.1
2. O_2	1.1.3; 1.2.3; 1.4.3; 1.5.3; 1.7; 1.8; 1.10; 1.99.2
3. H_2O_2	1.4.3; 1.11
4. NH_3	1.4.1; 1.4.3; 1.7; 3.5; 4.2; 4.3; 6.3
5. CO_2	4.1.1; 6.4
6. ATP	2.7; 6.2; 6.3
7. PO_4^{3-}	6.2; 6.3; 6.4
8. Acid/Base	3.1.1–3.1.3; 3.1.6; 3.4; 3.5

readily available. Obviously the precise assignment of indicator species is perfectly arbitrary on our part and most certainly this group will expand with time as new chemical and physical measurement methods are introduced. As an example, 10 years ago, iodide ion was not a useful indicator species since there were few convenient techniques for measuring it under a wide variety of conditions. This situation has changed completely with the introduction of a highly selective solid-state iodide electrode (62). There has developed a high level of specialization in analytical chemistry, for example, a trace-metal analytical chemist leans naturally upon certain techniques, most particularly those of atomic spectroscopy. Clearly the stock-in-trade techniques of organic analytical chemists include gas chromatography and infrared and nuclear magnetic spectroscopy, as well as a variety of quantitative functional-group reactions. Precisely the same situation exists in the measurement of metabolites in biological matrices, that is, one tends to couple reactions together until some species which has proven to be particularly easy to measure in previous work is obtained.

The methods and techniques used in bioanalytical measurements share a number of common requirements. As is universally the case, the approaches need to be rapid, precise, and free from interferences. More specifically the methods should be compatible with water containing reasonable amounts of salt and buffer, should not be dependent on even high protein concentrations (serum contains approximately 70 g/liter protein), should not be influenced by turbidity (e.g., hyperlipemic sera are milky), should not interact substantially with the sample so as to alter a reaction rate, and should be compatible with flow systems or other types of automated analyzers. For these reasons bioanalytical chemists rely heavily on UV–visible absorption spectroscopy, fluorescence and, electroanalytical methods. Thermoanalytical methods offer enormous advantages but often lack the sensitivity and or high sample throughput required for measurement of many species at their normal levels in biological samples.

Table 3.10 gives a gross overview of measurement techniques and chemical methods frequently encountered for certain key species. The rest of this chapter is a brief description of the analytical chemistry involved in some of these approaches. The interested reader is referred to some of the excellent sources such as: Glick's *Methods of Biochemical Analysis* (in 25 volumes as of 1979), Methods in Enzymology (63 volumes), Bergmeyer's *Methods of Enzymatic Analysis* (4 volumes) and Henry's *Clinical Chemistry: Principles and Techniques* for a general introduction to this area.

3.4.1 Measurement of Pyridine Nucleotides

Chemical and instrumental methods for the determination of the pyridine nucleotides are based upon the following properties of these materials:

1. Direct absorption of NADH and NADPH at 340 nm.
2. Direct fluorescence of NADH and NADPH when excited at 340 nm.
3. Chemical conversion of all the nucleotides to a very strongly fluorescent material upon treatment with strong base and hydrogen peroxide.
4. Reaction of the reduced nucleotides with various oxidizing agents to form compounds which absorb in the visible wavelength region.
5. Electrochemical activity of all nucleotides at solid and mercury electrodes.

As we have noted previously, the reduced form of the nucleotides absorb light at 340 nm very strongly ($\epsilon = 6.2 \times 10^6$ cm^2/mole) but the oxidized forms do not. The oxidized nucleotides are virtually transparent at 340 nm but do absorb even more strongly than the reduced nucleotides at 270 nm. The spectra of NADH and NADPH near 340 nm are essentially identical (53) (see Table 3.11). Note that the values in this table were measured with the closest lines or group of lines in a mercury lamp, which is the most common source of radiation for NADH/NADPH because of its high intensity near 340 nm. The spectral positions are slightly displaced with temperature; an increase in the molar absorptivity of about 1% per 10°C is a good average value. Very frequently the decrease in absorption of NADH is measured. Since in a 1 cm cuvette a 0.2 mM solution will have an absorbance of 1.24, some care must be paid to minimize stray light effects and scattering in the cuvette in order to obtain a linear calibration curve above a concentration of a few tenths millimolar. With fairly common equipment, linear calibration curves with NADH or NADPH can be obtained from 10–200 μM ($A = 0.01$ to 1.2). As little as 5 nanomoles

Table 3.10 Selected Analytical Methods for Indicator Species

Species	Method
1. Acid(H^+)—Base(OH^-)	a. Titration b. pH stat c. Glass electrode d. Colorimetric indicator dye e. Manometry in bicarbonate buffer
2. ATP	a. Bioluminescence with firefly extract b. Reaction with glucose to glucose-6-phosphate and coupling to NADP c. Deaminate enzymatically and measure NH_3 d. Ultraviolet absorption at 260 nm e. Hydrolysis and determination of inorganic phosphate
3. CO_2	a. Manometry b. Gas-diffusion electrode c. Determination as an acid, see 1.
4. H_2O_2	a. Oxidation of colorimetric or fluorometric indicator dye b. Direct amperometry on platinum or glassy carbon electrodes c. Oxidation of $Fe(CN)_6^{4-}$ and amperometry d. Coupling to I^-/I_3^- and potentiometry e. Chemiluminescence with luminol
5. PO_4^{3-}	a. Heteropolyblue colorimetry b. As a base
6. Pyridine nucleotides (NAD, NADH, NADP, NADPH)	a. Absorption at 340 nm b. Direct fluorescence c. Fluorescence in strong base d. Colorimetry with redox indicator e. Amperometrically on platinum or glassy carbon electrodes f. Concommitant acid or base production
7. NH_3/NH_4^+	a. As a base b. Ammonium ion selective electrodes (NH_4^+) c. Gas sensitive electrode (NH_3) d. Colorimetry (Berthelot, Nessler, ninhydrin) e. Couple to NADH with glutamate dehydrogenase f. Manometry g. Microdiffusion
8. O_2	a. Clark electrode amperometry b. Manometry

Table 3.11   Extinction Coefficients of Pyridine Nucleotides ($\times$ 10^{-6} cm^2/mole)

λ(nm)	NADH	NADPH
334	6.11	6.13
340	6.22	6.22
365	3.39	3.45

can be measured in a 1 cm, 0.5 ml cuvette. Detailed pH and temperature studies of the absorption spectra of NADH have been reported (63–65) and high-performance liquid chromatography has been applied to the preparation of very pure solutions of the nucleotides (65). It is very important to realize that NADH and NADPH are rather unstable and care must be exercised in storing the solids. NADH solutions must be used very shortly after their preparation, especially when they are exposed to air, light, or low pH. The reduced nucleotides may be stored at 4°C in the dark at pH 8.2 in ammonium carbonate buffer (0.01 to 0.10 M) for several days without significant decomposition (63).

McComb et al. (64) found that the absorbance of a 0.1 mM NADH solution in pH 7.80 (1M THAM buffer) at 25°C drifted at a rate of 0.0005 absorbance units/hr, the drift tripled at pH 6.8, and was six times faster in 0.09 M phosphate buffer (pH 7.8). These workers and others (65) have carried out extensive high-precision measurements of the spectroscopic (ϵ and λ_{max}) properties of NADH and NADPH as a function of temperature, pH, buffer composition, and ionic strength.

The second analytically useful property of the reduced nucleotide (but not the native oxidized forms) is that they fluoresce when excited at 340 nm. Emission occurs at a maximum of about 460 nm. The fluorescence spectra of NADH and NADPH are virtually identical. The typical linear working range of the native fluorescence is from 0.1 to 10 μM. Fluorescence spectra are independent of pH from 6–13 but the intensity decreases about 1.6% per °C. Extraneous materials, particularly cations and enzymes can alter the fluorescent intensity of the nucleotides. For example, 1 mM magnesium at pH 11.5 increases the fluorescence of NADH by 50% and that of NADPH by 300%. There is little effect below pH 10. The lower limit of detection is about 0.1 μM, which is the blank obtained with water in a simple fluorometer. Higher blanks indicate undue scattering, stray light, or impurities in the reagents.

The native fluorescence of the reduced nucleotides is very much less than that of a material obtained by treating the oxidized nucleotide with strong base (6 M sodium hydroxide). In order to selectively measure NAD or NADH by this method one must be able to eliminate their mutual interference. This is very easily done by converting NADH to an inert product by treatment with acid. At pH 2, NADH is 99% destroyed at 23°C in less than 2 min, whereas at pH 4 the reaction time is about 2 hr. NADPH is destroyed more rapidly. The reaction rate of NADH is 3–4 times faster at 37°C and nearly ?0 times faster at 60°C. NAD is very stable at low pH (less than 7) but is rapidly destroyed in weak base (0.04 M sodium hydroxide). Conditions can be achieved where NADH is not consumed at all while the NAD is totally destroyed (66). In very strong base (6 M sodium hydroxide) NAD is converted to a highly fluorescent form mentioned above. The reaction is complete within 1 hr, 30 min, and 10 min at 25°, 38°, and 60°C, respectively. NADH can be converted to this same material by first destroying excess NAD by treatment with dilute base followed by oxidation of NADH with H_2O_2 to NAD in strong base. The fluorescence of this product is about 10 times as strong as that of the native NADH. Thus the detection limit is about 0.01 μM for either NAD or NADH. Detailed procedures for all of these chemical processes can be found in the monograph by Lowry and Passonneau (4). Colorimetric methods for the pyridine nucleotides based on their reaction to product formazans [equation (3.48)] have been employed. A minor disadvantage of the preceeding methods for NAD(P)H is that they require the use of a near-UV measurement or a fluorometer. Lee and Tan (67) have developed a sensitive colorimetric reaction for the reduced nucleotides based on the reaction

(chloropromazine radical, colored)

$$2 \quad + \quad NADH \; + \; H^+ \qquad (3.68)$$

(colorless)

The reduced chloropromazine radical is stable and absorbs intensely (ϵ = 9.7 × 10^6 cm²/mole) at 530 nm. Since two molecules of the radical react with one of NADH, the sensitivity is approximately three times as great as direct measurement of the nucleotide at 340 nm.

Recently Williams and Seitz have shown that NADH can be monitored by chemiluminescence methods (68). The basis of their method is the reaction of NADH with oxidized methylene blue to form reduced methylene blue, which then reacts with oxygen to form hydrogen peroxide. The hydrogen peroxide can be measured by chemiluminescence techniques, for example, the luminol method (see below) or by the peroxy oxalate method (69, 70). Williams and Seitz could detect 0.2 μM NADH with a 2:1 signal-to-noise ratio in a continuous-flow analysis system. This approach was applied to the measurement of lactic acid dehydrogenase activity in serum and correlated acceptably with a spectrophotometric method. Seitz and Neary (71) have reviewed the use of bacterial bioluminescence for the detection of NADH and NADPH. When either of these species is added to an aerated buffered solution of luciferase, an aldehyde, and excess FMN, bioluminescence is observed as the flavine nucleotide is reduced. Commercial luciferase gives a detection limit of 10^{-15} moles of NADH; however, the use of purified enzyme can significantly improve detectability. This system has been adapted for measurement of glucose, maleate, and NAD (72).

When photometric or fluorometric interferences are present in excessive concentrations, it may be advisable to measure the pyridine nucleotides by amperometry. Electrochemical oxidation of NAD or NADP, or reduction of NADH or NADPH can circumvent photometric interferences at the cost of possibly encountering electroactive interferants. These techniques have the advantage that no reagents need to be added and that the sensor is a probe and can be immersed directly in the test solution. The potential of the half-cell reactions of the pyridine nucleotides is such that one can work with solid electrodes and need not remove oxygen from the solution.

A major advantage of amperometric methods is that calibration curves are usually linear over three or four decades in concentration. In contrast, scattered and stray light, or self-absorption usually limit the upper limit of linearity of photometric and fluorometric analyses.

The reduction of NAD at the dropping mercury electrode has been studied extensively, the results of these studies have been reviewed by Elving, O'Reilley, and Schmakel (73). The formal potential for the NAD/NADH half reaction, which was obtained from measurements of the equilibrium constants and not by potentiometry or polarography, is about −0.320 V versus the standard hydrogen electrode. Reduction of NAD does not begin until almost 1 V more negative than the formal potential. Similarly NADH is not oxidized until a potential approximately 1 volt more positive than the formal potential. The electron exchange reaction is quite slow as indicated by the magnitude of the overvoltages. The electrochemical reduction of NAD proceeds in several steps; a biologically inactive dimer is the initial product, but at very negative potential, NAD is ultimately reduced to NADH. The final polarographic wave of NADH is only observable on mercury electrodes in nonelectroactive buffers in which quaternary ammonium salts are used as the supporting electrolyte. The second wave is not well-defined but the first wave does have a flat plateau and can be used to measure the concentration of NAD.

In contrast to the reduction of NAD, the oxidation of NADH has only been studied recently. The shape of a voltammogram of NADH on a solid electrode depends upon the solid used (Pt, carbon paste, glassy carbon, and pyrolytic graphite), the nature of the chemical and electrochemical pretreatment of the electrode surface, and, very significantly, on the nature of the buffer employed.

Voltammograms of NADH on platinum are very ill-defined and poorly reproducible; the oxidation wave at stationary platinum electrodes often appears as a slight knee on a rapidly increasing current–voltage curve. Because platinum surfaces start to oxidize at a lower voltage than do carbon surfaces and the oxidation of NADH is electrochemically more facile on carbon, most workers in the area have avoided platinum surfaces (74–78). Blaedel and Jenkins (74, 75) used glassy carbon, while Thomas and Christian employed carbon paste (76). Reasonably well-defined waves are obtained with both surfaces. Such electrodes have been applied to the determination of lactate (74), ethanol with alcohol dehydrogenase, and to measurement of lactic acid dehydrogenase activity (76). Elevated levels of ascorbic and uric acid, which are easily oxidized electrochemically, did not interfere because kinetic analysis based on the rate of change of current with time was employed.

Blaedel and Jenkins used the technique of steady-state voltammetry at rotating glassy carbon disc electrodes. When the electrodes were preconditioned in O_2-free solutions, well-defined voltammograms were obtained. At low rotation rates (~200 rpm) the wave appears to be reversible, that is, governed by the Nernst equation, but it becomes distinctly non-Nernstian at high rates of rotation. The half-wave potential is a function of both rotation rate and pH; it shifts from + 0.279 to + 0.308 V compared with a 0.01 M potassium chloride-silver chloride electrode as the pH is changed from 7.8 to 6.1 in dilute phosphate buffer. The position and height of the wave was dependent upon the buffer composition; thus phosphate, carbonate, and THAM behave differently. THAM appeared to cause fouling of the electrode and inhibition of electron transfer. The system noise and background currents are such that a 10 μM solution of NADH can be measured with a precision of 1%.

In a more recent study, Blaedel (78) compared glassy carbon and pyrolytic carbon electrodes. Once again, chemical (dichromate) and electrochemical (rapid cycling) pretreatment had a marked effect on the slope of the current–voltage curve. They found that even though oxidation of NADH proceeds more easily on conventionally pretreated glassy carbon than on the pyrolytic carbon film, a very well-defined limiting current can be obtained. A very slow decay in current was observed and assigned to a potential-dependent adsorption of a reaction product.

Moiraux and Elving (77) carried out an extensive study of the mechanism of oxidation of NADH on both glassy carbon and pyrolytic graphite in a variety of buffers. The peculiar results found in borate buffer were attributed to interaction of borate with the ribose part of the nucleotide. Their results indicate that oxidation of NADH proceeds via two processes, one of which involves adsorption of the reaction product NAD^+ on the electrode. They did not observe electrode fouling.

NADH has been determined coulometrically (79) in 0.1 M pH 9 pyrophosphate buffer. The platinum anode employed must be exhaustively pretreated, both chemically and electrochemically, to reproducibly obtain 100% current efficiency.

One of the most promising approaches for producing well-defined and reversible NADH oxidation waves is the use of chemically modified electrodes. Tse and Kuwana have employed electrodes modified by covalent attachment of a variety of quinones for the electrochemical oxidation of NADH (80).

A final electrochemical method for NADH is based on the use of a Clark O_2 sensor (see below) (81). NADH can be quantitatively oxidized by O_2 to NAD^+ in the presence of horseradish peroxidase in pH 8.0 THAM–succinate buffer containing 0.1 mM Mn^{2+}. This technique avoids the complexities of the electrochemical oxidation of NADH, but at the cost of a much poorer limit of detection.

3.4.2 Measurement of ATP

Undoubtedly the most sensitive method and one of the simplest for quantitating ATP is its well-known chemiluminescence reaction. The enzyme luciferase, which can be obtained from several sources but is commonly a firefly lantern extract, induces the following sequence of reactions (82):

$$LH_2 + E + ATP + Mg^{2+} \rightarrow E \cdot LH_2 \cdot AMP + Mg \cdot PP_i \qquad (3.69)$$

$$E \cdot LH_2 \cdot AMP + O_2 \rightarrow [oxyluciferin]^* + AMP + CO_2 + H_2O \qquad (3.70)$$

$$[oxyluciferin]^* \rightarrow oxyluciferin + h\nu \quad (\lambda_{max} = 562 \text{ nm}) \qquad (3.71)$$

In this series of reactions LH_2 represents luciferin, E is the enzyme luciferase, and PP_i represents pyrophosphate. The structures of luciferin and oxyluciferin are as follows:

Luciferin   Oxyluciferin

The chemiluminescent spectra obtained when the above series of reactions takes place is identical with the fluorescence spectra of oxyluciferin. Because the quantum efficiency of biological luminescence reactions are high, that is, close to the theoretical limit of unity, this reaction is the basis of an extremely sensitive assay for ATP. Using crude extracts, one can measure 0.1 to 1.0 pmole of ATP and, with extreme care in reagent preparation, as little as 2×10^{-5} pmole of ATP can be measured although a detection limit of 10^{-14} moles is easily achieved. As Seitz and Neary point out, detection limits in chemiluminescent methods is more a function of reagent purity and background luminescence than of instrumental noise or sensitivity (82). Both cytidine-5′-triphosphate and inosine-5′-triphosphate will emit the same quantity of light under the same conditions as ATP; thus the enzyme assay is not perfectly selective. The reaction will proceed without any added metal but is much less rapid. Other metals may be used as catalysts (Mn^{2+}, Mg^{2+}, Co^{2+}, Zn^{2+}, Fe^{2+}, Cd^{2+}, Ni^{2+}, Ca^{2+}, Si^{2+}) but high concentrations (0.15 M) of group IA and IIA cations can suppress the signal by as much as 80%.

The optimum pH for the reaction is 7.8 (1 mg/ml enzyme, 0.05 M THAM, 0.01 M Mg^{2+}); a decrease in pH decreases the quantum efficiency, shifts the wavelength of the emission, and decreases the reaction rate. Use of metals other than magnesium can also shift the λ_{max}. The review of Strehler and Totter should be consulted for other techniques of ATP measurements and a variety of applications of the firefly enzyme system to the measurement of other important biochemical species (83).

3.4.3   Measurement of Oxygen

A variety of enzymes, most particularly many oxidases and all oxygenases, utilize dissolved oxygen (O_2) as a coreagent. Classical methods for the determination of O_2 are based on manometry (84) in which the O_2 is released and absorbed in a solution of a strong reductant. Gasometric techniques have been invaluable in the past for O_2 measurement; in fact a now-archaic enzyme activity unity, the "Q_{O_2}," was based on the number of $\mu l/min$ of O_2 used up in a reaction. The development of reliable amperometric oxygen sensors has essentially replaced manometry for the purpose of rapid, routine measurement of O_2.

Even though oxygen is electroactive at the dropping mercury electrode, producing two well-defined waves corresponding to the reactions

$$O_2 + 2H^+ + 2e = H_2O_2 \quad E_{1/2} = -0.1 \text{ V versus SCE} \quad (3.72)$$

$$H_2O_2 + 2H^+ + 2e^- = 2H_2O \quad E_{1/2} = -0.9 \text{ V versus SCE} \quad (3.73)$$

most amperometric measurements are carried out with the Clark oxygen electrode which is illustrated in Figure 3.11. The Clark electrode is comprised of a very small platinum electrode (~0.02 mm diameter) which is covered with a thin oxygen-permeable plastic membrane (Teflon, mylar, polyethylene, polypropylene, colloidion, silicone rubber, etc.). Both the platinum microelectrode which serves as the cathode for reduction of oxygen to peroxide and a silver auxilliary anode are placed on the same side of the membrane and are in electrical connection through an aqueous solution of a dilute potassium chloride–potassium hydroxide solution.

An external circuit holds the platinum electrode at a potential of about -0.8 V with respect to the silver anode so that virtually all of the oxygen at the sensor is electrochemically reduced. Current–voltage curves with Clark electrodes have the usual sigmoidal shape, but lack a well-defined, that is, flat, limiting current region (85). In general, the potential should be set so as to cause no reduction of water on the platinum via the reaction:

$$H_2O + 2e \leftrightharpoons H_2 + 2OH^- \quad (3.74)$$

The alkaline supporting electrolyte is used to suppress the electrolysis of water since it has a much lower overvoltage for hydrogen evolution on platinum than on mercury. Because all components of the cell are in contact behind the membrane, the Clark electrode, which is really a com-

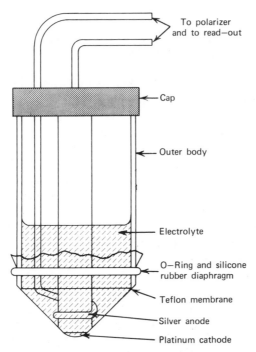

Fig. 3.11.   Diagram of a Clark Amperometric Oxygen Electrode and Associated Circuitry. (Reproduced from D. T. Sawyer and J. L. Roberts, *Experimental Electrochemistry for Chemists* (John Wiley and Sons, Inc., New York, 1974). Reprinted by permission of John Wiley and Sons, Inc.)

plete cell in itself, can be used to measure electroreducible gases (O_2, Cl_2, SO_2) either in solution or in the gas phase.

The three chief analytical characteristics of a Clark electrode are: sensitivity (current per unit partial pressure of O_2), the so-called "gas/water ratio" (see below) which should preferably be unity for ease of calibration, and response time. Jessen, Jacobsen, and Thomsen (86) have reviewed theoretical and experimental studies of these characteristics and devised a reasonably accurate two-dimensional diffusion model which predicts the sensitivity of micro Clark electrodes. Clearly the sensitivity will be controlled by the area of the active electrode, the O_2 permeability of the membrane, and the geometry of the electrode. In general, when the ratio of the electrode area to membrane area is low, the electrode will give the same response to a gaseous sample as to a solution equilibrated with that gas. Crude diffusion models do not successfully predict a steady-state current in unstirred solution, nor do they come anywhere near estimating the very high sensitivity of micro Clark electrodes. Silicone rubber membranes are frequently used in Clark electrodes due to the high O_2 solubility in this media and the concommitant high O_2 permeability of these materials.

A common electrode (Beckman Macro Electrode) has a sensitivity of 10^{-2} μA/mm Hg of O_2 (85). This is equivalent to a sensitivity of about 100 μA/mM of dissolved oxygen. Since currents of this magnitude are easily measured with conventional electronics, there is hardly any necessity for using bigger electrodes. Measurements precise to 1–2% can be obtained in stirred solutions (87) and compared within 1% with van Slyke O_2 measurements by manometry (88). The response times of commercial electrodes is typically well under one minute (89). The response time will depend upon the thickness and nature of the membrane, provided that the solution is well stirred. In unstirred solutions, the response time may be as long as 5 min. A device based on using a 6 μm Teflon membrane had a response time of less than 1 sec (89).

A detailed study of Clark electrodes indicates that the lower limit of detection, noise, and drift can be reduced to well less than the typical 100 pA (= 1 μM O_2) of commercial electrodes. The principal factors which increase the background or blank reading are: improperly cleaned electrode surfaces, non-uniform chloridation of the silver anode, poor pretreatment of the electrode body, and incorrect application of the membrane. The paper by Koch and Kruuv should be consulted for details (85) as to how to achieve detection corresponding to about 10 pA or about 0.1 μM O_2. Their results also indicate that the electrode is linear over a range of about 1000 in O_2 concentration.

The Clark electrode has a number of advantages over conventional polarographic measurement of O_2 at a dropping mercury electrode:

1. The sensitivity is rather independent of stirring.

2. The membrane improves selectivity so that only reducible gases (O_2, SO_2, Cl_2, etc.) are measured.

3. Proteins and other materials which can foul solid and mercury electrode surfaces do not contact the platinum electrode although the membrane can deteriorate in proteinaceous solution.

4. An external reference electrode is not needed therefore the entire sensor assembly can be immersed in the test solution as a single unit.

In addition to manometry and amperometry, dissolved oxygen can also be measured titrimetrically by the well-known Winkler method (89) by colorimetric methods (90). These techniques are generally more tedious and require more manipulations than does the use of the Clark electrode. An advantage of the titrimetric and manometric methods is that they can be calibrated in a fairly straightforward fashion. Calibration of the electrochemical method requires that the solution be equilibrated with a gas mixture of known oxygen content (91). Alternatively, the electrode can be calibrated in a closed cell by decomposing a known solution of hydrogen peroxide with catalase (87).

The oxygen electrode is extremely useful for the determination of glucose and has been extensively employed for this purpose in conjunction with glucose oxidase (88–99). It may be used by placing it in a reacting solution of glucose, oxygen, and glucose oxidase, as a detector at the end of a column containing the immobilized enzyme (97–99), or as an immobilized enzyme probe (96).

3.4.4 Measurement of Hydrogen Peroxide

Hydrogen peroxide is generated or consumed in a variety of enzymatic reactions including those of many oxidases and all peroxidases. Several very important clinical methods including: the determination of uric acid, cholesterol, and glucose involve detection of peroxide as an indicator species:

$$\text{uric acid} + O_2 \xrightarrow{\text{uricase}} \text{allantoin} + CO_2 + H_2O_2 \qquad (3.75)$$

$$\text{cholesterol} + O_2 \xrightarrow{\text{cholesterol oxidase}} \text{cholest-4-en-3-one} + H_2O_2 \qquad (3.76)$$

$$\text{glucose} + O_2 \xrightarrow{\text{glucose oxidase}} \text{glucono lactone} + H_2O_2 \qquad (3.77)$$

Because this species is so important, a considerable literature on its measurement by several very different techniques exists. Essentially five different types of methods have been described: colorimetric, fluorometric, chemiluminescent, amperometric, and potentiometric. At this point the colorimetric methods are probably the most popular. Both the colorimetric and fluorometric methods are based on the oxidation of an organic compound to a material which either absorbs or fluoresces:

$$\text{leuco dye} + H_2O_2 \rightarrow \text{absorber} + H_2O \qquad (3.78)$$

$$\text{nonfluorescent dye} + H_2O_2 \rightarrow \text{fluor} + H_2O \qquad (3.79)$$

Of the leuco dyes (see Table 3.12) used the most popular are: o-dianisidine (100, 101), benzidine (102), p,p' benzylidene bis-(N,N'-dimethyl aniline)(malachite green) (103), 2,4-dichlorophenol and 4-methoxy-1-napthol. In addition to direct oxidation to a colored dye, oxidative coupling of two materials, for example, of 4-amino antipyrine with phenol (104), or N,N'-

diethylaniline (105) has been employed:

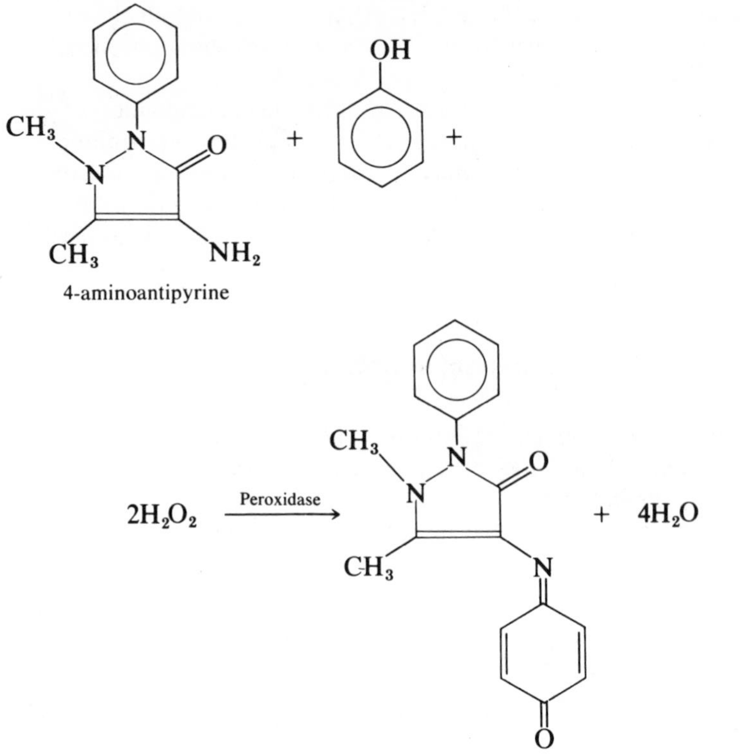

4-aminoantipyrine

$$2H_2O_2 \xrightarrow{\text{Peroxidase}} \qquad + \quad 4H_2O$$

(3.80)

quinoneimine
($\lambda_{max} = 500$ nm, $\epsilon = 5.3 \times 10^6$ cm^2/mole)

Fluorometric methods for peroxide have been reviewed by Guilbault (106). He states that homovanillic acid (4-hydroxy-3-methoxyphenyla-cetic acid) is an excellent reductant for H_2O_2. This reaction produces an intense fluor ($\lambda_{excitation} = 315$ nm, $\lambda_{emission} = 425$ nm). A wide variety of fluorogenic substrates have been tested as reagents for peroxide meas-urement (107, 108); p-hydroxyphenylacetic acid is quite stable and the product of its oxidation is a more intense fluor than that of oxidation of homovanillic acid. The homovanillic acid method has been used to detect cholesterol via cholesterol oxidase (109).

It has been known for some time that measurements of serum glucose via glucose oxidase and determination of H_2O_2 with a reducing dye are subject to interferences by uric acid. Blaedel and Ulh (100) carried out a detailed investigation of the effect of three types of interferants which were isolated from serum by gel permeation. All three interferences caused low recoveries of added glucose. They have shown that uric acid, which accounts for all of the poor recovery in the low molecular weight fraction, interferes by a peroxidase-catalyzed reaction of uric acid with H_2O_2. Normal levels of uric acid in serum easily caused 20% errors. Obviously any method for H_2O_2 based on a peroxidase will be subject to the same interference. It is very significant that Blaedel and Uhl observed no effect of any of the interferents on the rate of decrease in the oxygen concentration.

Chemiluminescence (CL) has proven to be an extremely sensitive approach to the measurement of hydrogen peroxide; as little as 10^{-12} moles in a 1 ml sample volume can be measured. The light-emitting reaction takes place in basic solution (pH $\cong$ 11) in the presence of an oxidant (H_2O_2, O_2, OCl^-, MnO_4^-, I_2, etc.) and a suitable catalyst. When H_2O_2 is the oxidant, ferricyanide is analytically the best catalyst (110). The overall mechanism or light emission is complex and depends upon the nature of both the oxidant and catalyst. For peroxide reactions, the overall process is probably as follows:

$$(\lambda_{max} = 425 \text{ nm}) \qquad\qquad (3.81)$$

The CL spectrum of the reaction matches the fluorescence spectrum of the aminophthalate product. The background light emission is often more than two orders of magnitude greater than the photomultiplier tube dark current thus improvements in reagent purity can significantly decrease the detection limits (82).

Bostick and Hercules have developed a continuous-flow analyzer for

Table 3.12   Indicator Dyes for Peroxide

Dye	λ_{max} (nm)	ϵ (cm^2/mole)
4-Methoxy-1-napthol	620	1.8×10^7
4-Aminoantipyrine + phenol	500	5.3×10^6
4-Aminoantipyrine + N,N-diethylaniline	553	1.2×10^7
Leucomalachite green	620	1.4×10^7
Benzidine	610	4.3×10^7
o-dianisidine	530	8.6×10^7
Tetraguaiacal	436 (Hg line)	2.5×10^7

H_2O_2 which is also suitable for analysis of glucose via immobilized glucose oxidase (110). In their optimization of peroxide measurement, they found that ferricyanide gives good sensitivity and reproducible results. The light intensity is a strong function of all reaction conditions. Optimum signal-to-noise ratios were obtained under the following conditions: 10^{-2} M $K_3Fe(CN)_6$, pH 10.8 (0.1 M borate), and 0.2 mM luminol. The relative peak heights are a strong nonlinear function of flow rate. Due to the back pressure of the glucose oxidase column, the actual measurements were carried out at a total liquid flow rate of about 6 ml/min. The limit of detection of the system for glucose is $2 \times 10^{-8}M$ and the calibration curve is linear up to $10^{-4}M$. The luminol system has been used to assay both amino acids and amino acid oxidase. A significant drawback of the luminol method for measuring enzyme activities is that its optimum light emission occurs at high pH; this is generally incompatible with enzymes. Recently Seitz and coworkers introduced a new reagent [bis-(2,4,6-trichloro-phenyl)oxalate] for peroxide measurement by chemiluminescence which can detect 10^{-8} M peroxide and can be used between pH 7 and 9 (69, 70).

A number of electrochemical methods for H_2O_2 have been developed. Even though peroxide is electroactive, its oxidation and reduction are very irreversible and its concentration cannot be determined by zero-current potentiometry at a noble-metal electrode. It has proven to be much more useful to chemically couple peroxide to a more reversible system (I_2/I^-, ferro/ferricyanide) for either potentiometric or ampero-metric detection. The most outstanding exception to this rule is Guil-bault's development of a biamperometric method in which the rate of change of the potential of a pair of polarized electrodes, that is, a small anode and cathode through which a current of 20 μA is flowing is used to quantitate the H_2O_2 concentration (111).

Direct potentiometry is usually based on the reaction of H_2O_2 with iodide

$$2H^+ + H_2O_2 + 3I^- \xrightarrow[\text{ammonium molybdate}]{\text{catalyst}} I_3^- + 2H_2O \qquad (3.82)$$

Since the iodide–tri-iodide half-reaction is freely reversible it can be detected potentiometrically at a platinum electrode (112). As indicated, the oxidation of iodide to tri-iodide is not particularly rapid; most investigators utilize molybdenum (VI) as the catalyst. The potential of the platinum electrode will be given by the Nerst equation:

$$E_{Pt} = E^0 + \frac{0.059}{2} \log \frac{[I_3^-]}{[I^-]^3} \qquad (3.83)$$

As is the case with all redox potential measurements, the presence of other more electroactive species in the solution may cause errors. With the availability of highly selective fast and sensitive solid-state ion-selective electrodes for detection of I^-, the redox-electrode approach to peroxide measurement is rapidly falling out of favor.

Papastathopoulos and Rechnitz indicate that a flow-through design of the iodide solid-state electrode can be used conveniently in conjunction with continuous-flow analysis (113). In their work, cholesterol was determined via cholesterol oxidase [reaction (3.76)] and peroxide oxidation of iodide. A 0.5 mM solution of potassium iodide in 1.6 M perchloric acid with 1 g/l of $(NH_4)_6Mo_7O_{24} \cdot 4H_2O$ catalyst was employed. The decrease in iodide concentration measured in millivolts of electrode potential was linear with cholesterol concentration from 80–420 mg/dl which encompasses the normal range.

Amperometric measurements of peroxide are problematic because its oxidation on noble-metal electrodes generally produces rather poorly defined current–voltage curves. Generally the peroxide can be measured through the reaction

$$2Fe(CN)_6^{4-} + H_2O_2 + 2H^+ \rightarrow 2Fe(CN)_6^{3-} + 2H_2O \qquad (3.84)$$

where the ferricyanide produced can be detected amperometrically at a platinum electrode (114, 115). Guilbault and Lubrano (116) have used the direct oxidation of peroxide on platinum surfaces for amperometric measurement of peroxide and glucose via glucose oxidase. The electrode must be pretreated before each run to provide reproducible surface conditions. Reasonably well-defined voltammograms in stirred solution were obtained from pH 4 to 11 and peroxide could be measured with a precision of 3% from 10 μM to 5 mM from the limiting current of the voltammogram. A commercial instrument developed by Leeds and Northrup for glucose analysis, which uses immobilized glucose oxidase, detects the generated peroxide by amperometry on glassy carbon electrodes. Although a flat limiting current plateau does not exist on this surface, very reproducible (0.2%) currents are obtained (117).

3.4.5 Measurement of Ammonia

Ammonia as NH_3 or NH_4^+ is a very common product of enzymatic reactions, particularly in the oxidation of nitrogeneous compounds (amino acids), and is a very important species in its own right. Some typical reactions which generate ammonia are:

L-amino acid + H_2O + O_2

$$\xrightarrow[\text{oxidase}]{\text{L-amino acid}} \text{2-oxo-acid} + NH_3 + H_2O_2 \tag{3.85}$$

D-amino acid + H_2O + O_2

$$\xrightarrow[\text{oxidase}]{\text{D-amino acid}} \text{2-oxo-acid} + NH_3 + H_2O_2 \tag{3.86}$$

diamine + H_2O + O_2

$$\xrightarrow{\text{diamine oxidase}} \text{aminoaldehyde} + NH_3 + H_2O_2 \tag{3.87}$$

L-glutamine + H_2O

$$\xrightarrow{\text{glutaminase}} \text{L-Glutamic Acid} + NH_3 \tag{3.88}$$

L-asparagine + H_2O

$$\xrightarrow{\text{asparaginase}} \text{L-Aspartic Acid} + NH_3 \tag{3.89}$$

L-glutamic acid + NAD^+ + H_2O

$$\xrightarrow{\text{glutamate dehydrogenase}} \text{2-Oxoglutarate} + NH_3 + \text{NADH} + H^+ \tag{3.90}$$

urea + H_2O

$$\xrightarrow{\text{urease}} 2NH_3 + CO_2 \tag{3.91}$$

By and large, methods for detection of ammonia are limited to colorimetry and electrochemical approaches. Since NH_3 is a base, it can also be detected with glass pH electrodes, indicator dyes, or titrated. Ammonia can be detected very specifically by microdiffusion methods which are essentially preconcentration techniques based on its volatility at high pH (118).

Colorimetric methods for NH_3 have been reviewed by Fleck (119) and fall into four distinct groups namely: modifications of Nessler's reaction to produce a colloid, reaction with ninhydrin to form a colored dye (a reaction shared by all amino acids and many other amino compounds), the Berthelot reaction in which ammonia and phenol are oxidatively coupled by hypochlorite, and coupling to NAD/NADH via reaction with α-ketoglutaric acid (see Table 3.13).

Both Nessler's method and the ninhydrin condensation reaction have many problems. For example, the Nessler reaction is very sensitive to temperature and the order and speed of reagent addition and thus it is not very reproducible; colloid formation is a very fast reaction. Colorimetry with ninhydrin is very nonspecific, usually requiring a preliminary separation of ammonia by distillation or diffusion (120) and requires prolonged heating at 95–100°C. The Berthelot reaction is reasonably fast and convenient, but there are several interferences from metal ions. A new, closely related method, based on the use of salicylate in place of phenol and dichloroisocyanurate in place of hypochlorite, overcomes many of these interference problems (121).

A variety of modifications of the Berthelot method have been described in the literature including: substitution of oxidants other than hypochlorite (e.g., chlorine water, chloramine T and B, and dichloroisocyanurate), various phenols (phenol, thymol, m-cresol, α-napthol, guaiacol, o-chlorophenol and 2-methyl-8-oxyquinoline) and many different catalysts (manganese dioxide, acetone, sodium nitroprusside). An examination of 65 phenols shows that an unsubstituted *para* position is necessary. A comparative study (122) on a half-dozen of the most popular methods indicates that the modification of Mann (123) yields the best results in terms of sensitivity ($\epsilon = 29.6 \times 10^6$ cm^2/mole), reproducibility (standard deviation $= 0.0017$ μg/ml), and color stability, although other modifications are simpler and faster (124). An excellent kinetic study of the Berthelot reaction should be consulted for details of the method as well as for the optimum conditions for its implementation (125).

A number of potentiometric methods for ammonium ions based on the use of ion-selective electrodes (glass and liquid ion-exchange membrane types) and gas sensing electrodes have been developed. The glass ion-selective electrodes are constructed physically in the same form as the conventional pH glass electrode. Chapter 5 should be consulted for the design of common ion-selective electrodes. It was, in fact, the development of a glass formulation with a useful selectivity for NH_4^+ which enabled Montalvo and Guilbault to construct the first urease electrode (126, 127). As the data in Table 3.14 indicates the so-called univalent-ion glass electrode has a fair selectivity over certain cations which permits the direct measurement of changes in ammonium-ion concentration. Certain cations (K^+, Na^+) are not tolerated by glass NH_4^+ sensors and must be excluded from solution. The selectivity is not sufficient to permit work

Table 3.13 Colorimetric Methods for Measurement of Ammonia

Method	Reaction	λ_{max} (nm)	ϵ (cm^2/mole)
Nessler's Method	$2\left[HgI_2 \cdot 2KI\right] + 2NH_3 \rightarrow NH_2Hg_2I_2 + 4KI + NH_4I$ (yellow colloid)	420–460[a]	3×10^6 [b]
Berthelot	$NH_3 + OCl^- + 2$ (phenol, OH) $\xrightarrow{\text{nitroprusside}}$ indophenol	614	2.0×10^7 [b]
Ninhydrin	$+ NH_3 \rightarrow$ $+ 4H_2O$	546	1.7×10^7 [b]
NADH/NAD	$NH_4^+ + H^+ + NADH + \alpha\text{-ketoglutarate} \xrightarrow{\text{glutarate dehydrogenase}}$ glutamate $+ NAD + H_2O$	340	6.2×10^6

[a] There is no λ_{max} since this is a light-scattering measurement. This is the usual wavelength employed.
[b] See reference 120.

138

Table 3.14   Selectivity Coefficients of Potentiometric Ammonium Sensors

			Electrode Type			
Ion	Univalent glass[a]	Univalent glass[b]	Modified glass[c]	Rubber impregnated nonactin[e]	Liquid exchanger[f]	Univalent glass[d]
H^+	2.1	2.5	2.4	—	0.18	3.5
Na^+	0.25	0.25	0.24	1.2×10^{-3}	0.14	0.38
K^+	2.4	2.3	2.4	1.5×10^{-3}	0.14	2.1
Li^+	1.2×10^{-2}	1.0×10^{-2}	1.5×10^{-2}	—	2×10^{-3}	$2.3 \cdot 10^{-2}$
Ca^{++}	—	—	N.R.[g]	—	4.2×10^{-4}	—
Mg^{++}	3.8×10^{-5}	8.1×10^{-4}	N.R.	—	—	1.0×10^{-3}
Ag^+	2.5	4.0	3.6	—	—	5.4

[a] Beckman 39137 glass; reference 128.
[b] Beckman 39047 glass; reference 128.
[c] Beckman 39137 electrode covered with 350 μm polyacrylamide containing urease, reference 129.
[d] A. H. Thomas electrode; reference 130.
[e] Reference 131.
[f] Philips IS 560. Also responds to many protonated amines and quaternary ammonium salts, reference 132.
[g] N.R. = not reported.

in buffers containing high concentration of these ions; thus THAM buffers (trishydroxymethylamino methane) are usually employed.

The selectivity coefficients in Table 3.14 are usually calculated from an equation of the form:

$$E = \text{const} + \frac{RT}{\beta F} \ln \left([NH_4] + K_i[M^+]\right) \qquad (3.92)$$

A potentiometric interferant invariably acts to decrease both the total potential change and the rate of potential change, thus the results will always be low when such ions are present. Equations of the form given above are actually quite poor descriptions of the actual interference; thus they should not be used to correct data except under extreme duress. Rather, they should be used as a guide to whether or not the electrode can function without error in the sample matrix of the expected concentration of foreign ions.

The sensitivity of glass and liquid-exchange electrodes to H^+ ions determines the ultimate limit of detection of ammonium ion. As shown in Figure 3.12 the lower limit of linearity for NH_4^+ moves toward higher concentration as the pH is decreased.

Because the selectivity of ion-selective electrodes is never ideally suited to measurements under all circumstances, considerable care must be

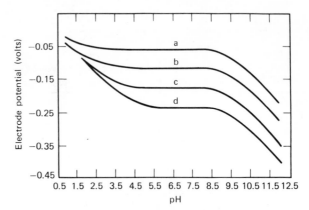

Fig. 3.12. Theoretical effect of pH on the response of ion-selective and gas-sensitive ammonia electrodes. All curves were computed from equation (3.92) with $K_i = 2.4$. Curve a, 0.1 M NH_4^+; curve b, 0.01 M NH_4^+; curve c, 0.01 M NH_4^+; curve d, 0.01 M NH_4^+.

taken to optimize the solution conditions. This is particularly true of the measurement of ammonium ions, since at high pH the ion is transformed to an uncharged molecule which is not detected by the electrode. The exact opposite is true of a gas-sensing electrode. In a very real sense, the ammonium-ion glass and liquid-exchange electrodes operate between "a rock and a hard place," that is, at low pH the activity of the hydrogen ion dominates in equation (3.92) and the sensor does not respond well to NH_4^+; at the other extreme, the equilibrium reaction below dominates and the response falls off:

$$NH_4^+ \leftrightharpoons NH_3 + H^+; \quad pK_a = 9.2 \qquad (3.93)$$

These problems are best illustrated in terms of a plot of electrode potential versus pH at a given formal ammonium-ion concentration. Under ideal circumstances, the electrode potential will be independent of pH. In reality (see Figure 3.12), at any given level of ammonium ion, the electrode potential becomes more negative as the pH is increased, flattens out, and then decreases again. The initial decrease is due to a drop in the activity of the hydrogen-ion activity which the electrode senses; at a sufficiently high pH the ammonium ion dominates and, starting at about 2 pH units below the pK_a, the ammonium ion is gradually converted to an inert form, thereby decreasing its effective activity. Since the gas sensor does not detect H^+ ions and does detect ammonia, its response is different and gradually improves until all of the ammonium ion in a sample is deprotonated.

   The pH dependence can serve to complicate an enzymatic analysis if the enzyme is not active near the optimum pH range of the electrode employed, that is, pH 7 for the glass and liquid exchanger and above pH 10 for the gas sensor. The data presented in Table 3.15, which is from the work of Guilbault, Smith, and Montalvo (128), was carried out by the slope method of kinetic analysis with a Beckman 39137 electrode in pH 7.0 THAM buffer (0.1 M). This indicates that the pH restriction is not severe for the glass electrode since it works well at pH 7, which is near the pH optima of many enzymes. In contrast it is often necessary with a gas sensor to carry out the enzyme reaction at one pH and then add base to obtain the active form for the sensor (NH_3).

The response time of the liquid ion-exchange electrode for NH_4^+ is about 30 sec for a change from 10^{-3} to 10^{-2} M (132). The gas-type sensors are somewhat slower; at concentrations greater than about 0.1 mM, about 60 sec are required, but at 30 μM, the steady-state response time exceeds 2 min; furthermore the pH must be above 11 for fast response. A precision of 2–5% has been attained for the measurement of ammonia in seawater at the 0.05–60 μM level (133).

The detection limit of most ion-selective electrodes of glass and liquid-exchanger or rubber-impregnated types are for all practical purposes in excess of 10^{-6} M and may be much worse depending on the composition and pH of the test solution. In the linear concentration range, where the electrode potential is proportional to the logarithm of the concentration of ammonium ion, the precision is set by the reproducibility of the elec-

Table 3.15   Substrates and Enzymes Measured
with Ammonia-Ion Electrode[a]

Substance	Concentration range	Relative standard deviation (%)
Urea	0.5–100 μg/ml	2.0
Urease	1–40 milliunits/ml	2.5
Glutamine	0.5–10 μg/ml	5.0
Glutaminase	1–400 milliunits/ml	5.0
Asparagine	0.5–10 μg/ml	3.0
Asparaginase	1–400 units/ml	3.0
L-Tyrosine	1–40 μg/ml	2.0
L-Leucine	1–40 μg/ml	2.5
D-Tyrosine	10–100 μg/ml	3.5
D-Methionine	10–100 μg/ml	4.0
L-amino acid oxidase	10–200 milliunits/ml	2.5
D-amino acid oxidase	0.05–1.0 units/ml	3.5

[a] From reference 128.

trode potential. It is easy to show that the relative uncertainty in concentration will be:

$$\frac{\%\sigma_{NH^+}}{[NH_4^+]} = \frac{\beta F}{RT}\sigma_E = 3.91\ \beta\cdot\sigma_E \tag{3.94}$$

where $\sigma_{NH_4^+}$ is the uncertainty in concentration (moles/liter) and σ_E is the uncertainty of the cell potential (millivolts). In the case of a univalent ion ($n = 1$), a 1 mV error causes about a 4% error in concentration. The major causes of imprecision with modern electronics is not electrical noise, but uncertainty and drift in temperature, liquid-junction, and reference-electrode potentials. The latter two have little effect on kinetic analyses since the absolute value of the potential is unimportant and these factors should be relatively constant, that is, independent of time, during a run.

Many of the same considerations discussed above apply to the so-called gas-sensor and air-gap electrodes. There has been a flurry of interest in these devices which have been used to detect both CO_2 and NH_3 (134–140). Their importance lies in their near-perfect selectivity. In the case of CO_2, only volatile acids can interfere, whereas for NH_3, only volatile bases can constitute a selectivity problem. Such devices utilize either a hydrophobic gas-permeable membrane or an air gap to separate an ordinary pH-sensitive glass electrode from an analyte solution. Upon contacting the analyte, a gas (e.g., NH_3) will diffuse into an internal electrolyte solution (a dilute alkaline buffer for NH_3 detection) which wets the internal glass electrode. Ultimately the partial pressure of the analyte (NH_3) is equal everywhere and, obviously, the pH of the internal electrolyte must reflect the level of ammonia in the test solution. The internal electrode contains a reasonable level of NH_4^+ and consequently the pH is directly related to the ammonia concentration.

A number of attempts to estimate the response characteristics of gas-sensor potentiometric electrodes have appeared (136–138). Recently, von der Pol has obtained an excellent model which predicts realistic detection limits and response times (138). Basically, the best detection limit and fastest response is achieved by use of very dilute internal buffers (0.001 M in ammonium chloride).

A very recent development is a gas-sensor conductance electrode (140). This device is very similar to the potentiometric system but the change in conductance of an internal electrolyte is detected.

Measurements with ion-selective electrodes can be implemented through the use of calibration curves as well as standard additions techniques. The excellent review of Covington on ion-selective electrodes as well as several recent monographs should be consulted (141–146); these contain detailed considerations for the theory of operation and use of ion-selective electrodes.

References

1. M. Dixon and E. C. Webb, *Enzymes*, 2nd ed., (Academic, New York, 1964), pp. 210–215.
2. J. I. Peterson and D. S. Young, *Anal. Biochem.* **23**, 301 (1968).
3. A. L. Chaney and E. P. Marbach, *Clin. Chem.* **8**, 131 (1962).
4. H. L. Pardue, *Clin. Chem.* **23**, 2189 (1977).
5. O. H. Lowry and J. V. Passonneau, *A Flexible System of Enzymatic Analysis* (Academic, New York, 1972).
6. G. Benzonana and P. Desneulles, *Biochim. Biophys. Acta* **164**, 47 (1964).
7. O. H. Lowry and J. V. Passonneau, *A Flexible System of Enzymatic Analysis* (Academic, New York, 1972), p. 30.
8. H. V. Malmstadt and S. R. Crouch, *J. Chem. Ed.* **43**, 340 (1966).
9. H. L. Pardue, C. S. Frings, and C. J. Delaney, *Anal. Chem.* **37**, 1426 (1965).
10. R. H. Callicott and P. W. Carr, *Anal. Chem.* **46**, 1840 (1974).
11. W. J. Blaedel and C. Olson, *Anal. Chem.* **36**, 343 (1964).
12. G. E. James and H. L. Pardue, *Anal. Chem.* **41**, 1618 (1969).
13. W. J. Blaedel and G. P. Hicks, "Analytical Applications of Enzyme Catalyzed Reactions," in *Advances in Analytical Chemistry and Instrumentation,* C. N. Reilley, Ed., Vol. 3 (Interscience, New York, 1964).
14. H. L. Pardue, "Application of Kinetics to Automated Quantitative Analysis," in *Advances in Analytical Chemistry and Instrumentation,* C. N. Reilley and F. W. McLafferty, Eds., Vol. 7 (Interscience, New York, 1969), p. 141.
15. R. H. Stehl, D. W. Margerum, and J. J. Latterell, *Anal. Chem.* **39**, 1346 (1967).
16. G. E. James and H. L. Pardue, *Anal. Chem.* **40**, 796 (1968).
17. R. A. Parker, H. L. Pardue, and B. G. Willis, *Anal. Chem.* **42**, 56 (1970).
18. S. R. Crouch, *Anal. Chem.* **41**, 880 (1969).
19. J. D. Ingle, Jr., and S. R. Crouch, *Anal. Chem.* **43**, 697 (1971).
20. H. L. Pardue, T. E. Hewitt, and M. J. Milano, *Clin. Chem.* **20**, 1028 (1974).
21. H. V. Malmstadt, C. J. Delaney, and E. A. Cordos, *Crit. Rev. Anal. Chem.* **2**, 559 (1972).
22. J. D. Ingle, Jr., and S. R. Crouch, *Anal. Chem.* **45**, 333 (1973).
23. E. M. Cordos, S. R. Crouch, and H. V. Malmstadt, *Anal. Chem.* **40**, 1812 (1968).
24. A. C. Javier, S. R. Crouch, and H. V. Malmstadt, *Anal. Chem.* **41**, 239 (1969).
25. J. D. Ingle, Jr., and S. R. Crouch, *Anal. Chem.* **42**, 1055 (1970).
26. E. S. Iracki and H. V. Malmstadt, *Anal. Chem.* **45**, 1766 (1973).
27. L. C. Thomas and G. D. Christian, *Anal. Chim. Acta* **77**, 153 (1975).
28. H. Weisz and H. Ludwig, *Anal. Chim. Acta* **55**, 303 (1971).

29. H. Weisz and K. Rothmaier, *Anal. Chim. Acta* **68**, 93 (1974).
30. S. Pantel and H. Weisz, Anal. Chim. Acta **68**, 311 (1974).
31. H. Ludwig and H. Weisz, *Anal. Chim. Acta* **72**, 315 (1974).
32. S. Pantel and H. Weisz, *Anal. Chim. Acta* **70**, 391 (1974).
33. H. Weisz, K. Rothmaier, and H. Ludwig, *Anal. Chim. Acta* **73**, 224 (1974).
34. S. Pantel and H. Weisz, *Anal. Chim. Acta* **74**, 275 (1975).
35. H. Weisz and K. Rothmaier, *Anal. Chim. Acta* **75**, 119 (1975).
36. L. C. Thomas, G. D. Christian, and J. D. S. Danielson, *Anal. Chim. Acta* **77**, 163 (1975).
37. H. V. Malmstadt and E. H. Peipmeier, *Anal. Chem.* **37**, 34 (1965).
38. R. E. Karcher and H. L. Pardue, *Clin. Chem.* **17**, 214 (1971).
39. R. E. Adams, S. R. Betso, and P. W. Carr, *Anal. Chem.* **48**, 1989 (1976).
40. C. F. Jacobsen, J. Leonis, K. Linderstrom-Lang, and M. Ottsen, "The pH-Stat and Its Use in Biochemistry," in *Methods of Biochemical Analysis,* D. Glick, Ed., Vol. 4, (Interscience, New York, 1957).
41. G. E. Mieling and H. L. Pardue, *Anal. Chem.* **50**, 1611 (1978).
42. P. W. Carr, *Anal. Chem.* **50**, 1602 (1978).
43. J. G. Atwood and J. L. Di Cesare, *Clin. Chem.* **21**, 1263 (1975).
44. J. E. Davis and B. Reno, *Anal. Chem.,* **51**, 529 (1979).
45. T. E. Hewitt and H. L. Pardue, *Clin. Chem.* **19**, 1128 (1973).
46. J. B. Landis, M. Rebec, and H. L. Pardue, *Anal. Chem.* **49**, 785 (1977).
47. P. Seelig and H. N. Blount, *Anal. Chem.* **48**, 252 (1976).
48. H. V. Malmstadt, C. J. Delaney, and E. A. Cordos, *Anal. Chem.* **44**, 79A (1972).
49. H. A. Mottola, *Crit. Rev. Anal. Chem.* **4**, 229 (1975).
50. G. G. Guilbault, *Enzymatic Methods of Analysis* (Pergamon Press, Oxford, 1970).
51. H. B. Mark and G. A. Rechnitz, *Kinetics in Analytical Chemistry* (Interscience, New York, 1968).
52. E. Maclin, D. Rohlfing, and M. Ansour, *Clin. Chem.* **19**, 832 (1973).
53. H. U. Bergmeyer, Ed., *Methods of Enzymatic Analysis,* Vol. 1–4 (Academic, New York, 1974).
54. N. O. Kaplan, M. M. Ciotti, and F. E. Stolzenbach, *J. Biol. Chem.* **221**, 833 (1956).
55. E. Negelein and H. Brömel, *Z. Biochem.* **301**, 135 (1939).
56. S. R. Goode and R. J. Mathews, *Anal. Chem.* **50**, 1608 (1978).
57. P. Talalay, "Enzymic Analysis of Steroid Hormones," in *Methods of Biochemical Analysis,* D. Glick, ed., Vol. 8 (Interscience, New York, 1960), p. 119.
58. G. Michal, "Enzymatic Analysis," in *Methodicum Chimicum,* F. Korte, ed., Vol. 1 (Academic, New York, 1974), p. 1070.
59. H. U. Bergmeyer, Ed., Methods of Enzymatic Analysis, Vol. 1 (Academic, New York, 1974), p. 111.
60. W. J. Blaedel and R. C. Boguslaski, *Anal. Chem.* **50**, 1026 (1978).
61. D. E. Leyden, W. K. Nonidez, and P. W. Carr, *Anal. Chem.* **47**, 1449 (1975).

62. R. A. Llenado and G. A. Rechnitz, *Anal. Chem.* **45**, 2165 (1973).
63. J. Ziegenhorn, M. Senn, and T. Bücher, *Clin. Chem.* **22**, 151 (1976).
64. R. B. McComb, L. W. Bard, and R. W. Burnett, *Clin. Chem.* **22**, 141 (1976).
65. S. A. Margolis, B. F. Howell, and R. Schaffer, *Clin. Chem.* **22**, 1323 (1976).
66. O. H. Lowry and J. V. Passonneau, *A Flexible System of Enzymatic Analysis* (Academic, New York, 1972), p. 15.
67. K. T. Lee and I. K. Tan, *Microchim. Acta* **II**, 139 (1975).
68. D. C. Williams and W. R. Seitz, *Anal. Chem.* **48**, 1478 (1976).
69. D. C. Williams, G. F. Huff, and W. R. Seitz, *Anal. Chem.* **48**, 1003 (1976).
70. H. R. Schroder and F. M. Yeager, *Anal. Chem.* **50**, 1114 (1978).
71. W. R. Seitz and M. P. Neary, in *Methods of Biochemical Analysis*, D. Glick, Ed., Vol. 23 (Wiley, New York, 1976).
72. S. E. Brolin, E. Borgland, L. Tegner, and G. Wettermark, *Anal. Biochem.* **42**, 124 (1971).
73. P. J. Elving, J. E. O'Reilly, and C. O. Schmakel, in *Methods of Biochemical Analysis*, D. Glick, Ed., Vol. 21 (Interscience, New York, 1973).
74. W. J. Blaedel and R. A. Jenkins, *Anal. Chem.* **47**, 1337 (1975).
75. W. J. Blaedel and R. A. Jenkins, *Anal. Chem.* **48**, 1240 (1976).
76. L. C. Thomas and G. D. Christian, *Anal. Chim. Acta* **78**, 271 (1975).
77. J. Moiroux and P. J. Elving, *Anal. Chem.* **50**, 1056 (1978).
78. W. J. Blaedel and G. A. Mabbott, *Anal. Chem.* **50**, 933 (1978).
79. H. Jaegfeldt, A. Torstensson, and G. Johansson, *Anal. Chim. Acta* **97**, 221 (1978).
80. D. C. S. Tse and T. Kuwana, *Anal. Chem.* **50**, 1315 (1978).
81. F. S. Cheng and G. D. Christian, *Anal. Chem.* **49**, 1785 (1977).
82. W. R. Seitz and M. P. Neary, *Anal. Chem.* **46**, 188A (1974).
83. B. L. Strehler and J. R. Totter in *Methods of Biochemical Analysis*, D. Glick, Ed., Vol. 1 (Interscience, New York, 1954).
84. O. Warburg, "Biological Manometry," in *Methods of Enzymatic Analysis*, H. U. Bergmeyer, Ed., Vol. 1 (Academic, New York, 1974), p. 248.
85. C. J. Koch and J. Kruuv, *Anal. Chem.* **44**, 1258 (1972).
86. O. J. Jensen, T. Jacobsen, an K. Thomsen, *J. Electroanal. Chem.* **87**, 203 (1978).
87. C. J. Koch and J. Kruuv, *Anal. Chem.* **44**, 1258 (1972).
88. W. J. Wingo and G. M. Emerson, *Anal. Chem.* **47**, 351 (1975).
89. K. Damaschke and E. Saling, *Klin. Wochenschr.* **39**, 265 (1961).
90. J. Ikuda and I. Miwa, in *Methods of Biochemical Analysis*, D. Glick, Ed., Vol. 21, (Interscience, New York, 1973), p. 158.
91. F. Kruezer, G. A. Rogeness, and P. Barnstein, *J. Appl. Physiol.* **15**, 1157 (1960).
92. E. C. Potter and J. F. White, *J. Appl. Chem., London* **7**, 309, 317, 459 (1957).
93. P. A. Hamlin and J. L. Lombart, *Anal. Chem.* **43**, 618 (1971).
94. A. H. Veefkind, R. A. M. Van den Camp, and A. H. J. Mass, *Clin. Chem.* **21**, 685 (1975).
95. H. J. Kuntz and M. Stasny, *Clin. Chem.*, **20**, 1018 (1974).

96. S. J. Updike and G. P. Hicks, *Nature* **214**, 986 (1967).
97. M. K. Weibel, W. Dritschilo, H. J. Bright, and A. E. Humphrey, *Anal. Biochem.* **52**, 402 (1973).
98. W. Dritschilo and M. K. Weibel, *Biochem. Med.* **11**, 242 (1974).
99. M. Notin, R. Guillien, and P. Nabet, *Anal. Biol. Clin.* **30**, 193 (1972).
100. W. J. Blaedel and J. M. Uhl, *Clin. Chem.* **21**, 119 (1975).
101. P. N. Tarbutton and C. R. Gunter, *Clin. Chem.* **20**, 724 (1974).
102. D. A. Ahlquist and S. Schwartz, *Clin. Chem.* **21**, 362 (1975).
103. G. G. Guilbault, P. Brignac, and M. Juneau, *Anal. Chem.* **40**, 1256 (1968).
104. C. C. Allain, L. S. Poon, C. S. Chan, W. Richmond, and P. C. Fu, *Clin. Chem.* **20**, 470 (1974).
105. P. Kabasakalian, S. Kalliney, and A. Westcott, *Clin. Chem.* **20**, 606 (1974).
106. G. G. Guilbault, *Enzymatic Methods of Analysis* (Pergamon Press, Oxford, 1970), p. 86.
107. G. G. Guilbault, P. Brignac, and M. Juneau, *Anal. Chem.* **40**, 1256 (1968).
108. G. G. Guilbault and D. N. Kramer, *Anal. Chem.* **36**, 2494 (1964).
109. H. S. Huang, J. W. Kuan, and G. G. Guilbault, *Clin. Chem.* **21**, 1605 (1975).
110. D. T. Bostick and D. M. Hercules, *Anal. Chem.* **47**, 447 (1975).
111. G. G. Guilbault, *Anal. Biochem.* **14**, 61 (1966).
112. H. V. Malmstadt and H. L. Pardue, *Clin. Chem.* **8**, 606 (1962).
113. D. S. Papastathopoulos and G. A. Rechnitz, *Anal. Chem.* **47**, 1792 (1975).
114. W. J. Blaedel and C. L. Olson, *Anal. Chem.* **36**, 343 (1964).
115. H. Pardue and R. Simon, *Anal. Biochem.* **9**, 204 (1964).
116. G. G. Guilbault and G. J. Lubrano, *Anal. Chim. Acta* **64**, 439 (1973).
117. J. A. Burns, *Cereal Food World* **21**, 594 (1976).
118. J. S. Annino and R. W. Giese, *Clinical Chemistry: Principles and Procedures*, 4th ed. (Little, Brown 1976), pp. 162–164.
119. A. Fleck, *Crit. Rev. Anal. Chem.* 3141 (1974).
120. R. Richterich, *Clinical Chemistry: Theory and Practice* (Academic, New York, 1969), p. 257.
121. R. V. E. Pym and J. P. Milham, *Anal. Chem.* **48**, 1413 (1976).
122. L. I. Glebko, Zh. I. Ulkina, and E. M. Lognenko, *Mikrochim. Acta* **11**, 641 (1975).
123. L. T. Mann, *Anal. Chem.* **35**, 2179 (1963).
124. Y. Morita and Y. Kogure, *J. Chem. Soc. Japan* 84, 816 (1963).
125. C. J. Patton and S. R. Crouch, *Anal. Chem.* **49**, 464 (1977).
126. G. G. Guilbault and J. Montalvo, *J. Amer. Chem. Soc.* **91**, 2164 (1969).
127. G. G. Guilbault and J. Montalvo, *J. Amer. Chem. Soc.* **92**, 2533 (1969).
128. G. G. Guilbault, R. K. Smith, and J. G. Montalvo, Jr., *Anal. Chem.* **41**, 600 (1969).
129. J. Montalvo, Jr., and G. G. Guilbault, *Anal. Chem.* **41**, 1897 (1969).
130. G. G. Guilbault and F. R. Shu, *Anal. Chim. Acta* **56**, 333 (1971).
131. G. G. Guilbault and G. Nagy, *Anal. Chem.* **45**, 417 (1973).
132. R. Dewolfs, G. Broddin, H. Glysters, and H. Deelstra, *Z. Anal. Chem.* **275**, 337 (1975).
133. T. R. Gilbert and A. M. Clay, *Anal. Chem.* **45**, 1757 (1973).

134. J. Ruzicka and E. H. Hansen, *Anal. Chim. Acta* **69**, 129 (1974).
135. E. H. Hansen and J. Ruzicka, *Anal. Chim. Acta* **72**, 353 (1974).
136. E. H. Hansen and N. R. Larsen, *Anal. Chim. Acta* **78**, 159 (1978).
137. P. L. Bailey and M. Riley, *Analyst* **102**, 213 (1977).
138. F. von der Pol, *Anal. Chim. Acta* **97**, 245 (1978).
139. M. Mascini and C. Cremisini, *Anal. Chim. Acta* **97**, 237 (1978).
140. H. A. Mimpler, S. F. Brank and M. J. D. Brand, *Anal. Chem.* **50**, 1623 (1978).
141. A. K. Covington, *Crit. Rev. Anal. Chem.* **4**, 335 (1974).
142. R. A. Durst, *Ion Selective Electrodes*, National Bureau of Standards Special Publication 314 (U.S. Government Printing Office, Washington, 1969).
143. G. J. Moody and J. D. R. Thomas, *Selective Ion Sensitive Electrodes* (Merrow Publishing Co., Watford, Harts., England, 1971).
144. J. Koryta, *Ion Selective Electrodes*, Cambridge Monographs in Physical Chemistry. No. 2 (Cambridge University Press, New York, 1975).
145. R. P. Buck, *Anal. Chem.* **50**, 17R (1978); **48**, 23R (1976); **46**, 28R (1974).
146. M. Kessler, Ed., *Ion and Enzyme Electrodes in Biology and Medicine* (University Park Press, Baltimore, MD, 1976).

IMMOBILIZATION OF ACTIVE BIOCHEMICALS

Enzymes and other biological agents can be physically localized in or on a variety of insoluble matrices by an ever-increasing number of techniques with the concommitant retention of biological activity. This area of technology has proliferated rapidly with nearly 1000 papers in the literature. A number of reviews have appeared in the last eight years which discuss various aspects of the field (1–6). To date, much of the immobilized-enzyme literature which has appeared has dealt with the investigation of a new support or immobilization reaction. This first stage of immobilized-enzyme technology is rapidly drawing to a close. It is now important to consider the advantages of immobilized enzymes in analysis, industrial production, biomedical engineering, and other applications. In order to evaluate immobilized enzymes effectively, the stability, ease of preparation, and usefulness of various physical forms must be considered.

The usefulness of enzymes as analytical reagents has been discussed in Chapter 3. The immobilization of an enzyme results in several advantages. First, there is frequently an increase in the stability of the enzyme. Second, the enzyme–support system is easily removed from solution without contamination of the reaction mixture by the contents of the enzyme preparation. Third, because of the stability of the immobilized enzyme, a single aliquot of enzyme can be used repetitively to achieve hundreds or thousands more analyses than could be performed with the same amount of enzyme in solution. Conversely, in many cases we could economically load large amounts of enzyme into a reactor for equilibrium analyses. The long, predictable half-lives of the decay of enzyme activity allow the immobilized enzyme to become part of the analytical instrumentation. Thus, calibration of the device includes calibration of the enzyme reactor. Finally, it is possible to prepare unstable, sensitive, or expensive reagents using an immobilized enzyme, an advantage which has not been exploited significantly.

In any enzyme immobilization, two of the major considerations are the recovery of activity and the stability of the preparation. It is well-known that there is a loss of enzyme activity in the immobilization process. Since this recovery is related to the severity of the protein modification, the interaction of the enzyme and the matrix, and other less rigorously defined parameters, it should be a factor in the choice of an immobilization scheme. It must be remembered that the economic feasibility of an application must be measured against the *original* amount of enzyme used. Thus for an expensive enzyme which exhibits large losses in the attachment step, an application may be untenable although the preparation can be used for a large number of analyses. The stability of the preparation is a rather difficult parameter to assess. As with all of the parameters involving matrix-bound enzymes, both increases and decreases in this property have been observed. In addition, several types of stability can be considered: thermal, storage, and operational. Thermal stability is a reflection of the ability of the enzyme–support conjugate to withstand higher temperature before denaturation occurs. In many cases, if the thermal stability is increased, the enzyme may be used at elevated temperatures while maintaining some degree of efficiency. Storage stability is frequently reported in the literature and is simply the ability of the preparation to retain its activity under some specified storage conditions. The main value of this parameter is that it can provide the investigator with some idea of the shelf life of this reagent. Probably the most important stability parameter, and the one least frequently reported, is the operational stability. The operational stability of the preparation is not only a function of the enzyme, but also a function of the carrier durability, inhibitor concentrations in the analyte solution (in particular irreversible inhibitors), and other considerations such as particulate matter which may clog an immobilized-enzyme reactor.

It is beyond the scope of this monograph to survey the multitude of enzymes which have been immobilized. A relatively complete, but outdated, listing can be found in either Zaborsky's book (1) or a compendium published by the Corning Glass Works (2). In addition, it is impossible to discuss all of the immobilization procedures which have appeared. It is our intent to discuss the major types of immobilization processes with regard to their chemistry and their advantages and limitations. We will place particular emphasis on schemes which have been useful analytically and on the recent advances.

The methods used for the immobilization itself can be classified systematically and their merits and disadvantages discussed in general terms. Although any system of classification is rather subjective, the grouping employed by Zaborsky (1) seems to be quite adequate and is given in Table 4.1. A single modification has been the addition of a "hybrid methods" classification for those methods which involve both physical and chemical modification of the protein. The first basic classification, physical methods, includes any methods which do not involve the formation of covalent bonds. Chemical methods involve the formation of at least one covalent bond between the enzyme and a functionalized insoluble

Table 4.1   Classification of Enzyme-Immobilization Methods

_____

Physical Methods

Adsorption of enzyme onto water-insoluble matrix
Entrapment of enzyme inside water-insoluble polymer lattice
Entrapment of enzyme within a semipermeable microcapsule

Chemical Methods

Attachment of enzyme to derivatized water-insoluble matrix
Intermolecular crosslinking of enzyme molecules

Hybrid Methods

_____

carrier, or between two protein molecules. The main difference between the two types of methods are the rigor of the formation of the enzyme–matrix conjugate.

In the remainder of this chapter, we will discuss the role of the immobilization scheme in an analytical setting. The largest portion of work done thus far has been on covalent coupling of the enzyme to a solid support. Since the carrier has an important role in the mode in which the enzyme is used, support characteristics will be addressed in the section on covalent binding. The primary emphasis of this chapter will be to review methods of immobilization which are rapid, inexpensive, and widely applicable to analytical applications. In addition, some new techniques which show promise have been presented. In many cases, sufficient detail has been included so that the reader can prepare the immobilized-enzyme system.

4.1   EFFECT OF IMMOBILIZATION ON ENZYME STABILITY

One of the major goals of enzyme immobilization, particularly for analytical purposes, is an increased lifespan of the reagent. Unfortunately, there appears to be little predictive value in most of the work reported to date. This is true in part because of the complex nature of the problem. Our knowledge of the role of enzyme structure and the forces affecting it is far from complete. As pointed out in Chapter 2, species such as polyelectrolytes or sulfonic acid zwitterions appear to stabilize some enzymes in solution. Similar effects have been observed with covalent binding, where polyanionic and polycationic matrices seem to increase the stability of the enzyme in alkaline and acid solutions respectively (17, 18).

This appears to be due to a microenvironmental "buffering." In addition, the use of polylysine, polyornithine, or polyvinylamine as a mediator between the support and the enzyme seems to increase stability (19). This area has not been extensively studied and general theories are not available. If the decrease in enzyme activity is primarily due to denaturation of the protein, a better understanding of how and why denaturation occurs would be valuable in increasing enzyme stability. Once again, since different proteins display different denaturation and refolding characteristics, it has been difficult to postulate a general theory of denaturation. Despite this absence of theoretical framework regarding conformational, microenvironmental, and inactivation effects, we can garner some information from the literature about the role of immobilization in increased enzyme stability.

The storage stability of 50 immobilized-enzyme systems has been reviewed (20). Of these, 12 systems showed no enhancement of stability relative to their solution-phase counterparts and 30 exhibited greater storage stability. The remaining eight showed decreased stability. Thus, in a majority of cases, increased storage stability can be expected. The entire stability issue has been confused somewhat by the use of a variety of assay conditions. If, a large excess of enzyme is used under conditions of complete conversion, the rate of decrease in enzyme activity may not be indicated with fidelity.

Since denaturation is thought to be the main cause of the loss of enzyme activity, a technique which prevents this process would be expected to increase enzyme stability. If unfolding of the protein is the major mechanism of denaturation, then the immobilization of the enzyme in such a way that conformational transitions leading to denaturation are prevented while those necessary for activity are allowed should lead to a more stable preparation. As shown in Figure 4.1, a number of stereochemical arrangements of an enzyme and matrix are possible. As the local bonding activity of the matrix increases, multiple matrix–enzyme linkages are possible. Although the bonds of Figure 4.1 are illustrated as covalent linkages, any enzyme–matrix interaction may assist in stabilization. It has been shown that the greater the number of bonds formed between trypsin and Sephadex, the less susceptible the enzyme is to denaturation by urea (21). Other studies of thermal stability as a reflection of resistance to denaturation have confirmed these observations. A more recent study in which the amine functions of several enzymes were modified by a monomeric species, then incorporated into a polymer matrix found a 1000-fold increase in thermal stability (22). It should be apparent, however, that a change in conformation caused by the immobilization process, excessive bond formation restricting conformational changes required for

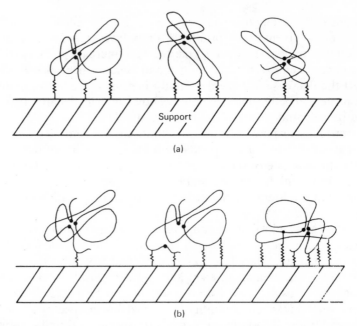

Fig. 4.1.   Possible random orientations of enzymes bound to a solid matrix. Heterogeneity not only arises from orientation (a) but also from the rigidity of the binding (b).

active-site formation, or stereochemical hindrances caused by the matrix may result in an inactive molecule being bound to the matrix (see Figure 4.1). A study of noncovalent interactions has shown that these forces can also stabilize an enzyme. Polyacrylamide, which has no charged groups in the matrix, shows no stabilization of trypsin in the range 0–50% gel. Polymethacrylate, on the other hand, showed a 10,000-fold decrease in the thermal decomposition rate at greater than 30% gel. Further study showed that transitional motion of the entrapped enzyme was decreased 1000-fold and that the rotational motion was also dramatically reduced. Modification of the protein indicated that specific hydrogen-bonding interactions were responsible for the stabilization (23). From this work it might be anticipated that large proteins would show greater thermal stabilization than small ones under entrapment procedures, and this appears to be true at least for glucose oxidase and peroxidase relative to acetylcholinesterase in a silastic matrix. It is anticipated that more supporting evidence will be presented.

It is of some interest to consider how many bonds actually form when an enzyme is attached to a solid support. Immobilization of trypsin on arylamine CPG was shown to have caused modification of an average of

11 amino acid residues (24). The use of the enzyme–substrate complex for immobilization, a relatively common practice to increase activity recovery and stability, increased the number of bonds formed to 14, thus lending some credence to this maneuver. In the immobilization of leucine aminopeptidase to a cyanogen bromide activated agarose, it was found that 22 bonds were formed between the matrix and the enzyme (25). It should be apparent from these results that the coupling of an enzyme to an insoluble matrix results in a multiplicity of bonds and environments.

As mentioned above, a significant cause of decreased enzyme activity is denaturation, but elution of the enzyme from the carrier matrix also contributes to this decrease. This is particularly true when the substrate and/or matrix is of the charged polyelectrolyte type. This phenomenon has been observed even with covalently bound enzyme (25). This slow leakage from cyanogen bromide activated agarose was significant above pH 5.0 and was increased by the presence of nucleophilic agents. The decay of activity is not simple, probably due to the multiplicity of bonds and microenvironments. The amount of elution which occurs is a function of the immobilization method, as will be discussed in later sections.

Some interesting relationships have been observed between stability and the enzyme–matrix system used. Using the same matrix and coupling reaction, differences in stability were exhibited by the various isoenzymes of lactate dehydrogenase and glucose isomerase. The recovery of activity from the isoenzymes was also significantly different. No explanation has been forthcoming regarding this phenomenon. Significant differences in the stability of lactate dehydrogenase bound to arylamine glass through diazo or glutaraldehyde linkages have been observed with the glutaraldehyde method being more stable (26). As we will see, the carrier itself plays a significant role in the stability of the preparation due to its hydrophilicity, its charge, and other properties. One final point about the interaction of the enzyme and immobilization scheme concerns the reaction conditions under which coupling occurs. It has often been observed that increasing the amount of protein bound to the matrix increases the stability of the product (27–30). It is interesting to note that the actual protein used is unimportant. Thus albumin can be mixed with the protein of interest to achieve greater stability. In addition, it has been claimed that increased time of contact between the aqueous solution of enzyme and the matrix results in increased stability (29).

Enzyme immobilization may result in the occurence of new factors which influence stability. In the broadest sense, compaction of an organic ion-exchange resin at high ionic strength or dissolution of the silica matrix at high pH both cause a loss of enzyme activity. Microbial attack on the matrix or the enzyme or plugging of a reactor by uncontrolled microbial growth both present new but surmountable difficulties. A more significant problem can arise because of the mass-transfer limitations of relatively large porous particles (see Section 5.2). Although this condition may result in some beneficial kinetic properties, it also increases the microenvironmental concentration of product. In the case where the product is harmful to the enzyme activity, this situation causes rapid deterioration of the catalytic capacity of the preparation. A good example is glucose oxidase where a reaction product, H_2O_2, was found to deactivate the enzyme (31). The diffusional restrictions imposed by the support significantly increased the deactivation rate constant. Operationally, this problem can be overcome by co-immobilizing the enzyme catalase which decomposes hydrogen peroxide. Another example is the extraction of the loosely bound cofactor FAD from amino acid oxidase by the constant flow of buffer through the matrix, resulting eventually in complete deactivation. Once again, the operational solution is to add a small amount of FAD to the flow stream. It should be apparent that although immobilization makes enzymes more appealing reagents, it does not exempt one from learning enzymology.

4.2   EFFECTS OF IMMOBILIZATION ON ENZYME KINETICS

It was observed relatively early in the investigation of immobilized enzymes that the kinetic properties of an insolubilized enzyme often did not agree very closely with those of the soluble species. Further investigation has to a large extent identified the sources of these differences. The effects can be classified into essentially three types. First, the limitations imposed by slow mass transfer of substrate or product to or from the site of the enzyme can seriously affect the observed kinetics. Since mass transfer through the unstirred Nernst layer is a first-order process, the expected, and observed, effect would be to extend the apparent first-order kinetic region. The effects of internal and external mass transfer are considered in detail in Chapter 7.

Second, one must consider the steric orientation of the enzyme. As illustrated in Figure 4.1, the enzyme may be coupled to the carrier in such a way that the active site is inaccessible or only partially accessible to the substrate, or that the enzyme is held in an inactive conformation such that the catalytic function is partially or totally blocked. It is possible that these situations contribute to the kinetic constant of the overall system although this situation has not been well studied.

Finally, the microenvironment of the enzyme, as controlled by the support itself, has a considerable influence on the kinetic parameters. There are two discernable situations under which these effects might be observed. First, a partitioning of a kinetic effector which might be a result of the chemical affinity of the effector for the support would give rise to an environment for the enzyme which is not representative of the bulk, and therefore measurable, concentrations. There can also arise microenvironmental effects which might best be envisioned as an interaction of the support with the enzyme in such a way that the mechanism of catalysis is altered. One example of this type of behavior would be an interaction of the support with the active site itself as has been proposed for chymotrypsin bound to ethylene–maleic acid copolymers through polyornithyl or polyglutamyl side chains (32).

Since any of these modifications might result in the same observed kinetic behavior, a nomenclature to distinguish the various types of kinetic behavior has been developed. In the event that mass-transfer effects are rate-limiting, the observed rate and kinetic parameters are generally referred to as the *effective* rate (see Chapter 7). In the absence of diffusional limitations, the kinetic parameters are referred to as the *inherent* rate parameters to distinguish them from the *intrinsic* rate parameters observed for the enzyme in solution. A difference in the latter two properties would arise from the microenvironmental parameters which are discussed in greater detail below.

Of the two effects of the microenvironment on enzyme kinetics, partitioning of the various species between the bulk of solution and the local area of the support has been studied in the greatest detail. One may represent the distribution of any species between the bulk and the matrix as a partition coefficient, ρ, where $\rho = C_i/C_b$, and C_i and C_b represent the matrix and bulk concentration, respectively, of a charged substituent. If the distribution can be represented as a Boltzmann distribution, $C_i = C_b \cdot \exp x$, then for an ionic species the partition coefficient can be written

$$\rho = \exp(-ze\psi/kT)$$

where z is the charge on the species, e is the charge on the electron, ψ is the electrostatic potential, k is the Boltzmann constant, and T is the temperature (33–37). If we now consider the pH gradient between the bulk and the interior of the support,

$$\Delta pH = -\log [H^+]_i + \log\{\exp(-ze\psi/kT)\} [H]_i$$

$$= \log\{\exp(+ze\psi/kT)\}$$

$$= 0.43 \, ze\psi/kT$$

It can be seen that for a negatively charged support ($\psi < 0$) the interior pH would be less than the bulk pH. As a result, one would expect to observe a shift in the pH optimum toward a more alkaline pH to compensate for the increased local hydrogen-ion concentration. One would expect the opposite effect for a positively charged ($\psi > 0$) support. Goldstein and coworkers have studied this effect and a sample of their results are shown in Figure 4.2.

Using the same logic, one can predict the change in kinetic properties of an enzyme reacting with a charged substrate.

$$V_{obs} = \frac{V_{max}[S_i]}{K_M + [S_i]} = \frac{V_{max}[S_b] \exp(-ze\psi/kT)}{K_M^{obs} + [S_b] \exp(-ze\psi/kT)}$$

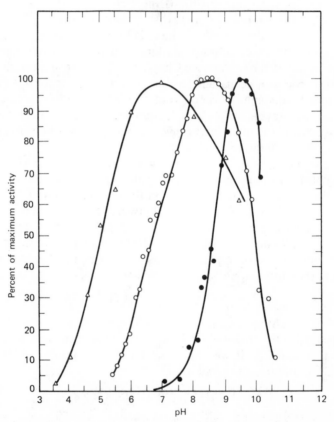

Fig. 4.2. pH–activity curves for ($\bigcirc$) chymotrypsin ($\bullet$) a polyanionic ethylammaleic acid (EMA) copolymer derivative of chymotrypsin (EMA–chymotrypsin), and ($\triangle$) a polycationic polyornithyl derivative of chymotrypsin. [Reproduced with permission from R. Goldman, L. Goldstein, and E. Katchalski, in *Biochemical Aspects of Reactions on Solid Supports,* G. R. Stark, Ed. (Academic, New York, 1971); Copyright by Academic Press]

From this relation, the K_M^{obs} can be shown to be given below and where S_i is the matrix substrate concentration and S_b is the bulk concentration:

$$K_M^{obs} = K_M \exp(ze\psi/kT)$$

Once again, one would expect that for a substrate with the same charge as the matrix, the observed K_M would be greater than that observed for the soluble enzyme. The converse would apply for substrates and matrices of opposite charge. Goldstein and coworkers have shown that for a polyanionic derivative of trypsin acting on a positively charged substrate at low ionic strength K_M^{obs} was over an order of magnitude lower than that observed for the soluble species.

The above effects arise from the fact that the support can be considered to be a polyelectrolyte. As such, the interactions that are responsible for the above changes in the kinetic properties are purely electrostatic. It would thus be expected that the ionic strength would have a substantial influence on the electrostatic interactions. Goldstein and other workers have shown that shifts in the kinetic parameters are abolished at high ionic strengths. Wharton and coworkers have presented an expression to explain quantitatively the effects of ionic strength of the apparent Michaelis constant (38). As the ionic strength of the solution was increased, the observed Michaelis constant increased in close agreement with the theoretical treatment. The observation that the apparent K_M at high ionic strength was lower than the K_M observed for the soluble enzyme appears to indicate that electrostatic interactions are not the only forces in effect. The effects of a hydrophobic matrix on kinetic parameters have recently been described (39). Two substrates of differing hydrophobic character were tested in conjunction with matrix-supported enzyme. As would be expected, increasing hydrophobicity of the matrix resulted in a greater change in the apparent K_M of the more hydrophobic substrate.

4.3 ADSORPTION TECHNIQUES

The adsorption of an enzyme onto a water-insoluble material is the simplest method for obtaining enzyme–support conjugates. Basically, it consists of placing an aqueous solution of enzyme in contact with an active material for some period of time after which the excess enzyme is washed off the insoluble matrix. Numerous surface-active materials have been used in the preparation of enzyme-adsorption complexes, some of the most popular being anion and cation exchange resins, activated charcoal, silica gel, diatomaceous earths and clays, alumina, and controlled-porosity glasses and ceramics.

The adsorption process is extremely complex. Depending on the nature of the surface, the enzyme binding may be the result of ionic interactions, physical adsorption, hydrophobic bonding, or van der Waals interactions. The characteristics of these forces are such that changes in the pH, temperature, ionic strength, concentration of enzyme and adsorbant, and the presence of a solvent may have a significant effect on the efficiency of the immobilization. Zittle (40) has reviewed the importance of these factors in enzyme adsorption.

Adsorption on controlled-porosity glasses and ceramics has been investigated in detail by Messing (41). Two different reaction rates were observed. The first rate was proportional to the isoelectric point of the protein, while the second was inversely related to the molecular weight of the macromolecule. Messing attributed the former to the formation of ionic amine–silanol bonds on the surface. The latter slower rate was thought to be due to diffusion of the enzyme into the pores of the material where most of the surface area is located. The fact that urea or mineral acids did not remove the protein indicates that some other type of interaction beside the ionic bond is present, possibly hydrogen bonding. Two other general considerations which have come from the work on controlled-porosity materials are: (1) the activity is proportional to the surface area of the adsorbant and (2) in order for the enzyme to enter the matrix, the pore diameter of the support must be at least twice its longest axis of rotation. For other matrices which have internal surface area, such as Sephadex, the pore size must also be considered.

The advantages in using adsorption to fix an enzyme to a carrier arise from its simplicity, the wide variety of adsorbents which can be used, and the reusability of the adsorbent. The manipulations required for immobilization are quite simple after the appropriate conditions are developed. The nature of the interactions are such that a high level of activity is retained by the preparation. A large number of adsorbents with a variety of physical shapes have been used, that is, sheets, fibers, beads, etc. In some cases, the adsorbent can be reused after the inactivation of a previous enzyme preparation, often without removing the adsorbant from its environment (43).

The major difficulties in using adsorbed enzymes are the empirical process required in optimizing the system and the reversibility of the adsorption process. The selection of an adsorbent should involve the following considerations: (1) the adsorbent should have a high affinity and capacity for the enzyme; (2) the enzyme must retain a high percentage of its activity in the adsorbed state; (3) the adsorbent should not adsorb the reaction product or enzyme inhibitors. The optimal values of pH, temperature, and ionic strength to use with a given support and enzyme system must be empirically determined to achieve good activity and stability. It has been found that in many cases, maximal adsorption occurs near the isoelectric point of the protein.

The adsorption of a protein onto a surface should, in theory, be a reversible process. A change in the pH, ionic strength, temperature, etc., should cause a breakdown of the enzyme–matrix bonds, with a resultant release of the enzyme into the aqueous phase. If "leakage" of the enzyme is detrimental to the overall process, either contaminating the product or allowing the catalytic process to proceed, this reversible binding may be a serious drawback. Systems which desorb under conditions of changing substrate concentration (44, 45), pH (46) and ionic strength (47) either individually or in combination have been reported. Irreversible behavior has been reported, however, for a number of adsorption systems (1, 48). An example of the scheme used for the adsorption of glucoamylase on activated charcoal is given in Figure 4.3.

Another type of charge interaction adsorption is the binding of an enzyme by an ion-exchange resin. As in ion-exchange chromatography, the binding is dependent on the pH and the ionic strength of the surrounding medium. The technique is only applicable where low ionic strength and an appropriate pH can be used. The requirement for low ionic strength includes the substrate concentration if the substrate or product are ionic species. It is possible to crosslink the enzyme molecules after adsorption to remove some of these objectionable characteristics but one does so at the expense of carrier regenerability.

Recently, Porath and his coworkers have reported a hydrophobic adsorption technique as a result of observations on the interaction of proteins with detergents such as Triton X-100 and with hydrophobic arms

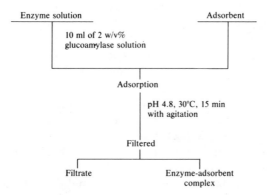

Fig. 4.3. Reaction scheme for the adsorption of glucoamylase onto activated charcoal. [Data taken from reference (49).]

in bioaffinity chromatography. Although more systematic studies are required, it is postulated that the length of the hydrophobic arm from the support can act as a probe of the exposed hydrophobic regions of the enzyme molecule in "hydrophobic chromatography." An excellent example can be found in the studies of β-galactosidase by bioaffinity chromatography (50). Porath and his coworkers have taken the reversible adsorption due to hydrophobic interactions one step farther and used it to prepare a simple reversible immobilization technique (51–54). Using β-amylase and amyloglucosidase for their initial studies, a hexyl substituted crosslinked Sepharose was prepared as an adsorbant. As in hydrophobic chromatography, the length of the hydrocarbon arm has a significant effect, with long chains resulting in denaturation and extensive nonspecific binding. The immobilization of β-amylase onto the substituted Sepharose was rapid, requiring less than 5 min, and had good loading capacity, binding 35 mg of protein onto 1 ml of the packed gel (53). In contrast to β-amylase, amyloglucosidase adsorption was significantly influenced by the salt concentration to 3 M, equilibrium adsorption conditions were reduced from 6 hr to 5 min in 0.02 M acetate buffer at pH 4.8.

As with any adsorption system, the effects of changing ionic strength, particularly with the amyloglucosidase, are of concern. In their studies, these authors observed that although elution of some protein occurred on washing with a salt-free buffer, the final product after the wash retained more activity than the preparation made in salt-free buffer. After about 15 volumes of wash, the adsorbate was stable.

As mentioned above, hydrophobic adsorption is dependent on the hydrocarbon chain length and on the density of substitution on the solid support (55, 56). These authors report a heterogeneous distribution of hydrocarbon on the surface of the polymer. As a result, several types of binding sites with varying binding affinities will result. This will probably effect the immobilization properties. Further studies involved the relationship of the degree of substitution of the solid support and the amount of enzyme recovered. As one might expect, the amount of protein bound increases monotonically with the degree of substitution. The activity recovered as a function of the number of moles of hydrocarbon per gram of support, however, seems to show a distinct maxima. This property of the immobilization bears a strong resemblance to the phenomenon often observed in covalent-attachment schemes.

A second type of hydrophobic binding carrier has been produced by the reaction of phenoxyacetyl chloride with cellulose (57). This particular matrix is limited mainly by the relatively rapid degradation of the ester bond. Thus the enzyme slowly leaks from the resin. This material is commercially available from Regis.

This method of immobilization would seem to have a number of advantages, particularly for lipophilic enzymes. First, it is a reversible adsorption technique, desorption being accomplished through the use of organic solvents such as butanol or by detergents. The hydrophobic nature of the matrix may assist in the interaction of lipophilic substrates and enzymes. Disadvantages include selecting the appropriate chain length for optimal immobilization, the potential denaturation of the protein, and the difficulties encountered in obtaining a homogeneous surface on which to adsorb the protein. It should be borne in mind that all of the problems and benefits associated with charge-interaction adsorption will also be true of hydrophobic adsorption.

4.4 GEL ENTRAPMENT TECHNIQUES

The use of physical entrapment of an enzyme within the interstitial spaces of a polymeric matrix was first employed by Bernfield and Wan in 1963. Because of their size, enzymes which are entrapped within the polymeric lattice during the crosslinking process should not diffuse out, while appropriately sized substrate or product molecules can. The most popular matrices for gel entrapment include polyacrylamide, silicone rubber, polyvinyl alcohol, starch, and silica gel.

Polyacrylamide gel is the most frequently used method of entrapping enzymes. The formation of polyacrylamide is performed by mixing acrylamide and N,N'-methylbisacrylamide in the presence of a catalyst. Chrambach and Rodbard have discussed the chemistry of the acrylamide polymerization reaction (58). The pore size of the gel and its mechanical properties are determined by the total and relative concentrations of acrylamide and N,N'-methylbisacrylamide. Hicks and Updike studied the operational stability of gels prepared with various amounts of acrylamide and its crosslinking agent (59). They found that at a fixed total acrylamide and crosslinking-agent concentration, increasing rigidity was observed with lower relative crosslinking-agent concentration. Decreased relative retention of the enzyme activity was also observed under these conditions. These investigators also found that useful beads could be made from the gel and could be stored in a lypholized state with very little loss in activity.

The agent used to catalyze the polymerization process is also an important consideration. Hicks and Updike reported that riboflavin and a photocatalyst work well with high concentrations of crosslinking agent, while high monomer concentrations require potassium persulfate. This author has had good success using a mixture of both catalysts as suggested by the above authors. Other chemical methods and X-ray irradiation have been used as catalysts. It is important, since radical formation is involved in the polymerization, that oxygen be rigorously excluded from the reaction vessel. The method of Hicks and Updike is summarized in Figure 4.4.

Silicone rubber has also been used as an enzyme-trapping matrix. Silastic resin® (Dow Chemical Company) contains silyl ether chains on the order of 10,000 Si groups in length. An immobilized enzyme is prepared by adding aqueous enzyme to a silastic resin solution, after which the catalyst, stannous octoate, is added. Guilbault and Das have reported that a relatively low yield of active enzyme was obtained due to the "rigorous" polymerization conditions (60). In addition, the hydrophobic nature of the polymer requires that the rubber be stored in a hydrated state. Other silicone rubber systems, however, have been used with good activity retention (61).

Polyvinyl alcohol has also been used to entrap enzymes by Boguslaski, Blaedel, and coworkers (62, 63). A polyvinyl alcohol solution was mixed with a buffered solution of enzyme after which glycerine was added as a plasticizer and Black BS salt was added as a crosslinking agent. The solution was then spread in a thin layer and allowed to dry. Further polymerization was accomplished by exposure of the resulting membrane to ultraviolet irradiation for a short period of time. The dry membrane could then either be stored or moistened and used. The present authors

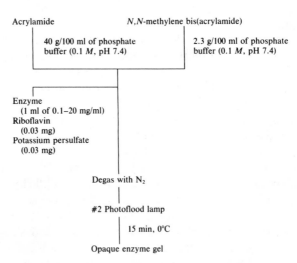

Fig. 4.4. An enzyme immobilization technique using Polyacrylamide. [Data taken from reference (59).]

found this method to yield mechanically durable membranes with activity yields of about 15% with the enzyme urease.

The entrapment of enzymes in urethane foam-supported starch gels has been used by Guilbault and coworkers in the development of entrapped enzyme pads. Buffer, glycerine, and starch were boiled, then cooled to 47°C after which the enzyme solution is added. The resulting solution is poured into a urethane foam form and cooled at 4°C for several hours. The pad could then be cut to the appropriate size for use. Variations of this procedure have been used for a variety of enzymes (64).

Silica gel has been used to successfully entrap an enzyme by Johnson and Whately (65). A solution of silica gel was adjusted to pH 7 with sodium hydroxide, enzyme was added, the ionic strength was adjusted, and the solution was set aside to allow condensation of the silicic acid groups to form a "hydrogel." The solution was then lypholized into a "xerogel" which contained the entrapped enzyme.

The advantages of gel entrapment include the experimental simplicity, the relatively mild conditions of preparation, and the wide variety of forms in which the gels can be used. The disadvantages include the control of numerous experimental factors, the possible inactivation of the enzyme by the radicals necessary for polymer formation, and as alluded to earlier, the restriction of the size of the substrate and products to relatively small species. Another problem is the leakage of the enzyme from the matrix. This occurs as a result of the problems encountered in trying to achieve homogeneous lattice size. Due to this nonhomogeneity, enzyme tends to leak from the gel, particularly during the early stages of use.

4.5   INTERMOLECULAR CROSSLINKING TECHNIQUES

The preparation of insoluble enzyme derivatives through the use of low molecular weight bifunctional reagents has been accomplished in several ways. The use of bifunctional reagents such as glutaraldehyde to crosslink proteins has been used extensively for biochemical studies of proteins and biomembranes (66, 67). Two separate methods have been used to make enzymes insoluble: the use of only the enzyme and the use of the enzyme and a carrier protein such as albumin. Another use of the bifunctional reagents has been in the preparation of "hybrid" immobilized-enzyme systems, where the coupling occurs after adsorption onto a water-insoluble matrix or as a functionalization step for a preformed water-insoluble polymer. Reagents have also been used which cause enzymatic condensation without incorporating the reagent into the polymer (68).

The polymerization of the enzyme directly with bifunctional reagents suffers from a lack of selectivity. That is, it is exceedingly difficult to control intramolecular crosslinking while obtaining a high degree of intermolecular crosslinking. Because intramolecular crosslinking is favored by low enzyme and reagent concentrations, it is advantageous to increase the overall protein concentration with the use of a second "carrier" protein. Control of the pH, ionic strength, temperature, and reaction time are all important in determining whether a water-soluble monomer or oligomer or a water-insoluble polymer is obtained. These factors also affect the degree of intramolecular crosslinking and the mechanical strength of the insoluble polymer. Determination of the optimal conditions for retention of activity is a trial and error proposition.

Because of the chemical disadvantages listed above, intermolecular crosslinking has not been widely used in analytical applications. The single advantage of the method is that a single reagent can be used to prepare numerous enzyme derivatives. The adsorption or entrapment of an enzyme onto silica or cellophane followed by crosslinking has achieved greater popularity. Although these techniques result in high enzyme recovery, a major drawback arises from the fact that two sets of optimal conditions must be found for the immobilization. The most popular immobilization method involving the use of bifunctional reagents has been the derivatization of insoluble polymers, which will be discussed in the next section. A compilation of crosslinking agents is shown in Figure 4.5.

4.6   COVALENT ATTACHMENT TECHNIQUES

The chemical modification of a protein and/or matrix with an ensuing covalent attachment of the protein is the method of localization which comes to mind most frequently in association with the term "immobilized enzyme." This is due in part to the fact that the greatest amount of work has been done on covalent binding, either to the surface directly or by crosslinking after adsorption. There are several methods of preparing covalent conjugates, but the most popular techniques involve the reaction of an aqueous solution of enzyme with an activated, functionalized water-insoluble support or the copolymerization of an enzyme with a reactive monomer. The expansion of covalent-conjugate technology has resulted in an impressive array of reactions and supports, a detailed discussion of which is beyond the scope of this monograph. A partial list of support media is given in Table 4.2.

Chemical modification of an enzyme is a relatively rigorous process which should involve only those amino acid residues which are not involved in the active site. Groups which appear to be chemically reactive

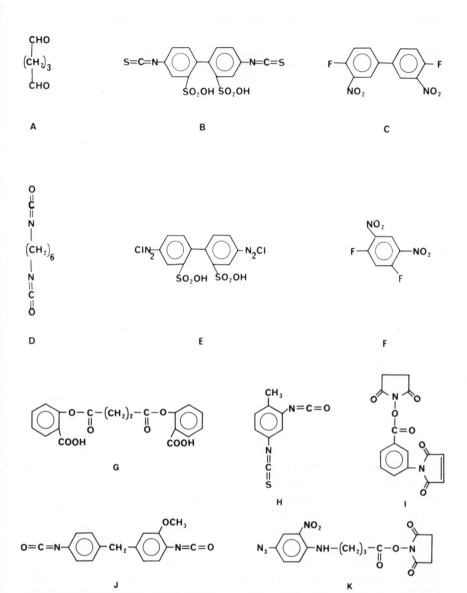

Fig. 4.5. Multifunctional reagents employed for crosslinking proteins. Homogeneous functional group reagents include: (A) glutaraldehyde (69); (B) 4,4'-diisothiocyanatobiphenyl-2,2'-disulfonic acid (70); (C) 4,4'-difluoro-3,3'-dinitrobiphenyl (71); (D) hexamethylenediisocyanate (72); (E) bis-diazobenzene-2,2'-disulfonic acid (73); (F) 1,5-difluoro-2,4-dinitrobenzene (74); and (G) succinyldisalicylate (75). Heterogeneous functional group reagents include: (H) toluene-2-isocyanato-4-isothiocyanate (76); (I) maleimidobenzoyl-N-hydrozysuccinimide ester (77); (J) 3-methoxydiphenylmethyl-4-4'-diisocyanate (76); and (K) N-succinylimidyl-6(4'-azido-2'-nitrophenylamino) propionate (78).

165

Table 4.2 Support Materials for the
Immobilization of Enzymes

Material	Reference
Synthetic	
Acrylamide based polymers	100–102
Maleic anhydride based polymers	145
Methacrylic acid based polymers	146
Polypeptides	147
Styrene based polymers	148
Acrylonitrile based polymers	149
Dacron	150
Nylon	91, 92, 98
Dialdehyde starch–methylenedianiline	151
Natural	
Agarose	84
Dextran	84
Cellulose	82
Glass	83, 115, 119
Nickel oxide	152
Titanium	115
Manganese oxide	153
Collagen	154
Magnetite	155
Alumina	104, 158

include α- and ϵ-amino groups, α-, β-, and γ-carboxyl groups, the phenolic portion of tyrosine, sulfhydryl groups, and the imidazole group of histidine. Mainly the first three groups of reactive sites have been used extensively in the immobilizations. The use of sulfhydryl groups is increasing in popularity due to the ease with which the coupling can be reversed. It should be apparent from the discussion of enzyme mechanisms that reaction of a residue involved in catalysis, such as serine in α-chymotrypsin, would seriously reduce the overall recovery of activity. The degree of the reduction would be related to the overall number of reactive serine residues. Several techniques have been used to protect the active site by blocking it through combination of a competitive inhibitor or substrate with the enzyme, by combination of the enzyme with a reversible covalently-linked inhibitor such as *para*chloromercuribenzoate with urease, by the addition of amino acids to the protein chain through which the covalent linkage can be made, or through the use of a zymogen which is the biological precursor of the enzyme of interest. It should be pointed out that the technique of chemical modification is not universally applicable, nor does it always result in increased recovery of enzyme activity.

There are essentially three steps in an immobilization scheme: activation of the support, enzyme coupling, and removal of loosely bound enzyme. Activation of the support is an important aspect of the process. It can be a simple, one step process, such as the CNBr activation of Sepharose, or a multistep process such as the activation of nylon or the inorganic matrices. It should be pointed out that the coupling reactions outlined in the following sections are not matrix-specific. For example, an alkyl amine derivative of CNBr activated Sepharose can be handled in the same manner as the γ-aminopropyltriethoxysilane derivative of controlled-pore glass. The degree of activation has a significant affect on the recovery of enzyme activity. As pointed out in the hydrophobic adsorption section, the greater the substitution, the greater the possibility of multiple-bond formation to the enzyme, possibly resulting in restriction of molecular orientation and a loss of activity. It is in general easier to control the substitution by varying the concentration of the activation reagent than by manipulating reagents during the coupling step. The concentration of enzyme used during the coupling step is also a consideration. In general, the use of 50–100 mg of enzyme per gram of support will saturate the support. We have always attempted to use the smallest volume of buffer which would cover the support to achieve the highest enzyme concentration possible. We have found that decreases in the enzyme concentration do affect the amount of protein bound although the degree of decrease in activity recovered is difficult to predict.

4.6.1 Support Considerations

The proliferation of viable support media is a result of the search for the perfect matrix. Obviously the carrier used in any application is important since its interaction with the enzyme may have an influence on stability and kinetics. Among the other factors which must be considered are the capacity of the carrier to bind protein, the surface charge and hydrophilicity, the dimensional and chemical stability, the ease of activation, the interaction of the support with the analyte or sample matrix, and the cost, regenerability, and availability.

The capacity of the carrier to bind protein is an important parameter since relatively high activity per unit volume of support is advantageous. If we can immobilize 100 I.U./g of support, we can drive many reactions to completion in a small reactor. As we will see, the loading of the support with enzyme is a major factor in determining the linearity and the limit of detection of methods using immobilized enzymes. A summary of the key factors in support selection, including binding capacity, can be found in Table 4.3. The binding capacity is a function of the number of binding

Table 4.3   Comparison of Characteristics of Some Common Support
Materials[a]

Support	Capacity	Flow characteristics	Stability	Ease of activation	Cost
Silica	+	+ + +	+ +	+	0
Titania	+	+ + +	+ + +	+	0
Zirconia	+	+ + +	+ + +	+	0
Alumina	+	+ + +	+ + +	+	0
Polyacrylamide	+ + +	+	+ +	+ +	+ +
Sephadex	+ + +	0	0	+ + +	+ + +
Agarose	+ + +	0	0	+ + +	+ +
Cellulose	+ +	+	+	+ +	+ + +

[a] + + + Excellent
+ +   Very good
+   Good
0   Poor

sites which can be activated. The polysaccharide-based supports such as
Sephadex, Agarose, and cellulose bind the largest amount of protein.
From the studies of Messing, it is apparent that the surface area, both
internal and external, and the pore size of the support materials are important (41). The open nature and various degrees of crosslinking available
in these supports make them ideal with respect to binding capacity. Some
of the moderately to highly crosslinked polymers such as polyacrylamide,
polyvinyl pyrrolidone, and nylon lose these advantages to some degree
depending on the amount of crosslinking. The inorganic supports such
as common glass, silica, and alumina have essentially no pores of the
appropriate dimensions and bind almost no enzyme. An intermediate case
can be found in the porous ceramics used for filtration. Alumina, which
has 0.5 μm pores, for example, can be obtained from producers of scientific labware. This material has been used successfully in enzyme immobilization. By far the best materials however, are the controlled-pore
glass and ceramics which were produced by Messing and Weetall at Corning Glass works. Due to the narrow pore-size distributions achieved with
these supports, both optimum pore size and surface area are achieved.
As a result, some of the disadvantages of inorganic supports have been
removed. One remaining disadvantage is the strong interaction between
proteins and silica causing denaturation of the enzyme. Hydrophilicity
is a highly desirable characteristic. The surface charge is also important
as was described in Section 4.2.

The mechanical and chemical stability of the support are also important,
particularly in applications using packed-bed reactors. The highly porous

structure of Sephadex, Agarose, polyacrylamide, and other weakly cross-linked polymers is easily deformed under conditions of flow. Flow rates of more than a few tenths of a milliliter per minute per cm^2 cause compaction of the bed into an impermeable plug and cessation of flow. Cellulose has somewhat greater dimensional stability, particularly the macroporous bead type. The development of organic polymer-based supports for high-speed gel-permeation chromatography may yield results beneficial to the immobilized enzyme field. The inorganic supports with their rigid macroporous structure are nearly ideal for packed-bed reactor applications. Also to be considered is the size of the particles to be used and its relationship to pressure drop across the reactor. This will be discussed in detail in Chapter 6. An attempt to achieve both the hydrophilicity and binding capacity of the organic polymers and the dimensional stability of the inorganic supports has been made by crosslinking polyethylene imine (79) or maleic anhydride ethylene copolymer (80) on the surface of a nonporous particle to make a "pellicular" support. In applications other than reactors, such as enzyme electrodes and membranes, the formability of the organic polymers is an advantage.

The chemical stability of the support is also important. Polysaccharide-based supports are subject under some conditions to microbial attack. The structural integrity of silica above pH 7 is not good. With the relatively large particles used in enzyme immobilization one would not expect bed collapse, but these conditions may increase the rate of enzyme leakage. The buffer type and particle size have also been shown to make a difference in the useful lifetime of the support (81). If the pH maximum of the enzyme of interest is in the range 3.5–9.0, one of the controlled-pore ceramics would be the matrix of choice.

The ease with which the support can be prepared for coupling is important, since economical preparation of the enzyme is essential to a useful system. Support activation is the subject of the remainder of this chapter.

The interaction of the support with the analyte or sample matrix is often overlooked. The effects on enzyme kinetics have been discussed. In a clinical setting, it is well known that substances such as wool can cause an immunological response if used in vivo. The use of controlled-pore glass in the analysis of plasma or whole blood would be disadvantageous due to the well-known catalysis of the coagulation reactions by glass surfaces. This latter effect has not been a problem in at least one application because of the addition of large amounts of heparin (159).

The initial cost, the regenerability and the availability, are interrelated factors. The cost factor favors the use of organic polymers if possible. Controlled-porosity inorganic supports are on the order of four to five times as expensive. Regeneration of the support would be useful in view of their expense. It is possible to regenerate the inorganic supports by pyrrolizing the carbonaceous material. The organic carriers cannot be reused unless adsorption was used as the coupling method. The availability of the organic polymers is excellent. Silica supports for gel-permeation chromatography will suffice as a support. Alumina and other ceramics used for small-particle filtration also can be used. At this time, however, Corning Glass Works is the only manufacturer of controlled-pore titanium oxide and zirconium oxide supports and these are produced in industrial particle sizes and quantities. Some polymer-coated porous silica particles which may be useful in the alkaline range have recently been introduced.

4.6.2   Activation of Organic Supports

Cellulose

Cellulose has been used extensively in immobilizing active biochemicals of many types. A wide variety of functional derivatives of cellulose can be used ranging from the azide formed from carboxylmethyl (CM)-cellulose to the isothiocyanato-derivative of 3-(*p*-aminophenoxy)-2-hydroxypropyl ether cellulose. A definitive review of the derivatization of cellulose was presented by Weliky and Weetall (82). Several methods of derivatization deserve mention due to their popularity. The activation of cellulose with cyanuric chloride or one of its derivatives allows the simple coupling of the cellulose to a primary amino group of the enzyme (83). A number of commercial preparations have appeared using this activation method due to its simplicity. More recently, cellulose has been used as a support for other polymers which are actually involved in the enzyme immobilization. It should be pointed out that many of the derivatization reactions which are used on cellulose can be produced on other supports which have a similar reactive group on the surface, that is, a hydroxyl or amine function. A list of functional groups for binding enzymes to cellulose is given in Figure 4.6.

Agarose and Dextran

The use of *Agarose* and *dextran* in enzyme coupling has also been quite common. Axen, Porath, and Ernback originally described the cyanogen halide activation of cross-linked polysaccharides such as Sephadex and Sepharose (84). The activated matrix appears to contain either an iminocarbonate or an isourea derivative. The pH optimum for this reaction is in the range of 11–12.5. Since hydroxide ions are consumed in the reaction, it is necessary to continually add base to maintain the proper pH in the absence of strong buffer. Operationally simple cyanogen bromide coupling schemes using concentrated phosphate (85) or carbonate (86) buffer to maintain the pH have been introduced. The coupling of the enzyme to the activated Sepharose occurs through the amine functions of the protein. The extent and rate of reaction are very pH-dependent and are most favorable when the amine is not protonated, that is, pH 9–10 for aliphatic amines.

There are other factors which must be considered in CNBr activation. The degree of activation is a serious consideration. If the degree of activation is high, there is increased probability of multiple-site binding with the protein. This quite frequently results in inactivation. It is not generally useful to obtain the preparation by vigorous activation followed by the use of low protein concentration or less than optimal pH conditions and subsequent blocking of the excess reactive sites with a small ligand. A general approach would be to use a weight ratio of CNBr to dry Agarose of about 1:10. Secondly, vigorous activation of Sepharose with CNBr results in cross-linking of the gel which, while improving its mechanical strength, does decrease the pore size of the gel.

It has been found that crosslinking of Agarose beads is beneficial in many respects. More suitable crosslinking schemes have been developed using either epichlorohydrin or 2,3-dibromopropanol (87). Another useful derivatization of CNBr activated agarose has been the addition of various diamines (88). This allows coupling to the Agarose with any of the primary amine reactions given in Figure 4.6. A number of other activation schemes for Agarose can be found in the literature (89) and their utility in the analysis should be assessed by the individual.

Nylon

The immobilization of enzymes on nylon has become quite widely used, particularly among those using immobilized-enzyme tubes with continuous-flow analyzers. There have been two types of activation used with nylon. The earlier methods involved the hydrolysis of the peptide bonds of the matrix. Hydrolysis of the secondary amide groups with acid results in the liberation of equimolar amounts of amino and carboxyl groups. The activation of either the carboxyl group through benzidine or hydrazine using carbodiimide (90), or the amino group through any of the aliphatic amine schemes (91, 92) can be used to prepare the enzyme–nylon conjugate. A second approach involved cleavage of the nylon with N,N-dimethyl-1,3-diaminopropane to obtain free amino and amide groups (93). This latter procedure results in a positively charged tube which may ex-

Support Function	Coupling Agent	Active Intermediate	Activation Conditions	Coupling Conditions	Major Reacting Groups on Proteins
$\sim$-OH $\sim$-OH	CNBr	$\sim$-O, O–C=NH (cyclic)	pH 11–12.5, $2M$ carbonate	pH 9–10, 24 hr at 4°C	—NH₂
$\sim$-OH or $\sim$-NH₂	2,4,6-trichlorotriazine, R=Cl, NH₂, OCH₂COOH, or NHCH₂COOH	triazine ring with Cl, R, O-linked	Benzene, 2 hr at 50°C	pH 8, 12 hr at 4°C, 0.1M phosphate	—NH₂
$\sim$-NH₂	$\overset{S}{\underset{\parallel}{Cl-C-Cl}}$	$\sim$-N=C=S	10% thiophosgene/CHCl₃, reflux reaction	pH 9–10, 0.05M HCO₃⁻, 2 hr at 25°C	
$\sim$-NH₂	$\overset{O}{\underset{\parallel}{Cl-C-Cl}}$	$\sim$-N=C=O	Same as isothiocyanate	Same as isothiocyanate	
$\sim$-NH₂	$\overset{O}{\underset{\parallel}{HC(CH_2)_3CH}}\!=\!O$	$\sim$-N=C(H)-(CH₂)₃-CH=O	2.5% Glutaraldehyde in pH 7.0, 0.1M PO₄	pH 5–7, 0.05 M phosphate, 3 hr at R.T.	—NH₂ HO— phenol ring
$\sim$-NH₂	succinic anhydride (H₂C, H₂C ring with two C=O and O)	$\sim$-NH-C(=O)-(CH₂)₂-C(=O)-OH	1% succinic anhydride, pH 6	See carboxyl derivatives	

172

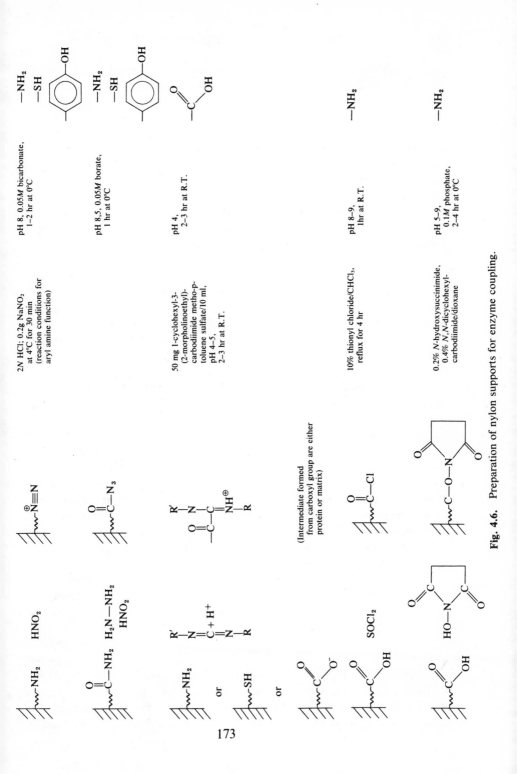

Fig. 4.6. Preparation of nylon supports for enzyme coupling.

173

hibit disadvantageous electrostatic interactions with proteins at neutral pH. All of these procedures allow only limited amounts of enzyme to be immobilized since the surface area of the tubing is rather small. Inman and Hornby used a mixture of 18.6% $CaCl_2$ and 18.6% water in methanol (50°C for 20 min) to remove amorphous areas of the nylon (94). Horvath and Soloman used the reaction of the nylon with polyethyleneimine or fumed silica to yield a wall-supported porous layer to achieve increased surface area (91). In both cases, significant increases in the enzyme activity bound were observed and in the latter case resulted in a diffusion-controlled reaction (see Chapter 7). It should be noted, however, that hydrolysis of the nylon backbone or removal of amorphous regions may compromise the mechanical strength of the tubing.

A second approach to activation of the nylon has recently been reported. The activation of nylon by O-alkylation of the secondary amine has been accomplished by Campbell, Morris, and Hornby. Initially, the derivatization was accomplished by incubating the tube with dimethyl sulfate at 100°C (95). The reactive secondary imidate groups can also be introduced with triethyloxonium tetrafluoroborate at room temperature (96). The latter procedure is more favorable since the reagents are less caustic and toxic; the reaction is effected at room temperature and the activation is more easily controlled. It has recently been reported that the imidoester formed in this reaction is reversible, leading to enzyme "bleed" from the reactor (160).

The application of the Passerini reaction to the activation of nylon (97, 98) and polyester (99) has been reported. In this reaction, a four-component mixture of an amine, a carboxyl group, isocyanide, and an aldehyde condense to form an N-substituted amide. Nylon was hydrolyzed to provide the amine and carboxyl groups, which were then reconnected by the condensation with acetaldehyde and 1,6-di-isocyanohexane. The result is shown in Figure 4.7. Enzymes can then provide either the amine or carboxyl function in a further four-component condensation in which the isocyanide component is supplied by the activated support. The isocyanide can also be derivatized in other more conventional ways, such as addition of diaminodiphenylmethane to form an arylamine. A summary of the nylon activation schemes is shown in Figure 4.7.

Polyacrylamide

A series of modifications of *polyacrylamide gels* has been described using diamines to facilitate an amine exchange (100). The use of hydrazine gives a simple acid azide while ethylenediamine yields an aminoethyl derivative. These groups can then be further modified in the manner of

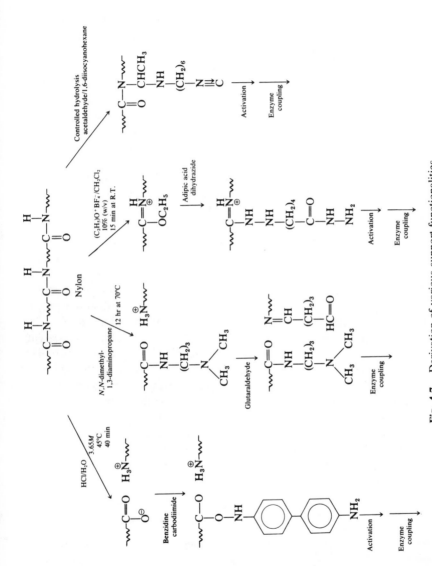

Fig. 4.7. Derivation of various support functionalities.

the above primary amines. Another activation process involves the use of glutaraldehyde, which presumably reacts with the free amino and amide groups on the polyacrylamide matrix, resulting in a Schiff's base formation and the presence of an aldehyde function for protein binding (101). More recently, aminoethyl polyacrylamide beads were reacted with glyceraldehyde to form the 2,3-dihydroxypropyl derivative (102). Aqueous *meta*-periodate was used to produce the aldehyde derivative of the gel, which was then coupled to the protein by reductive amination with sodium cyanoborohydride. Sodium cyanoborohydride was found to be preferable to sodium borohydride due to the milder reduction conditions and lack of evolution of gas that would block the pores. In addition, it appears that for the system studied, the milder reduction resulted in greater adduct stability.

A new group of matrices called *Enzacryl®* gel have become available for enzyme immobilization. The resin is patented by Koch-Light Ltd. and is distributed in the United States by Aldrich Chemical Company. The support is a hydrophilic copolymer of *N,N'*-methylenediacrylamide and various derivatives of acrylamide. Five functional groups are available: aromatic amine (AA); acid hydrazide (AH); sulfhydryl groups (Polythiol); thiolactone rings (Polythiolactone); and acetal functions (Polyacetal). The first two groups can be activated in the same manner as has been shown above in Figure 4.6. The polythiol function reacts with the disulfide groups of the protein, a reaction which can be reversed by the addition of sulfhydryl compounds. Because of its reversibility, the thiol linkage is receiving increase study in conjunction with other matrices. Enzacryl Polythiolactone reacts with the ε-amino groups of lysine as well as the hydroxyl groups of serine and tyrosine. The polyacetal group becomes an aldehyde function in acid media, forming a Schiff's base with the amino groups of the protein. This reaction can be reversed with some enzymes acting on macromolecular substrates such as the α-amylase–starch system (103). The reversible nature of the binding is a potentially useful feature, allowing regeneration of the support after the immobilized enzyme activity has decreased (103).

4.6.3 Activation of Inorganic Supports

Substantial changes have come about recently in the immobilization of enzymes on inorganic supports. Unlike the organic matrices discussed above, in the vast majority of cases the surface of the glass beads must be derivatized before it can be activated. An organosilane is used to bridge the inorganic-to-organic interface. The inorganic functional groups, which can be esters, halides, or silanols, condense with hydroxyl groups on the surface. The mechanism of the condensation is dependent on the reaction solvent used. The reaction of organo-silane coupling reagents occurs with a number of inorganic materials including alumina, aluminum silicate, clays, glass, nickel oxide, quartz, silica, talc, and titania (104, 105).

The technology for the preparation of organic derivatives of inorganic supports has been developed in great measure by the chromatography establishment. The preparation of bonded-phase supports has been a great advance, especially in high performance liquid chromatography. Research has shown that two types of silane derivatives can be prepared: a thin monomeric bonded phase or a thicker polymeric bonded phase. Kirkland (106–108) is generally given credit for preparation of a number of polymeric bonded phases for chromatographic use. In the reaction, trialkoxysilanes hydrolyze, then couple to the support and polymerize, after which heating of the product to 120°C for a few hours completes the polymerization. Two things should be noted. First, solution-phase polymerization of the silane molecules competes with surface bonding. Second, due to the interference of steric factors, not all of the silanol groups polymerize, leaving free hydroxyl groups on the surface of the support. The thickness of the polymer can be controlled by varying the silanol concentration, the reaction solvent, the pH, the reaction time, and the surface area of the matrix (107, 109).

There are essentially three approaches that have been used to prepare organo–silane derivatives of inorganic supports. In all cases, it is necessary to begin by cleaning the surface of the support. Several techniques have been used: heating to 500°C (which causes formation of siloxane groups and eliminates water), washing with organic solvents, and heating in 5% nitric acid. The first approach entails the aqueous silanization of the support. A 10% aqueous solution of a silane such as γ-aminopropyltriethoxysilane is added to an aliquot of clean glass. The pH of the solution is then adjusted to pH 3 and incubated at 75°C for 2 hr. The product is then washed, and polymerization at 120°C for at least 4 hr follows. Although this type of silanization results in the lowest amount of bonded material, the aqueous procedure appears to yield good durability in enzyme preparations (110).

The second approach involves the use of organic solvents in the silanization reaction (111, 112). As originally proposed, a portion of clean glass was refluxed with a triethoxysilane in toluene for 12 to 36 hr. The preparation was then washed and baked at 120°C. The yield of amino groups from one procedure was found to be 30–35 μmoles/g of support.

The final approach is an evaporative technique involving a 1% solution (v/v) of silane in acetone (113, 114). A 1 g portion of clean glass is placed in contact with the acetone solution. The acetone is evaporated after which the glass is heated at 120°C for about 12 hr. This approach in our hands has yielded the highest loadings of functional binding groups. This does not, however, imply that the enzyme activity on these support materials is significantly higher. Although silanization reactions are extremely popular, recent reports in the literature indicate that direct CNBr activation of inorganic matrices (115), the use of Si–N bridges (116), and the use of metallic bridging techniques (117) may also be useful techniques.

The use of glutaraldehyde to activate alkylamine glass has recently been investigated (118). The finding that the carbon content of the activated glass is much greater than that expected for a monomeric coverage seems to support the theory that aldol condensation products bind to the amine groups on the surface. Above pH 9, the aldol condensation rate increases and the polymer formed does not bind enzyme very well. This may be due to plugging of the silica pores. The amount of glutaraldehyde bound increases with increasing concentration or increasing contact time. Optimal trypsin binding occurred at higher concentrations, pH's, and contact times, but the specific activity decreased after 2% glutaraldehyde at pH 8.5 and 1 hr of activation.

One of the most recent advances in the area of inorganic supports has been the introduction of a hydrophilic organic monolayer covalently bound to the glass surface. Thus, the mechanical durability and precise pore morphology of the glass support are effectively combined with the hydrophilic nature and easy derivatization of polysaccharide carriers. Glycidoxypropyltrimethoxysilane (119) and dextran derivatives of the inorganic carriers have eliminated to a large extent the adsorption and denaturation observed on the normally polar glass surface. The glycolpropyl derivative is available commercially under the name Glycophase®G (Corning Glass Works) controlled porosity glass (CPG). Preparation of the matrix for an immobilization consists of simply oxidizing the glycerol group with sodium metaperiodate. Exposure of a solution of enzyme to the oxidized glass results in a Schiff's base formation between the aldehyde function of the glass and the amino groups on the protein (120). In studies in our laboratories, we found that for the enzyme hexokinase, the use of Glycophase supports gave the highest yield of enzyme activity (121). Some of the data are shown in Table 4.4.

Other methods have been used for coupling an enzyme to derivatized CPG. The alkyl amine group has been used directly with the methods discussed above. A more popular approach has been the preparation of an aryl amine from the alkyl group (110) followed by further activation of the aromatic group. This practice makes the matrix somewhat more difficult to prepare for coupling than the organic carriers. In spite of this

**Table 4.4 Comparison of Specific Activities for
Hexokinase Immobilized onto Controlled-Pore
Glass by Different Reactions**

Coupling method	Specific activity[a]	Activity (%)[b]
Glutaraldehyde[c]	50	0.5
Glutaraldehyde and NaBH₄[c]	90	3
Dextran[c]	100	1
Cyanogen bromide[d]	239	2
Cyanogen bromide[e]	90	1
Glycophase oxidation[d]	430	15
Glycophase oxidation[e]	90	5

[a] Units/g glass.
[b] Percent of initial enzyme activity in contact with glass.
[c] 200–400 mesh, 500 Å pore diameter.
[d] 120–200 mesh, 250 Å pore diameter.
[e] 120–200 mesh, 550 Å pore diameter.

disadvantage, CPG has been extensively used in analytical applications due to its mechanical durability.

There are advantages and disadvantages to the use of covalently bound enzymes. The chemical modification of an enzyme is relatively simple, requiring at most several steps to activate the support, followed by reaction with the enzyme and removal of any excess protein by washing. One significant advantage of covalent coupling is the large number of examples in the literature which may be directly applicable to a given analytical system. Another advantage of the chemical modification of an enzyme is that it may yield an enzyme which is chemically superior to the soluble enzyme. This fact, combined with the fact that these materials can be made in a number of physical forms, that is, sheets, beads, or membranes, make this method of enzyme insolubilization very useful for industrial and analytical applications. Of course, covalently bound enzymes exhibit the same ease of handling present in almost all immobilized-enzyme preparations.

Of the several disadvantages of the technique, the most problematic is the relatively low recovery of enzyme activity observed after covalent modification. One important piece of information that should be obtained is the amino acid composition of the active center of the enzyme. As mentioned in Chapter 1, the chemical reactivity of the amino acids in the active site is the source of the enzymatic catalysis. Obviously, chemical coupling of the enzyme through these residues would yield no activity. It is important to realize that binding in such a way that the enzyme cannot move into the proper orientation for catalysis, as could be the case for carboxypeptidase A (see Chapter 1), the activity could be impaired. Unfortunately, only a few active sites have been characterized and thus many immobilizations have been "shot in the dark" approaches which have not been very efficient. A second problem with covalent coupling is the lack of activity exhibited by enzymes acting on large substrates, presumably due to steric hindrance of the matrix.

4.6.4 Miscellaneous Methods

We have included this section to discuss methods which differ in approach from those in the previous sections or which are novel methods. Each of these methods may fill a specific need and some may herald a new direction in immobilization technology.

The first method is a covalent-attachment technique which uses activation and coupling of the enzyme through carbohydrate residues naturally bound to the enzyme. Prior to the work of Zaborsky and Olgetree (122), only covalent modification of the amino acid residues was used in immobilization. It is well known that carbohydrate moieties play an important role in many proteins. If the sugar residues do not take part in the catalytic function of an enzyme, coupling through these groups may eliminate the possibility of modification of the active-site region and produce better recoveries of enzyme activity. The procedure involves the periodate oxidation of the carbohydrate residues and subsequent coupling via a Schiff's base reaction to an amine function on a support. In work with glucose oxidase, these authors describe a 100% recovery of enzyme specific activity on coupling through the carbohydrate moieties. Peroxidase, in contradistinction, gave low yields, presumably because of the interaction of the carbohydrate groups with the active site.

One limitation of immobilized enzyme technology has been its "requirement" for organic reactions which are compatible with aqueous media. Recently it has been found that proteins can be activated and coupled in organic solvents (123). There are several advantages to organic coupling including an increase in the available reactive sites on the protein, a wider variety of supports due to solubility and reactive capabilities, and use of the solvent to control the protein configuration during the coupling process. The use of organic solvents for immobilization on various types of supports may be of great value in improving the recovery of enzyme activity.

There are, of course, numerous other successful immobilization techniques in the literature. To review all of them would make this a mammoth volume. The reader who finds that his needs cannot be served by any of the more common techniques is referred to either Zaborsky (1), Weetall (10), or Mosbach (124). It should be recognized that there is no "best" method of immobilization. Thus, at present, one is resigned to empirical observations of the most suitable approach for the specific application.

Before moving to a consideration of the various immobilization techniques, it seems important to touch on the area of enzyme distribution in the support. As will be seen, diffusional limitations can control the reaction rate. Thus for theoretical reasons, it is important to know whether the enzyme immobilized on, for example, Agarose is dispersed throughout the particle or is in a thin layer on the surface. A recent study using electron microscopy has shown that the appearance of the protein–support conjugate is dependent on the reaction conditions (125). Using Sepharose as a support matrix, ferritin was coupled to the matrix which had been activated using various amounts of cyanogen bromide per ml of gel. If less than 50% of the ferritin (2 mg/ml gel) was bound, the preparation was uniformly distributed through the beads. If, on the other hand, over 90% of the protein was bound or if the activation was accomplished with greater than 50 mg/ml of cyanogen bromide, the ferritin was found to be in a superficial layer with no protein at the core of the bead. Other studies in which an antigen was bound to CNBr activated Sepharose (30 mg/ml gel) and subsequently reacted with ferritin-labeled antibody showed homogeneous spacing. Other authors have reported homogeneous distributions using bound fluoroscein (126).

The use of immobilized enzymes in packed-bed reactors can be accomplished by using either pellicular or small totally porous particles. The major differences between the two approaches is the capacity of the support and the pressure drop across the column. A paper has appeared which describes the preparation of pellicular immobilized enzymes (80). The important point here, however, is that inappropriate reaction conditions may lead to a "pellicular" preparation when the desired result is a totally porous particle with homogeneous activity.

4.7   COMPARISON OF IMMOBILIZATION TECHNIQUES

As mentioned earlier, many of the properties of immobilized enzymes are quite variable. The recovery of activity, the pH optimum, the K_M and the stability of the immobilized enzyme may be increased or decreased from those of the soluble species depending on the particular enzyme, the support material, the type of enzyme–support interaction, and possibly on the experimenter himself. At this time, there is no way to predict the results that will be obtained with any particular technique. In the remainder of this chapter, we will try to survey the various methods of attachment of several enzymes to a solid support, approaching the subject from the determinant end—the enzyme.

4.7.1 Urease

The popularity of urease in immobilized-enzyme techniques stems from its analytical usefulness in the determination of urea, its specificity, and its stability. One of the earliest immobilization methods was adsorption of the enzyme onto kaolinite (48). The method involved the addition of 12 mg of urease to 200 mg of clay in a total reaction volume of 7.5 ml of Tris–KCl–EDTA buffer. The adsorption process was pH-dependent and at pH 7.0, a total of 3.3 mg of protein was bound to 100 mg of clay. The activity of the preparation was not mentioned. Since the authors were particularly interested in the kinetics of the system, no determination of the stability was reported. It is interesting, however, that the pH optimum of the immobilized enzyme was the same as that of the soluble species, while the kinetics were somewhat changed. The V_{max}, corrected for protein concentration, was consistently higher for the immobilized enzyme while the apparent K_M was slightly higher at all pH values. Some of the data is given in Table 4.5.

A hybrid method has been used to immobilize urease, in which the adsorption of the enzyme onto fumed silica was followed by glutaraldehyde crosslinking of the protein (127). Eventually, the silica particles were encapsulated for use as a prosthetic device. It was found that the glutaraldehyde concentration and the length of the crosslinking incubation were critical parameters in obtaining high yields of enzyme activity. As can be seen from Table 4.5, 80–90% retention of activity was obtained. The storage stability of the preparation was very good with no appreciable losses at 22°C after 39 days. At 38°C, the immobilized enzyme retained 42% of its activity after 15 days—a 400% increase over the soluble species. Microencapsulation of the conjugate caused a loss of about 80% of the activity of the immobilized preparation, probably due to the organic solvents used in the encapsulation process.

A number of authors have used gel entrapment of urease in conjunction with an electrochemical sensor for the measurement of urea. Guilbault and Das presented an interesting study of the entrapment of urease in several types of matrices (60). Although no data were presented on the pH optimum, these authors found that the apparent K_M approximated that of the soluble species rather closely. In order to determine the activity retained by the entrapment, each of the matrices was leached of enzyme activity and a solution phase determination made. Of the matrices studies,

Table 4.5 Comparison of Immobilization Methods for Urease

Support/Method of attachment	Amount of enzyme bound	Recovery	Stability	K_M	pH Optimum	Comments	Reference
Soluble	—	—	—	4 mM	6.7–7.6		
Kaolinite/Adsorption	33 mg/g support	ND[a]	ND	3 mM at pH 6.6	6.6		32
Starch gel/Gel entrapment	100–120 mg/6 in. × 6 in. pad	100%	About 90 days storage at 4°C	3.75 mM	ND		44
Polyacrylamide/Gel entrapment		80%		5 mM			44
Silicon rubber/Gel entrapment		20%					
Fumed silica/Adsorption and crosslinking	ND	80–90%	100% after 39 days at 22°C	ND	ND	10 μm particles	105
p-Amino-DL-phenylalanine and amino acid copolymer/covalent (diazo)	2 mg/ml support	6%	ND			glycine	106
	1.8 mg/ml	42%	ND	ND	ND	L-alanine	
	1.6 mg/ml	47–78%	60% after 5 months			L-leucine	
Controlled-pore glass/covalent (diazo)	1 mg/g glass (from specific activity)	ND	No loss over 30 days	ND	6	1550 I.U./g glass	107
Controlled-pore glass/covalent (glutaraldehyde)	1 mg/g glass (from specific activity)	32%	60 days operational stability	10 mM	7		unpublished data
Nylon/Colvalent (glutaraldehyde)	62 μg/m tubing (2.55 units/μg)	ND		3.5 mM	7	81% conversion of urea at 2 ml/min flow rate	108

[a] ND = No data.

183

starch gel retained the most activity through the entrapment step with a 100% yield. This was followed by an 80% retention by polyacrylamide gel and a 20% recovery with silicon rubber, although the more rigorous leeching techniques employed with the latter two supports may have influenced the activity measurement. Although the recovery from the entrapment step is of interest, the important feature is the amount of enzyme which can be used for analysis or study. It was found that 25% of the activity of the starch-gel preparation could be washed off the gel with the first 20 ml of solution passed through the enzyme pad, while under the same conditions only 4% of the polyacrylamide-gel activity was lost. The activity losses were ascribed to adsorption of the enzyme to the gel which was easily disrupted. Both types of gel pads were found to have excellent storage stability, essentially retaining constant activity over 90 days of storage at 4°C.

Some of the most interesting studies on immobilized urease have been on the covalently modified enzyme. Riesel and Katchalski presented an extremely interesting study on the activity of urease bound to diazotized copolymers of p-amino-DL-phenylalanine and L-alanine, L-glycine, and L-leucine (128). A portion of a table from their work is shown in Table 4.6. There are several interesting features in the table. First, the two preparations at the top of the table give results when an increased amount of protein is offered to the polymer. Not only is the absolute amount of bound protein increased, but the recovery of enzyme is increased as well. The percent of the offered protein which is bound, however, is decreased, indicating the loss of some enzyme in an effort to obtain a high activity on the support. This phenomenon is quite common when trying to couple as much enzyme as possible to a matrix. Second, the nature of the support

Table 4.6   Immobilization of Urease on Amino Acid Copolymer[a,b]

Preparation	Protein offered (mg)	Protein bound (mg)	% Activity[d]
pApa/leucine	25	20	38
pApa/leucine	108	51	78
1:1 pApa/glycine	28	20	6
1:1 pApa/alanine	23	18	42
1:2.5 pApa/leucine[e]	20	16	46

[a] Data taken from reference 128.
[b] p-Amino-DL-phenylalanine and L-leucine or L-alanine or L-glycine copolymer with diazo coupling reaction.
[c] Calculated by difference from protein remaining in supernatant.
[d] insoluble units/(specific activity × mg bound) × 100.
[e] Inactivated with p-chloromercuribenzoate during immobilization.

plays an important role in the enzyme recovery as seen in the remainder of the table. Most notably, the glycine-containing polymer has a significantly lower activity recovery than any of the other preparations. This is apparently due to a problem with the activity of the enzyme such as inactivation or steric hindrance of the substrate, since the amount of protein bound to the glycine copolymer is comparable to the other derivatives. Finally, the leucine copolymer was found to have the best storage stability, losing only 40% of its activity after 5 months of storage at 4°C. This can be compared to a 90% loss after 1 month at 4°C for the glycine copolymer derivative.

A number of techniques have been used to attach urease to CPG. Weetall and Hersh (129) used a diazonium salt derivative to achieve a preparation which contained 1 mg of active urease per gram of support (1550 I.U./g CPG). The amount of protein offered was not given. The CPG-bound urease was found to exhibit a pH optimum at pH 6.0, slightly more acidic than the soluble species. We have used glutaraldehyde activation of the γ-aminopropyltriethoxysilane derivatized glass with good results. Covering the glass with 1 ml of a concentrated solution of urease (25 mg/ml), we were able to achieve 50% binding of the protein with recovery of over 10% of the activity of the coupled enzyme. The glutaraldehyde preparation exhibited no measurable protein loss in use and retained 50% of its activity in storage at 4°C as a wet cake for 3 months, at which time measurements were discontinued.

A method of considerable interest to continuous-flow automated-analyzer users was reported by Inman and Hornby (130). Urease was covalently bound to the inner surface of a nylon tube which had been hydrolyzed with HCl and derivatized with glutaraldehyde. Although the yield of enzyme activity was not reported, it is of interest that the K_M' for the system was 3.5×10^{-3} moles/l, considerably larger than the solution-phase K_M. This is due to the diffusional limitations of crossing the laminar flow lines of the stream to reach the enzyme. The operational lifetime of these enzyme tubes was studied and over 5000 analyses were run over a 30-day period with no loss of activity. It is of interest to note that the amount of enzyme used in the immobilization would have been sufficient for only 1100 solution-phase determinations.

4.7.2 Glucose Oxidase

Another enzyme which has been extensively studied in the immobilized state is glucose oxidase. The specificity and ease with which this enzyme can be used for blood glucose measurements has increased the popularity of the enzyme. It was the first enzyme to be entrapped and used in an "enzyme electrode" (131). Although no yield information was reported, gels were prepared which contained up to 420 mg of glucose oxidase per 100 ml of gel. In lyophilyzed form, the preparation showed no loss of activity over 3 months, while in the hydrated state the glucose oxidase gel lost only about 5% of its activity in the refrigerator during a 6-week period. One of the difficulties of comparing enzyme preparations has been illustrated by a similar glucose oxidase gel (132). It was found that in the unbroken form, the gel exhibited a 10% recovery of activity. When the gel was cut into 8 mm squares or ground into a fine powder, the same preparation exhibited activity recoveries of 12 and 15%, respectively. This is the result of the diffusion of the substrate into the gel matrix. At some point in the gel, the immobilized enzyme no longer is under zero-order kinetics and thus the measurement is not a true measurement of the activity. As the size of the particle decreases, the amount of enzyme in the matrix under zero-order kinetics increases with a concomitant increase in the observed activity. A similar effect has been observed for determination of the Michaelis constant.

Adsorption techniques have also been used in conjunction with glucose oxidase. The adsorption of the enzyme onto CPG has been reported (133). No mention of the amount of protein offered was made, but the highest yield of bound protein was 0.12 mg/gram of support. The amount of protein was calculated from the activity of the preparation. Adsorption onto a cellophane membrane followed by glutaraldehyde crosslinking, however, achieved an 80% recovery of enzyme activity. The capacity of the cellophane is limited to fairly small amounts of protein. The same technique was used for urease, trypsin, and α-amylase with activity yields of 60, 20, and 10%, respectively (134).

A number of interesting studies have been undertaken with covalently bound glucose oxidase. The activity of immobilized glucose oxidase was localized in the matrix by a study of the reduction of FAD by glucose (112). It was found that 6% of the glucose-reducible protein was kinetically active, most of it on the surface of the beads. This again illlustrates the importance of diffusional effects in immobilized-enzyme preparations. Probably the most reasonable comparative study of immobilized enzymes was presented by Weetall (135). A number of inorganic carriers and cellulose-bound glucose oxidase were compared to the soluble enzyme as shown in Table 4.7. Several facts are evident from the data in the table. First, the attachment of glucose oxidase to alumina, hydroxyapatite, and controlled-porosity glass results in a stable preparation regardless of the storage conditions. It also appears that the azo linkage is more stable than coupling with the isothiocyanato-derivative. Finally, it is interesting to

Table 4.7   Stability of Immobilized Glucose Oxidase Derivatives[a]

Derivative	Original activity (mg/g)[b]	Conditions of storage	Storage time (days)	Percentage activity
Alumina-CPG-	6.0	5°C	68	100
isothiocyanate		5°C, in water	68	100
		23°C	68	100
		23°C, in water	68	100
Hydroxyapatite-	10.7	5°C	68	100
isothiocyanato		5°C, in water	68	65
		23°C	68	100
		23°C, in water	68	100
CPG-isothiocyanate	11.8	5°C	68	81
		5°C, in water	68	73
		23°	68	73
		23°C, in water	68	77
CPG, azo	10.1	5°C	68	100
		5°C, in water	68	100
		23°C	68	100
		23°C, in water	68	100
Cellulose-azo	6.1	5°C	68	69
		5°C, in water	68	16
		23°C	68	57
		23°C, in water	68	9
Soluble enzyme		5°C	60	100
		23°C	60	75

[a] Data taken from reference 135.
[b] Activity per gram of support material was calculated assuming that the immobilized enzyme retains the same specific activity as the soluble enzyme.

note that the cellulose preparation is less stable than the soluble preparation during storage.

4.7.3   Trypsin

Although the previous examples of immobilized enzymes have been well characterized, both of them involve small substrate molecules. As mentioned earlier, the matrix and the immobilization technique have an affect on the recovery of enzyme activity as well as on its characteristics. Trypsin, as a protease and/or esterase, works on macromolecular substrates such as casein as well as other smaller esters. Thus, trypsin may be illustrative of some of the problems encountered with immobilized enzymes working on macromolecules.

Adsorption and gel entrapment of trypsin have met with varied success. No tryptic activity was observed for adsorption onto a glass support (136). Gel entrapment with polyacrylamide yielded a 2% yield of activity using α-N-benzoyl-L-arginine methyl ester (BAME) as the substrate (137). No activity was observed toward casein. Similar observations were made with a sialic acid entrapped preparation (65). A 34% recovery was obtained initially using α-N-benzoyl-L-arginine ethyl ester (BAEE). This preparation exhibited good storage stability retaining 90% of its activity after 75 days at 4°C. An excellent recovery of trypsin activity (85%) was obtained using the fumed-silica-adsorption–glutaraldehyde-crosslinking technique (138).

The covalent attachment of trypsin to a support has been quite extensively studied. As shown in Table 4.8, a number of carriers have been used with a number of coupling reactions. Several conclusions can again be drawn from the table. First, the "tightness" of the matrix has a large effect on both the amount of protein bound and the activity recovery, allowing larger amounts of protein to be bound to the tight support while retaining lower activity (139). It is also interesting to note the relationship between the activity of the agarose gel and the pH optimum of the preparation. The reaction of the trypsin with the substrate ester yields a hydrogen ion which in turn affects the local pH in the matrix. Thus, one would expect a direct relationship between the amount of enzyme bound to the matrix and the alkaline shift of the pH optimum. This is observed. It is also interesting that the preparation exhibits very little activity toward casein, probably as a result of steric exclusion of the substrate from the enzyme. Weetall also studied the storage stability of trypsin bound to various carriers (135). From Table 4.9, the disparity between the capacities of the various carriers can be readily seen. Once again, the greater

Table 4.8   Immobilization of Trypsin onto Polysaccharide Carriers[a]

Polymer	Amount of bound enzyme (mg/g)	Esterase activity		Caseinolytic activity	
		Activity ratio[b]	pH optimum	Activity ratio[b]	pH optimum
CNBr–cellulose	205	4	9.7	0	—
CNBr–Sephadex	78	12	9.6	3	8.0
CNBr–Agarose	30	30	9.5	25	7.9
	108	45	9.7	35	7.6
Soluble	—	100	8.2	100	7.6

[a] Data taken from reference 139.
[b] The activity ratio is calculated from the activity per mg of protein bound to the conjugate divided by the activity per mg protein for the soluble species.

Table 4.9 Stability of Immobilized Trypsin Derivatives[a]

Derivative	Original activity (mg/g)[b]	Condition of storage	Storage time (days)	Percentage activity
Azo–CPG	1.42	5°C	72	79
		23°C	72	89
Isothiocyano–CPG	0.97	5°C	68	0
		23°C	68	0
Azide–cellulose	3.82	5°C	42	60
		23°C	63	60
Maleic-anhydride–[c]ethylene–trypsin (polymer = protein)	6.50	5°C	22	100
		23°C	42	100
		5°C, in water	30	0
Maleic-anhydride–[c]ethylene-trypsin (polymer = 4 protein)	26.0	5°C	63	48
		23°C	36	84
Maleic-anhydride–[c]ethylene–trypsin (4 polymer = protein)	41.0	5°C	63	19.5
		23°C	63	13
Isothiocyano–fumed silica (4 μm diameter)	1.32	23°C	42	58

[a] Data taken from reference 135.
[b] Activity per gram of support calculated assuming that the enzyme retains the same specific activity on the support.
[c] The ratio of the weight of polymer to the weight of crystalline enzyme is given in parentheses.

stability of the azo linkage with respect to the sulfonamide linkage is observed. The most interesting data from the table, however, is the stability of the maleic-anhydride–ethylene copolymer preparations and the inverse relationship between the amount of enzyme bound and its stability on the copolymer carriers. An insufficient number of studies have been performed to know whether this is a general trend, but it is an interesting phenomenon. Data which is not presented in the table indicated that a significant difference existed between the organic carrier-bound enzyme and that attached to the inorganic matrix with respect to casein activity. The controlled pore size of the CPG allowed diffusion of the casein into the matrix and thus the preparation showed greater caseinolytic activity.

4.7.4 α-Amylase

Although α-amylase has not been immobilized as frequently as the other enzymes discussed above, it provides another opportunity to study an enzyme which acts on a macromolecular substrate. Some data on the activity recovery on immobilization on several different matrices is given in Table 4.10. As can be seen, the recoveries are comparable to those obtained with other immobilization techniques. The interesting point about α-amylase is that immobilization changes its functional specificity. The degree of attack of α-amylase can be studied since the combination of starch with iodine only occurs if the starch chain contains 25 residues or more. Another technique can be used to determine the number of free chain ends. In its normal solution behavior, α-amylase attacks the starch chain in a random manner, resulting in an almost linear relationship between the number of chain ends and the reduction of the starch iodine color. In its immobilized state, however, the enzyme appears to change its degree of attack and cleave the starch in a sequential manner, binding to one end of the chain and moving toward the other end by several residues at a jump (143). This is an interesting phenomenon, and has been reported for other enzymes acting on macromolecular species (144).

Although comparisons of immobilized enzymes are difficult, there are several conclusions that can be drawn from the above discussion. First, the amount of protein that can be immobilized by any method is limited by the capacity of the carrier. It was apparent that in general the organic matrices have a greater coupling capacity than do the inorganic matrices. The relationship between the amount of protein coupled to the matrix and the amount of activity observed per gram of the carrier has been studied by several authors. The observed activities varied widely with the amount

Table 4.10 Comparison of Immobilization Techniques for α-Amylase

Support/Method of Attachment	Recovery	Stability	Comments	Reference
Cellulose/diazo or isothiocyanato	ND[a]	50% loss after 5 uses		157
Microcrystalline cellulose/	6.0%	ND	1 mg protein/ 100 mg carrier	143
CM cellulose/	2.1%	ND		143
p-Aminobenzoate cellulose	1.5%	ND		143
Polystyrene/	4.0%	ND		143
Enzacryl AA or AH®/diazo	6.1%	73 % after 3 months		156
/isothiocyanote	9.5%	67% after 3 months		
AH/azide	16.0%	85% after 3 months		

[a] ND = no data.

of enzyme coupled. It would be of assistance to know the number of functioning active sites bound to the matrix so that the affect of mass transfer could be assessed. Another difficulty is the number of methods used to determine the amount of protein bound to the support. Although the determination of protein in the supernatant after the coupling reaction is often used, other authors base the amount of protein bound to the support on the observed bound activity assuming that the specific activity of the enzyme has not changed. Thus, it is difficult to make a conclusive comparison.

4.8 CONCLUSION

In summary, the final decision concerning an immobilized enzyme/coupling technique/support system must include a large number of considerations. The choice of a support matrix must be based on the system of which the enzyme will be a part. The capacity, mechanical and chemical stability, expense, availability, ease of preparation, and flow characteristics are all potential considerations in an analytical application. In selecting a technique for coupling the enzyme to the support, it must be emphasized that there is no "best" technique. Economic factors may be a significant part of the decision. Irreversible adsorption may yield reproducible and inexpensive preparations of the immobilized enzyme of interest. In situations in which the user might not wish to be burdened with the chemical manipulations of immobilization, storage stability may be a primary consideration. Last, but certainly not least, the characteristics of the enzyme itself must be satisfactory. It should be understood that many of the above factors are not predictable. There are, however, logical approaches to immobilization based on a knowledge of enzyme and active-site structure, and an understanding of the coupling chemistry. The above discussion should serve as a useful guide in a systematic approach to immobilization.

References

1. O. R. Zaborsky, *Immobilized Enzymes* (CRC Press, Cleveland, Ohio, 1973).
2. H. H. Weetall, *Immobilized Enzymes–A Compendium of References from the Recent Literature*, (Corning Glass Works, Corning, NY, and New England Research Applications Center, University of Connecticut, Storrs, C., 1973).
3. R. A. Messing, Ed., *Immobilized Enzymes for Industrial Reactors* (Academic, New York, 1975).

4. R. B. Dunlap, Ed., *Immobilized Biochemicals and Affinity Chromatography* (Plenum Press, New York, 1974).
5. M. Salomona, C. Saronia, and S. Garattini, *Insolubilized Enzymes*, (Raven Press, New York, 1973).
6. E. K. Pye and L. B. Wingard, Eds., *Enzyme Engineering*, Vol. II (Plenum Press, New York, 1973).
7. C. R. Lowe and P. D. G. Dean, *Affinity Chromatography* (Wiley, New York, 1971).
8. G. R. Stark, *Biochemical Aspects of Reactions on Solid Supports* (Academic Press, New York, 1971).
9. G. E. Means and R. E. Feeney, *Chemical Modification of Proteins*, (Holden–Day, San Francisco, 1971).
10. H. H. Weetall, *Immobilized Enzymes, Antigens, Antibodies and Peptides* (Marcel Dekker, New York, 1975).
11. H. H. Weetall and S. Suzuki, *Immobilized Enzyme Technology–Research and Applications* (Plenum Press, New York, 1975).
12. D. Thomas and J. P. Kernevez, Eds., *Analysis and Control of Immobilized Enzyme Systems, Proc. Intl. Symp.*, (American Elsevier, New York, 1976).
13. H. H. Weetall, *Anal. Chem.* **46**, 602A (1974).
14. L. D. Bowers and P. W. Carr, *Anal. Chem.* **48**, 544A (1976).
15. I. Karube and S. Suzuki, *Kagaku (Kyoto)* **68**, 177 (1976).
16. I. V. Berezin and A. A. Klesov, *Zh. Anal. Khim.* **31**, 786 (1976).
17. L. Goldstein, *Meth. Enz.* **19**, 935 (1970).
18. Y. Levin, M. Pecht, L. Goldstein, and E. Katchalski, *Biochemistry* **3**, 1905 (1964).
19. M. Wilchek, *FEBS Letters* **33**, 70 (1973).
20. G. J. H. Melrose, *Rev. Pure Appl. Chem.* **21**, 83 (1971).
21. D. Gabel, *Eur. J. Biochem.* **33**, 348 (1973).
22. K. Martinek, A. M. Klimbov, V. S. Goldmacher, and I. V. Berezin, *Biochim. Biophys. Acta* **485**, 1 (1977).
23. K. Martinek, A. M. Klimbov, V. S. Goldmacher, A. V. Tchernysheva, V. V. Mozhaev, I. V. Berezin, and B. O. Glotov, *Biochim. Biophys. Acta* **485**, 13 (1977).
24. G. P. Royer and R. Uy, *J. Biol. Chem.* **248**, 2627 (1973).
25. J. Lasch and R. Koelsch, *Eur. J. Biochem.* **82**, 181 (1978).
26. J. D. Dixon, F. E. Stolzenbach, J. A. Berenson, and N. O. Kaplan, *Biochim. Biophys. Res. Commun.* **52**, 905 (1973).
27. W. M. Herring, R. L. Laurence, and J. R. Kittrel, *Biotechnol. Bioeng.* **14**, 975 (1975).
28. D. L. Morris, J. Campbell, and W. E. Hornby, *Biochem J.* **147**, 593 (1975).
29. D. B. Johnson and M. P. Coughlin, *Biotechnol. Bioeng.* **20**, 1085 (1979).
30. D. B. Johnson, *Biotechnol. Bioeng.* **20**, 1117 (1978).
31. S. Krishnaswamg and J. R. Kittrell, *Biotechnol. Bioeng.* **20**, 821 (1978).
32. L. Goldstein, *Biochemistry*, **11**, 4072 (1972).
33. R. Goldstein, L. Goldstein and E. Katchalski, in *Biochemical Aspects of Reactions on Solid Supports*, G. R. Stark, Ed. (Academic, New York, 1971), p. 1.

34. L. Goldstein and E. Katchalski, *Z. Anal. Chem.* **243**, 375 (1968).
35. E. Katchalski, I. Silman and R. Goldman, *Adv. Enz.* **34**, 445 (1971).
36. L. Goldstein, *Meth. Enz.*, **44**, 397 (1976).
37. L. Goldstein, Y. Levin and E. Katchalski, *Biochemistry* **3**, 1913 (1964).
38. C. M. Wharton, E. M. Crook, and K. Brocklehurst, *Eur. J. Biochem.* **6**, 572 (1968).
39. A. C. Johansson and K. Mosbach, *Biochim. Biophys. Acta* **370**, 348 (1974).
40. C. A. Zittle, *Adv. Enz. Rel. Areas Mol. Biol.* **14**, 319 (1953).
41. R. A. Messing, *J. Non-Cryst. Solids* **19**, 277 (1975).
42. R. A. Messing, *Research/Development* **25**, 32 (1974).
43. I. Chibata, T. Tosa, T. Sato, T. Mori, and Y. Matsuo, *Proc. IV IFS Fermentation Technology Today,* 383 (1972).
44. T. Tosa, T. Mori, N. Fusi, and I. Chibata, *Biotechnol. Bioeng.* **9**, 603 (1967).
45. S. Usami and N. Taketomi, *Hakko Kyokaishi* **23**, 267 (1965).
46. R. A. Messing, *Enzymologia* **38**, 39 (1970).
47. M. A. Mitz, *Science* **125**, 1076 (1956).
48. P. V. Sundaram and E. M. Crook, *Can. J. Biochem.* **49**, 1388 (1971).
49. S. Usami and S. Inoue, *Asahi Garasy Kogyo Gijitsu Shorei-kai Kenkyu Hokoku* **25**, 39 (1974).
50. G. Baum and S. J. Wrobel in *Immobilization of Enzymes, Antigens, Antibodies, and Peptides,* H. H. Weetall, Ed. (Marcel Dekker, New York, 1976), p. 419.
51. K. D. Caldwell, R. Axen, and J. Porath, *Biotechnol. Bioeng.* **18**, 433 (1976).
52. K. D. Caldwell, R. Axen, M. Bergwall, and J. Porath, *Biotechol. Bioeng.* **18**, 1573 (1976).
53. K. D. Caldwell, R. Axen, M. Bergwall and J. Porath, *Biotechol. Bioeng.* **18**, 1589 (1976).
54. K. D. Caldwell, R. Axen, M. Bergwall, I. Olsson and J. Porath, *Biotechol. Bioeng.* **18**, 1605 (1976).
55. S. Shartiel, *Meth. Enz.* **34**, 126 (1974).
56. H. P. Jenissen and L. M. G. Heilmeyer, *Biochemistry* **14**, 754 (1975).
57. L. G. Butler, *Arch. Biochem. Biophys.* **171**, 645 (1975).
58. A. Chrambach and D. Rodbard, *Science* **172**, 440 (1970).
59. G. P. Hicks and S. J. Updike, *Anal. Chem.* **38**, 726 (1966).
60. G. G. Guilbault and J. Das, *Anal. Biochem.* **33**, 341 (1970).
61. H. D. Brown, A. B. Patel, and S. K. Chattopadhyay, *J. Biomed. Mat. Res.* **2**, 231 (1968).
62. W. J. Blaedel and R. C. Boguslaski, *Biochem. Biophys. Res. Commun.* **47**, 248 (1972).
63. W. J. Blaedel, T. R. Kissel, and R. C. Boguslaski, *Anal. Chem.* **44**, 2030 (1972).
64. H. K. Lau and G. G. Guilbault, *Clin. Chem.* **19**, 1045 (1973).
65. P. Johnson and T. L. Whately, *J. Colloid Interface Sci.* **37**, 557 (1971).
66. F. Wold, *Meth. Enz.* **11**, 617 (1967).
67. H. Fasold, J. Klappenberger, C. Meyer, and H. Remold, *Angew. Chem. Int. Ed., Eng.* **10**, 795 (1971).
68. S. Avrameas and T. Ternynch, *J. Biol. Chem.* **242**, 1651 (1967).

69. P. Monsan, G. Puzo, and H. Mazarguil, *Biochimie* **57**, 1281 (1975).
70. G. Manecke and G. Gunzel, *Naturwissenschaften* **54**, 647 (1967).
71. F. J. Wold, *J. Biol. Chem.* **236**, 106 (1961).
72. R. Goldman, L. H. Silman, S. R. Caplan, O. Kedem, and E. Katchalski, *Science* **150**, 758 (1965).
73. H. Ozawa, *J. Biochem.* **62**, 419 (1967).
74. H. Zahn and H. Meinenhofer, *Makromol. Chem.* **26**, 126, 153 (1968).
75. R. H. Zaugg, C. King, and I. M. Klotz, *Biochem. Biophys. Res. Commun.* **64**, 1192 (1975).
76. H. F. Schick and S. J. Singer, *J. Biol. Chem.* **236**, 2477 (1961).
77. T. Kitagawa and T. Aikawa, *J. Biochem.* **79**, 233 (1976).
78. P. Guire, D. Fliger, and J. Hodgson, *Pharmacol. Res. Commun.* **9**, 121 (1977).
79. G. P. Royer, *Chem. Technol.* **4**, 694 (1974).
80. C. Horvath, *Biochim. Biophys. Acta* **358**, 164 (1974).
81. A. M. Filbert, *Immobilized Enzymes for Industrial Reactors,* R. A. Messing, Ed., (Academic, New York, 1975).
82. N. Weliky and H. H. Weetall, *Immunochemistry* **2**, 293 (1965).
83. G. Kay and E. M. Crook, *Nature* **216**, 514 (1967).
84. R. Axen, J. Porath and S. Ernbach, *Nature* **215**, 1491 (1967).
85. J. Porath, K. Aspberg, H. Drevin, and R. Axen, *J. Chromatogr.* **86**, 53 (1973).
86. S. C. March, I. Parikh, and P. Cuatrecasas, *Anal. Biochem.* **60**, 149 (1974).
87. T. Kristiansen, *Biochim. Biophys. Acta* **362**, 567 (1972).
88. P. Cuatrecasas, *J. Biol. Chem.* **245**, 3059 (1970).
89. J. Porath and R. Axen, *Meth. Enz.* **44**, 19 (1976).
90. W. E. Hornby and H. Filippusson, *Biochim. Biophys. Acta* **220**, 343 (1970).
91. C. Horvath and B. Soloman, *Biotechnol. Bioeng.* **14**, 885 (1972).
92. P. V. Sundaram and W. E. Hornby, *FEBS Letter* **10**, 325 (1970).
93. W. E. Hornby, D. J. Inman, and A. McDonald, *FEBS Letter* **23**, 114 (1972).
94. D. J. Inman and W. E. Hornby, *Biochem. J.* **137**, 25 (1974).
95. J. Campbell, W. E. Hornby, and D. L. Morris, *Biochim. Biophys. Acta* **384**, 307 (1973).
96. D. L. Morris, J. Campbell, and W. E. Hornby, *Biochem. J.* **147**, 593 (1975).
97. L. Goldstein, A. Freeman, and M. Sokolovsky, *Biochem. J.* **143**, 497 (1974).
98. A. Freeman, R. Granot, M. Sokolovsky, and L. Goldstein, *J. Solid Phase Biochem.* **1**, 275 (1977).
99. D. Blassburger, A. Freeman, and L. Goldstein, *Biotechol. Bioeng.* **20**, 309 (1978).
100. J. K. Inman and H. M. Dintzis, *Biochemistry* **8**, 4074 (1969).
101. P. D. Weston and S. Avrameas, *Biochem. Biophys. Res. Commun.* **45**, 1574 (1971).
102. M. B. Fiddler and G. R. Gray, *Anal Biochem.* **86**, 716 (1978).
103. R. Epton and T. H. Thomas, *Aldrichemica Acta* **4**, 61 (1971).
104. B. M. Vanderbilt and J. J. Jaruzelski, *I and EC Product Res. Dev.* **1**, 188 (1962).
105. W. D. Bascom, *Macromolecules* **5**, 792 (1972).

106. J. J. Kirkland and P. C. Yates, U.S. Patent 3,722,181 (March 1973).
107. J. J. Kirkland, *J. Chromatogr. Sci.* **9**, 206 (1971).
108. J. J. Kirkland and J. J. DeStefano, *J. Chromatogr. Sci.* **8**, 309 (1970).
109. M. Lynn in *Immobilized Enzymes, Antigens and Antibodies*, H. H. Weetall, Ed. (Marcel Dekker, New York, 1976), p. 9.
110. H. H. Weetall, *Science* **166**, 616 (1969).
111. H. H. Weetall and L. S. Hersh, *Biochim. Biophys. Acta* **185**, 469 (1969).
112. M. K. Weibel, W. Dritschilo, H. L. Bright, and A. E. Humphrey, *Anal. Biochem.* **52**, 402 (1973).
113. P. J. Robinson, P. Dunnill, and M. D. Lilly, *Biochim. Biophys. Acta* **242**, 659 (1971).
114. E. C. Lee, G. F. Senyk, and W. F. Shipe, *J. Food Sci.* **39**, 927 (1974).
115. H. H. Weetall and C. C. Detar, *Biotechnol. Biogen.* **57**, 295 (1975).
116. R. A. Messing, L. F. Biolousz, and R. E. Lindner, *J. Solid Phase Biochem.* **1**, 263 (1977).
117. R. A. Messing, *Meth. Enz.* **47**, 166 (1976).
118. P. Monsan, *J. Mol. Catal.* **3**, 371 (1978).
119. F. E. Regnier and R. Noel, *J. Chromatogr. Sci.* **14**, 316 (1976).
120. G. P. Royer, F. A. Liberatore, and G. M. Green, *Biochem. Biophys. Res. Commun.* **64**, 478 (1975).
121. L. D. Bowers and P. W. Carr, *Biotechnol. Bioeng.* **18**, 1331 (1976).
122. O. R. Zaborsky and J. Ogletree, *Biochem. Biophys. Res. Commun.* **61**, 210 (1974).
123. H. D. Brown, G. J. Bartling, and S. K. Chattopadhyay, in *Enzyme Engineering*, E. K. Pye and L. B. Wingard, Jr., Eds, Vol. II (Plenum Press, New York, 1974), p. 83.
124. K. Mosbach, ed., *Methods in Enzymology*, Vol. XLVII (Academic, New York, 1976).
125. J. Lasch, M. Iwig, R. Koelsch, H. David, and I. Marx, *Eur. J. Biochem.* **60**, 163 (1975).
126. D. E. Stage and M. Mannik, *Biochim. Biophys. Acta* **343**, 382 (1974).
127. D. L. Gardner, R. D. Falb, B. C. Kim, and D. C. Emmerling, *Trans. Am. Soc. Art. Intern. Organs* **17**, 239 (1971).
128. E. Reisel and E. Katchalski, *J. Biol. Chem.* **239**, 1521 (1964).
129. H. H. Weetall and L. S. Hersh, *Biochim. Biophys. Acta* **185**, 464 (1969).
130. D. J. Inman and W. E. Hornby, *Biochem. J.* **129**, 255 (1972).
131. S. J. Updike and G. P. Hicks, *Nature* **214**, 986 (1967).
132. L. B. Wingard, C. C. Liu, and N. L. Nagda, *Biotechnol. Bioeng.* **13**, 629 (1971).
133. R. A. Messing, *Enzymologia* **39**, 12 (1970).
134. E. Selegny, G. Broun, and D. Thomas, *Physiol. Veg.* **9**, 25 (1971).
135. H. H. Weetall, *Biochim. Biophys. Acta* **212**, 1 (1970).
136. R. A. Messing, *Enzymologia* **38**, 370 (1970).
137. K. Mosbach and R. Mosbach, *Acta Chem. Scand.* **20**, 2807 (1966).
138. R. Haynes and K. A. Walsh, *Biochem. Biophys. Res. Commun.* **36**, 235 (1969).
139. R. Axen and S. Ernback, *Eur. J. Biochem.* **18**, 351 (1971).

140. K. Mosbach, *Acta Chem. Scand.* **24**, 2034 (1970).
141. S. A. Barker, P. J. Somers, R. Epton, and J. V. McLaren, *Carbohyd. Res.* **14**, 287 (1970).
142. S. A. Barker, P. J. Somers, and R. Epton, *Carbohyd. Res.* **14**, 323 (1970).
143. W. M. Ledingham and W. E. Hornby, *FEBS Letter* **5**, 118 (1969).
144. E. B. Ong, Y. Tsang, and G. E. Perlmann, *J. Biol. Chem.* **241**, 5661 (1966).
145. Y. Levin, M. Pecht, L. Goldstein, and E. Katchalski, *Biochemistry* **3**, 1913 (1964).
146. G. Manecke and S. Singer, *Makromol. Chem.* **39**, 13 (1960).
147. A. Bar-Eli and E. Katchalski, *Nature* **188**, 856 (1960).
148. N. Gurbhofer and L. Schleith, *Hoppe-Seyler's Z. Physiol. Chem.* **297**, 108 (1954).
149. P. A. Biondi, M. Pace, O. Brenna and P. G. Pietta, *Eur. J. Biochem.* **61**, 171 (1976).
150. R. Y. Ko and L. S. Hersh, *J. Biomed. Mat. Res.* **10**, 249 (1976).
151. L. Goldstein, M. Pecht, S. Blumberg, D. Atlas, and Y. Levin, *Biochemistry* **9**, 2322 (1970).
152. H. H. Weetall and L. S. Hersh, *Biochim. Biophys. Acta* **206**, 54 (1970).
153. Z. Dunvjak and M. D. Lilly, *Biotechol. Bioeng.* **18**, 737 (1976).
154. P. R. Coulet, J. H. Julliard, and D. C. Guatheron, *Biotechol. Bioeng.* **16**, 1055 (1974).
155. P. J. Robinson, P. Dunhill, and M. D. Lilly, *Biotechnol. Bioeng.* **15**, 603 (1973).
156. S. A. Barker, *Carbohyd. Res.* **14**, 287 (1970).
157. S. A. Barker, *Carbohyd. Res.* **14**, 323 (1970).
158. M. Keyes, U.S. Patent No. 2,933,589 (20 January 1976).
159. W. Dritschilo and M. K. Weibel, *Biochem. Med.* **9**, 32 (1974).
160. P. V. Sundaram, *Biochem. J.* **183**, 445 (1979).

THEORY AND APPLICATIONS
OF ENZYME ELECTRODES

An enzyme electrode may be considered as the combination of any type of *electrochemical* sensor and a small (thin) layer of enzyme which is used to measure the concentration of a substrate. As will be seen shortly, these devices inherently have many analytical advantages over the use of soluble enzymes. The basic features of an immobilized-enzyme electrode measurement system are simple. All that is absolutely required for operation is a thin enzyme layer $(10 - 200 \ \mu m)$ held in close proximity to the active surface of a transducer which might be either a potentiometric or an amperometric indicator electrode, an appropriate reference electrode, and a circuit for measuring either the potential difference between the two electrodes or the current which flows between them. The solution should be stirred to speed up mass transfer and provide fast response. The potentiometric sensor could be any one of a number of ion-selective sensors, for example, an ammonium ion sensitive glass electrode. Platinum amperometric sensors which can reduce O_2 or oxidize peroxide have also been employed as transducers for enzyme electrodes. When amperometric sensors are used the circuit generally contains some source of potential to drive current. A dual sensor electrode has been used so that a "blank" can be run and the difference current directly readout. A simple three-electrode polarographic system can be attached to the indicator and reference electrode, while a third auxiliary electrode is used to avoid polarizing the reference electrode.

The sample is assayed by placing it in a buffered solution. Since enzymes are pH-sensitive, some buffer is almost always needed. When a simple hydrolytic enzyme is used, for example urease, penicillinase, or β-glucosidase, no coreactants are required. When an oxidase or dehydrogenase is used, it will be necessary to add a coreactant, for example oxygen or NADH. The sensing electrodes are immersed and one then reads out the steady-state potential or current, which will be related to the analyte (substrate) concentration. This relationship is *logarithmic* for a potentiometric electrode and *linear* for an amperometric electrode. In some cases the rate of change of potential or current may be used. This generally tends to shorten the analysis time and may reduce interferences.

A bit more mechanistically, the conversion of substrate (S) to product (P) takes place only in the trapped enzyme layer.

$$S + R \xrightarrow{\text{enzyme}} P + R' \tag{5.1}$$

In this reaction R indicates any and all essential coreactants [O_2, NAD(H), ATP, water, H^+, OH^-, etc.]. Because the reaction occurs only in the trapped layer, the substrate concentration is lower and the product concentration higher in the vicinity of the surface than in the bulk of solution. As a consequence of the enzyme-generated concentration gradient, the substrate will diffuse from the bulk solution and the reaction will be self-sustaining. Ultimately (after 30 sec–10 min) a steady-state situation will occur in which the substrate diffuses in at a rate equal to that at which it is consumed by the chemical reaction. This steady state should not be considered as an *equilibrium* situation, since there is net consumption of substrate. Evidently the bulk substrate concentration drops continuously, but the rate of decrease is so low that it will be detectable only when the test volume is very small, mass transfer is fast, and the net enzyme activity is very high.

Potentiometric electrodes are usually chosen so as to detect the product of the enzyme reaction. Obviously, they detect the *product concentration only at their active surface*. Amperometric transducers measure the flux of the electroactive species which they are designed to detect. They may be used to measure the decrease in either the concentration of a coreactant (for example, O_2) or the increase in a product (for example, H_2O_2).

There is no *a priori* reason why this type of device should be restricted for use with electrochemical transducers. It should be possible to use a temperature transducer and measure the temperature change due to reaction in the enzyme layer. Similarly, the sensor might be replaced with a fiber-optic light pipe. Thus we should consider enzyme electrodes as one member of a larger class of "enzyme probes."

Clark and Lyons (1) were the first investigators to consider the possibility of making an enzyme electrode. They predicted in 1962 that it should be possible to use a pair of membranes to hold a thin layer of *soluble* enzyme in close proximity to an electrode and that one could use an ion selective (pH), amperometric (O_2), or conductivity electrode system to monitor the reaction in this layer. Clark and Lyons' paper preceeded the exponential growth of the use of immobilized enzymes. They assumed that the soluble enzyme would be trapped between dialysis membranes and that it could be held in place by the steric exclusion characteristic of a semipermeable membrane. The membrane would be selected to allow substrates (glucose) and coreactant (O_2) to diffuse into the enzyme layer and react. The membrane closest to the sensor could serve the double purpose of passing only O_2 to the amperometric sensor, as in a Clark O_2 electrode. Depletion of O_2 would then be measured to indicate the level of substrate in the external solution. Even in the face of the great advances in immobilized-enzyme technology which have taken place since the original paper of Clark and Lyons, their basic principle of trapping a soluble enzyme is still in use. They projected that it should be possible to mount an electrode behind two membranes which form a sandwich with the enzyme trapped between the membranes. This electrode is shown in Fig. 5.1(A).

The first *experimental* report on an immobilized-enzyme electrode is due to Updike and Hicks (2), who coined the term "enzyme electrode." Their device [Figure 5.1(B)] is quite sophisticated in that it incorporates

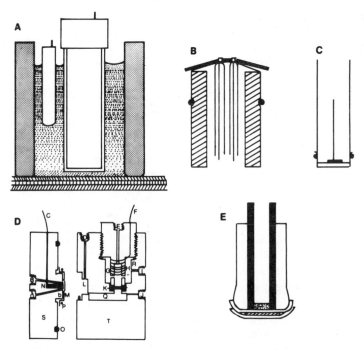

Fig. 5.1. Different types of amperometric enzyme electrodes. The component parts may be: (a) an amperometric sensor, (b) the trapped or immobilized enzyme, (c) a dialysis membrane, (d) gas-permeable membrane, (e) auxillary electrode. (A) Design of Clark and Lyons (redrawn from reference 1 courtesy of New York Academy of Science.) (B) Design of Updike and Hicks [redrawn from reference 2 courtesy of McMillan (Journals Ltd.)]. (C) Design of Guilbault (redrawn from reference 12 courtesy of Marcel Dekker, Inc.) (D) Design of Blaedel and Jenkins (redrawn from reference 38 with permission of the American Chemical Society). (E) Design of Clark with internal O_2 source (redrawn from reference 3 courtesy of John Wiley and Sons, Inc.).

two cathodes, one of which is in contact with active enzyme (glucose oxidase) and the other cathode is mounted behind a thermally denatured enzyme. This dual arrangement permits a direct measurement of the change in O_2 upon oxidation of glucose, as well as correction for changes in O_2 levels due to other reactions, for example, with uric and ascorbic acid. Their work was also the first report in which an enzyme electrode was made by immobilizing the enzyme (a polyacrylamide gel was used).

Tremendous analytical interest in enzyme electrodes has existed since their introduction. There are a number of reasons why these devices have such appeal:

1. Perhaps the most important reason is that they can provide convenient and straightforward measurement methods for species of biological and clinical interest.

2. Pretreatment of samples (add buffer and coreagent, dip, measure) is at a minimum. Prior removal of protein, cells, and other materials is often unnecessary. No incubation to develop a color or fluorescence, etc., is needed.

3. The devices are as easy to calibrate and use as glass pH electrodes.

4. The technique is virtually nondestructive. The sample can be tested for other constituents as long as the added buffer and coreactants do not interfere.

5. These sensors are often very sensitive; frequently detection limits of 10^{-5}–10^{-4} M can be obtained with potentiometric electrodes. Clark reports that his alcohol oxidase electrode can detect alcohol at 10^{-8} M levels (3).

6. These devices can be very fast; often a steady state will be achieved in 30 sec–2 min; seldom more than 10 min. Rate analyses require as little as 10–20 seconds. As will be seen the response time is very dependent upon the geometry of the probe.

7. They can be used in a small volume (2–10 ml), and therefore only a small volume of sample is needed. Llenado and Rechnitz reported an amygdalin electrode which can be used in a total volume of 0.2 ml (4).

8. An individual probe is very inexpensive. The cost of the few milligrams (10 mg) of enzyme needed to make one probe never exceeds the cost of the base transducer (glass, platinum electrode) and in most cases will only amount to a few dollars. Since a given electrode can be reused many times, the cost per sample is trivial.

9. The scope of application is in principle very broad. For example Clark (3) has listed 27 enzymes which are known to generate hydrogen peroxide and an additional 5 which consume oxygen without confirmation

of peroxide generation (see Table 5.1). Approximately 2000 enzymes had been identified by 1978 and the list is still growing exponentially; each of these could be the basis for an enzyme probe provided that a suitable transducer is available. Rechnitz has shown that whole bacterial cells can be trapped on an electrode and used directly (5). Thus one does not necessarily require a highly purified or even isolated enzyme.

10.   In principle microprobes could be developed to be placed in a catheter and used for continuous *in vivo* monitoring of the critically sick, as a warning device to be worn, for example by diabetics or in physiological studies of test animals in much the same way as are micro glass and oxygen electrodes. Ultramicroprobes could be designed to study responses of single cells to external stimuli as is presently being done with glass ion-selective electrodes (6). Clark has described a device which can take direct glucose readings on the surface of live rat brain tissue (7).

11.   A final reason for interest in enzyme electrodes has been the tremendous concurrent interest in the related area of ion-selective electrodes. There has been a remarkable cross-fertilization in these two areas. Although a wide variety of electrodes have been developed to detect cations and anions, with the exception of the acetylcholine electrode (8) and a methylamine sensor (9), very little success has been encountered in developing electrodes sensitive to molecular species. Potentiometric sensors based on pH glasses have been developed for ammonia, carbon dioxide and other volatile acids and bases. Basically, ion-selective electrodes can be used only to detect ionic species or molecules which decrease the activity of such species. Enzyme electrodes represent the major avenue of approach for using ion-selective electrodes to detect nonionic species.

The rapid growth in the number of enzyme electrodes over the past 10 years is related to advances in both immobilization technology and in the availability of novel, functional sensing electrodes. The first really stable enzyme electrode—the urea selective electrode of Guilbault and Montalvo (10) was based on the stabilization of urease in a polyacrylamide gel and the availability of a glass cation-selective electrode which is sensitive to ammonium ions. This device has been succeeded by several generations of new electrodes based on more selective and sensitive ammonium and ammonia electrodes as well as by carbon dioxide sensors.

Tables 5.2 and 5.3 present an overview of the types of potentiometric and amperometric electrodes which have been employed as sensors in immobilized-enzyme electrodes. To date, cation-selective glass electrodes, nonactin-impregnated silicon rubber, ammonia and carbon dioxide gas sensing, and cyanide and iodide solid-state electrodes have been used

Table 5.1 Enzymes Useful with O_2 Electrodes[a]

Enzyme	Number	Source	Typical substrates
Glycollate oxidase	1.1.3.1	spinach rat liver	glycollate L-lactate D-lactate (+)-mandalate
Lactate oxidase	1.1.3.2	*M. phlei*	L-lactate
Glucose oxidase	1.1.3.4	*Aspergillus niger* *Penicillium amagasakienses* honey (bee) *Penicillium notatum*	β-D-glucose 2-deoxy-D-glucose 6-deoxy-6-fluoro-D-glucose 6-methyl-D-glucose
Hexose oxidase	1.1.3.5		β-D-glucose D-galactose D-mannose
L-Gulonolactone oxidase	1.1.3.8	rat liver	L-gulono-λ-lactone L-galactonolactone D-manonolactone D-altronolactone
Galactose oxidase	1.1.3.9	*Dactylium dendroides* *Polyporus circinatus*	D-galactose stachyose lactose
L-2-Hydroxyacid oxidase	1.1.3.a	hog renal cortex	L-2-hydroxyacid
Aldehyde oxidase	1.2.3.1	rabbit liver pig liver	formaldehyde acetaldehyde
Xanthine oxidase	1.2.3.2	bovine milk porcine liver	purine hypoxanthine benzaldehyde xanthine
Pyruvate oxidase	1.2.3.3		pyruvate requires thiamine phosphate
Oxalate oxidase	1.2.3.4		oxalate
Dihydro-orotate-dehydrogenase	1.3.3.1	*Zymobacterium oroticum*	L-4,5-dihydro-orotate NAD
D-Aspartate oxidase	1.4.3.1	rabbit kidney	D-aspartate D-glutamate
L-Amino acid oxidase	1.4.3.2	diamond rattlesnake cotton mouth moccasin rat kidney	L-methionine L-phenylalanine 2-hydroxy acids L-lactate

202

Table 5.1 (*Continued*)

Enzyme	Number	Source	Typical substrates
D-Amino acid oxidase	1.4.3.3	hog kidney	D-alanine D-valine D-proline
Monoamine oxidase	1.4.3.4	beef plasma placenta	monoamine benzylamine octylamine
Pyridoxamine phosphate oxidase	1.4.3.5	rabbit liver	pyridoxamine phosphate
Diamine oxidase	1.4.3.6	bovine plasma pea seedlings procine plasma	diamines spermidine tyramine
Sarcosine oxidase	1.5.3.1	*Macaca mulatta* rat liver mitochondria	sarcosine
N-Methyl-L-amino acid oxidase	1.5.3.2		N-methyl-L-amino acids
Spermine oxidase	1.5.3.3.	*Neisseria perflava* *Serratia marcescens*	spermine spermidine
Nitroethane oxidase	1.7.3.1		nitroethane aliphatic nitro compounds
Urate oxidase	1.7.3.3	hog liver ox kidney	urate
Sulfite oxidase	1.8.3.1	beef liver	sulfite
Alcohol oxidase		*Basidiomycetes*	ethanol and methanol
Carbohydrate oxidase		*Basidiomycetes* *Polyporus obtusus*	D-glucose D-glucopyranose D-xylopyranose L-sorbose δ-D-gluconolactone
NADH oxidase		beef heart mitochondria	NADH
Malate oxidase	1.1.3.2		L-malate
Cholesterol oxidase	1.1.3.6	nocardia	cholesterol
N-Acetylinodoxyl oxidase	1.7.3.2		N-acetylindoxyl
Thiol oxidase	1.8.3.2		R: CR–SH
Ascorbate oxidase	1.10.3.3	squash	L-ascorbate

Table 5.2 Potentiometric Devices Used for Immobilized-Enzyme Electrodes

Sensor type	Species detected	Typical substrates
pH glass	H^+	penicillin, glucose, urea
Univalent cation glass	NH_4^+	urea, L-amino acids, D-amino acids, asparaginine, glutamine, glutamate, lactate
Nonactin–silicon rubber membrane	NH_4^+	urea, L-phenylalanine
Ammonia gas sensor	NH_3	L-asparagine, creatinine, urea, 5'-AMP, arginine
Carbon dioxide gas sensor	CO_2	urea, uric acid, phenylalanine, tyrosine, lysine
Iodide solid state	I^-	L-phenylalanine, glucose
Cyanide solid state	CN^-	amygdalin

in immobilized-enzyme electrodes. The physical designs of these potentiometric electrode configurations are shown in Figure 5.2. The similarity to the original Clark and Lyons design is striking.

It is of course possible to use any electrode with a soluble enzyme and implement either a bulk kinetic or equilibrium assay. When an enzyme can be immobilized or trapped with high efficiency in a stable form, a number of major analytical advantages will result in comparison to using a soluble enzyme in solution. First, the enzyme can be reused; sometimes as many as 1000 determinations can be carried out with just one electrode (11). Guilbault states that only about 10–20 units of enzyme/electrode are required and that the enzyme should have a specific activity of at least

Table 5.3 Amperometric Devices Used for Enzyme Electrodes

Electroactive species	Sensor electrode	Typical substrates
O_2	Pt, Clark electrode	L-amino acids, glucose, uric acid, alcohols, aldehydes, carboxylic acids, phosphate, lactate, diamines, peroxide, sucrose
H_2O_2	Pt	glucose, L-amino acids, alcohols
NADH	glassy carbon, Pt	lactate, ethanol
$Fe(CN)_6^{4-}$	Pt	lactate
I_2	Pt	glucose

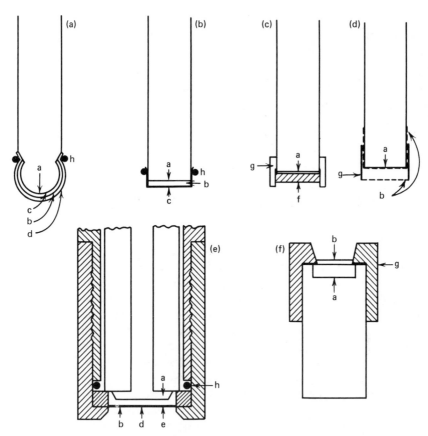

Fig. 5.2. Different types of poteniometric enzyme electrodes. The component parts may be: (a) a potentiometric sensor, (b) trapped or immobilized enzyme, (c) nylon spacer, (d) dialysis membrane, (e) gas-permeable membrane, (f) glass frit, (g) sleeve, (h) "O" ring seal. (A) Guilbault design with immobilized enzyme (redrawn from reference 37 courtesy of Marcel Dekker, Inc.). (B) Guilbault Design with Soluble Enzyme (redrawn from reference 37 courtesy of Marcel Dekker, Inc.). (C) Cullen Design (redrawn from reference 15 with permission of the American Chemical Society). (D) Immobilized substrate electrode (redrawn from reference 55 courtesy of Elsevier Scientific Publishing Co.). (E) Rechnitz design with gas sensor (redrawn from reference 5 with permission of the American Chemical Society.) (F) Microelectrode design (redrawn from reference 4 with permission of the American Chemical Society.)

1 unit/mg (12). This is not an illusory advantage, since a soluble enzyme is often the major cost component in a continuous-flow analysis. Second when the enzyme concentration is sufficiently high, *the electrode response will become independent of the amount of enzyme.* This has been demonstrated to be the case for both potentiometric and amperometric

electrodes (10, 13). This is *never true in a kinetic assay with a soluble enzyme*. Clearly, the reaction in the enzyme layer will be virtually complete if a vast kinetic excess of enzyme is present. Third, excess enzyme will decrease the sensitivity of the assay to all those variables, which influence the rate of enzyme reactions including pH, activator, and inhibitor concentration, temperature, and most significantly, loss of enzyme activity. Thus the enzyme electrode may well be more accurate than a kinetic analysis, with the corresponding soluble enzyme and, in the absence of problems with liquid-junction potentials and activity coefficient, it may be more precise.

5.1 PRINCIPLES OF OPERATION OF IMMOBILIZED-ENZYME ELECTRODES

A schematic drawing of an immobilized-enzyme probe and the sequence of steps leading to a response is shown in Figure 5.3. As indicated, the outer active surface of the transducer is placed in close contact with the thin layer of enzyme. No *mass* transfer is possible across this boundary. The outer edge of the enzyme is exposed to an analyte solution which is generally well stirred to minimize concentration gradients in it. In some instances a thin (20 μm) semipermeable membrane may be interposed between the enzyme and the electrode surface (see Figures 5.1 and 5.2). This was the case in the original design of Clark and Lyons and is still used when a gas-sensitive electrode is the sensor. This usually has the effect of increasing the response time of the electrode *unless the barrier is thin*. A second membrane which must be permeable to the sample and coreagents is placed over the probe assembly in order to hold the enzyme in place whenever a nonrigid support, such as a polyacrylamide or starch gel, or a spongy cross-linked immobilized enzyme is used. This membrane should be as thin as possible so that the response time is not affected. Ideally, the only slow steps in such an immobilized-enzyme probe will

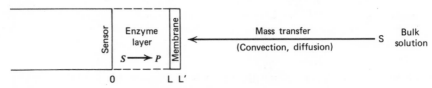

Fig. 5.3. Schematic diagram of the sequence of steps which take place in product generation in an enzyme probe. The substrate is transported from the bulk to the outer membrane surface located at L', diffuses through the membrane to point L and undergoes simultaneous reaction and diffusion in the enzyme layer.

be diffusion through the trapped layer and the chemical reaction. Clearly the primary sensor should be fast, the membranes ultrathin and the bulk solution well mixed. This is hardly ever the case in practice and these complications make a detailed theoretical model which closely emulates the behavior of a real electrode difficult to solve in closed form. What we will now attempt is a definition of various simplified models which allow estimation of the steady-state behavior of an enzyme electrode and then less rigorously, the transient behavior. Detailed considerations of both the time-dependent and steady-state behavior which include all major factors (membrane thickness, permeability, convection, kinetic nonlinearity, etc.) are not yet available.

A schematic diagram of the reaction sequence involved in the use of an enzyme electrode is shown in Figure 5.3. Obviously, the first step must be transport of the substrate, which is *assumed to be the rate-limiting species*, from the bulk to the outer layer at point L':

$$S_{in\ bulk} \xrightarrow{\text{external mass transfer}} S_{at\ L'} \tag{5.2}$$

the substrate must then diffuse through the outer membrane toward the enzyme layer at point L:

$$S_{at\ L'} \xrightarrow{\text{membrane diffusion}} S_{at\ L} \tag{5.3}$$

The substrate undergoes chemical conversion as it attempts to reach the sensor surface at point O. The chemical reaction rate between points O and L will in the simplest case be governed by Michaelis–Menten kinetics:

$$E + S \underset{k_1}{\overset{k_{-1}}{\rightleftharpoons}} ES \xrightarrow{k_2} E + P \tag{5.4}$$

The product formed will then diffuse back into the bulk solution. A final factor which should be considered is the possibility of partition equilibrium between the bulk phase and the enzyme layer. That is, the process:

$$S_{at\ L'} \rightleftharpoons S_{at\ L}; \quad K_d = \frac{[S]_{enzyme\ phase}}{[S]_{bulk\ phase}} \tag{5.5}$$

When the partition coefficient, K_d, is not unity, then even with no membrane present there will be a discontinuity in the substrate concentration at the outer surface of the enzyme. This is by no means a trivial factor. Montalvo and Guilbault have found that the presence of the enzyme in a polyacrylamide gel altered the detection limit of a cation-selective electrode for NH_4^+ (14). Papariello and coworkers (15) were forced to abandon the design of their original penicillin electrode (16), which used an ordinary pH-sensitive glass electrode, when they found that the presence of penicillinase in a polyacrylamide gel introduced sufficient ion-exchange sites to make the electrode very sensitive to sodium and potassium and less sensitive to hydrogen ions. If this is the case, then clearly the biochemical affinity of an enzyme for its substrate can act to generate a concentration difference between the bulk solution and the immobilized-enzyme layer, therefore *it is improbable that K_d will equal unity.*

5.1.1 External Mass-Transfer Model (Potentiometric Probes)

A somewhat simplified model, but one which incorporates effects due to enzyme activity, Michaelis constant, and mass transfer was developed by Racine and Mindt (17). Their work is important because it is the only one which permits a closed-form solution of the equations without any assumption with respect to the ratio of substrate concentration to the Michaelis constant. It allows us to see the effect of K_M on the linearity of a calibration curve. They assumed that the reaction rate is limited by diffusion through the membrane between L and L' and that the inner enzyme layer is so thin that diffusion within it is very fast, so that no concentration gradients of substrate exists between points O and L. Under these conditions, the rate of chemical reaction can be obtained from the Michaelis–Menten scheme [equation (5.4)]:

$$\frac{1}{V^*}\frac{dN_S}{dt} = \frac{d[S]}{dt} = \frac{-k_2[E][S]}{K_M + [S]} \quad 0 \le x \le L \qquad (5.6)$$

where V^* represents the volume of the trapped layer and N_S the number of moles of substrate in the layer. The rate at which substrate enters the enzyme layer is restricted by the membrane and will be given by

$$\frac{dN_S}{dt} = \phi_S([S]_{L'} - [S]) \qquad (5.7)$$

where ϕ_S is the permeability of the membrane which will depend upon its area, thickness and the ability of the substrate to pass through it. When the solution is well stirred, external mass transfer will be very fast and the substrate concentration at L' will be equal to the bulk concentrations denoted $[S]_0$. In the steady state the rate of the chemical reaction [equation (5.6)] and the rate of mass transfer are equal [equation (5.7)]. These equations can be solved simultaneously to determine the substrate concentration. Since a potentiometric electrode is sensitive to the product concentrations, we will calculate it rather than the substrate concentration. Assuming equal membrane permeabilities for the substrate and product

$(\phi_S = \phi_P = \phi)$, the following equations result

$$[P] = [S]_0 - [S] \tag{5.8}$$

$$[P] = \frac{1}{2}\left[\left([S]_0 + K_M + \frac{k_2[E]}{\phi}V^*\right)\right.$$
$$\left. - \sqrt{\left(K_M + \frac{k_2[E]V^*}{\phi} - [S]_0\right)^2 + 4K_M[S]_0}\right] \tag{5.9}$$

In the limit of a zero-order reaction ($K_M \ll [S]$) this becomes:

$$[P] = \frac{k_2[E]V^*}{\phi} \tag{5.10}$$

whereas for a first-order reaction ($K_M \gg [S]$)

$$[P] = [S]_0 \cdot \frac{k_2[E]V^*/K_M\phi}{1 + k_2[E]V^*/K_M\phi} \tag{5.11}$$

It is important to note the similarity of the term $k_2[E]V^*/K_M\phi$ to the enzyme loading factor considered in Chapters 6 and 7. The relationship between the product concentration in the enzyme layer and the bulk substrate concentration ($[S]_0$) that is, equation (5.9), is obviously nonlinear. A series of plots of the logarithm of the product concentration versus the logarithm of the bulk substrate concentration are shown in Figures 5.4 and 5.5 for several typical values of K_M and enzyme activities. Such log–log plots simulate the behavior of potentiometric calibration curves.

It is evident that the Michaelis constant has a significant effect on the linearity of an enzyme electrode. The sensitivity and linearity of the electrode is controlled by the relative ratio of the rate of chemical reaction to mass transfer. Equation (5.11) indicates that at high enzyme activity relative to the parameter $K_M \cdot \phi$, the product concentration in the trapped layer becomes equal to bulk substrate concentration. Obviously, if the rate of chemical reaction is slow, the product will escape into the bulk and its concentration in the trapped layer will be very low. Equation (5.11) also indicates that a low permeability will improve the sensitivity but it is obvious that this will also tend to increase the response time. A very important result is that these electrodes will give calibration curves which are linear at *bulk substrate concentrations which exceed the Michaelis constant* provided that the amount of enzyme is so high that the reaction rate becomes primarily *mass-transfer limited*. Linear calibration curves at $[S] \cong K_M$ are never observed in kinetic analyses with soluble enzymes.

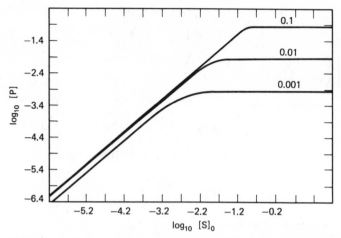

Fig. 5.4. Theoretical potentiometric enzyme electrode calibration curve based on external diffusion control of the reaction. A plot of $\log_{10}$ of the product concentration in the enzyme layer versus $\log_{10}$ of the bulk substrate concentration. All results were computed from equation (5.9) where $K_M = 10^{-3}$; the value of $k_2 [E]V^*/\emptyset$ is given on the curve.

Also note that as the enzyme activity of the electrode increases, the curves at low substrate concentrations ($< K_M$) become identical. This means that the product concentration is independent of the amount of enzyme on the electrode and therefore will become independent of those factors (pH, temperature, activators, inhibitors, denaturation) which alter the enzyme activity. Evidently the stability, precision, and accuracy of analyses based on the use of enzyme electrodes is in large part derived from rate control by mass transfer rather than chemical reaction kinetics.

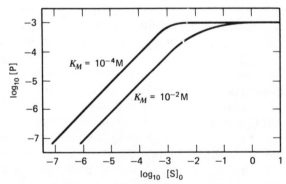

Fig. 5.5. Theoretical potentiometric enzyme electrode calibration curve based on external diffusion control of the reaction, where $k_2[E]V^*/\emptyset = 10^{-3}$; the value of K_M (M) is given on each curve.

In order to achieve these results, efficient immobilization techniques are essential. There is no question, however, that immobilization by occlusion or covalent binding can improve the inherent enzyme stability. This effect will be discussed later.

Since a closed-form solution can be obtained for both the substrate and product concentration, one can obtain an estimate of the maximum substrate concentration for which the electrode will provide linear response. If one assumes that the enzyme electrode will be linear as long as the substrate concentration in the trapped layer is less than about $0.1\ K_M$, then it is easy to show that this corresponds to

$$\frac{[S]_0}{K_M} \leq 0.1 \left(\frac{k_2[E]V^*}{\phi K_M} + 1 \right) \tag{5.12}$$

This result is in qualitative agreement with the data presented in Figures 5.4 and 5.5 and indicates that the upper limit of linearity increases with the amount of enzyme on the electrode. The linearity will be improved when the enzyme loading factor $(k_2[E]V^*/\phi K_M)$ approaches and exceeds unity. As Figure 5.4 indicates, the upper limit of linearity improves in proportion to the enzyme loading factor when this factor exceeds unity. The lower limit of detection will be discussed in a subsequent section.

5.1.2   Internal Mass-Transfer Model (Potentiometric Probes)

The next model to be considered is one in which all complicating factors outside the trapped layer are assumed to be negligible; we will disregard external mass transfer, membrane diffusion, and phase partitioning. All of these factors were included in the model of Blaedel and coworkers (18, 19), but their final equations are quite complicated. The results to be presented here are one limiting form of their equations in which *internal diffusion dominates* over all other factors.

The enzyme reaction in the trapped layer is described by Michaelis–Menten kinetics and their equations apply when the substrate concentration is a function of time and distance. Blaedel and coworkers have shown that the combined reaction rate and diffusion in the layer $(0 < x < L)$ can be accounted for by combining Fick's second law of diffusion and equation (5.6); thus:

$$\frac{\partial [S]}{\partial t} = D_S \frac{\partial^2 [S]}{\partial x^2} - \frac{k_2[E][S]}{K_M + [S]} \tag{5.13}$$

$$\frac{\partial [P]}{\partial t} = D_P \frac{\partial^2 [P]}{\partial x^2} + \frac{k_2[E][S]}{K_M + [S]} \tag{5.14}$$

Both the steady-state and transient response of the electrode will depend upon the nature of the boundary conditions applied to equations (5.13) and (5.14). When both slow mass transfer and phase partitioning effects are dismissed, and the product species is not present in the bulk of the test solution, the concentration of substrate at the outer boundary (L) of the sensor will be equal to the bulk concentration ($[S]_0$) and the product concentration will be zero at that point:

$$[S] = [S]_0; \quad [P] = 0 \text{ at } x = L \tag{5.15}$$

The boundary condition at the sensor surface ($x = 0$) depends upon the nature of the transducer. When none of the species are electrochemically consumed, for example, by electron transfer at an amperometric probe, by ion exchange or adsorption on a glass electrode, or diffusion through a gas-permeable sensor, then the flux of each species at the sensor surface will be zero and by Fick's first law of diffusion the concentration gradient of each species will be zero. Thus for potentiometric electrodes one can state that:

$$\frac{\partial [S]}{\partial x} = \frac{\partial [P]}{\partial x} = 0 \text{ at } x = 0 \tag{5.16}$$

The solution of the differential equations is also contingent upon the initial condition in the probe. In practice, a freshly washed electrode would be placed in a homogeneous solution of the sample or a sample would be added to a well-stirred buffer in which the electrode was previously immersed. We can assume that neither the sample nor the product is initially present in the trapped layer; thus the appropriate initial ($t = 0$) conditions for all probes are

$$[S] = [P] = 0 \quad 0 \leq x \leq L \text{ at } t = 0 \tag{5.17}$$

The steady-state reading of the electrode potential will occur when the product concentration in the probe is constant; this condition is satisfied by setting the *time* derivative in the differential equations equal to zero. After a slight transposition of terms we see that in the *steady state,* the supply of substrate by mass transfer *exactly counterbalances* its rate of removal by the enzyme reaction; similarly, the rate of product generation is counterbalanced by its rate of removal from the probe by diffusion:

$$D_S \frac{\partial^2 [S]}{\partial x^2} = \frac{k_2 [E][S]}{K_M + [S]} \tag{5.18}$$

$$D_P \frac{\partial^2 [P]}{\partial x^2} = -\frac{k_2 [E][S]}{K_M + [S]} \tag{5.19}$$

Note that the units of each term above are mole/sec·ml. The net reaction rate can be obtained by multiplying by the volume of the enzyme layer. Also note that the net reaction rate depends upon the substrate concentration in precisely the same way as does the rate of an enzyme process in solution, when it is understood that the substrate and product concentrations are both functions of distance.

This implies that the net reaction rate will become relatively independent of the substrate concentration as it approaches the Michaelis constant. When the rate of the chemical reaction becomes independent of the *bulk* substrate concentration, the electrode potential will become constant. Tran-Minh and Brown (20) calculated, by digital simulation, the concentration profile of both the substrate and product as a function of time and distance as, well as the steady-state product concentration at the sensor surface as a function of the bulk substrate concentration. Their results indicate that even when there is a substantial concentration gradient in the enzyme layer and the average substrate concentration is considerably less than the bulk substrate concentration, the product concentration at the sensor surface is not linearly related to the bulk substrate concentration when the bulk substrate concentration exceeds about $0.2 K_M$. The remarkable result of their theoretical calculation is that the *amount of enzyme on the electrode does not strongly influence the upper limit of linearity* even when the enzyme loading factor $(k_2[E]L^2/K_M D_s)$ is greater than unity. This stark contrast between the behavior of an electrode which is governed by external and internal mass transfer requires some explanation. When membrane diffusion and/or external mass transfer are slow, it is evident that they became rate-controlling in a conventional chemical sense since this process takes place in series with the chemical reaction. When internal mass transfer is considered, the chemical reaction and diffusion are coupled together in a very complicated fashion, since it is the chemical reaction which supplies the gradient which causes diffusion. In a sense we are dealing with processes which do not take place in series. Thus, as the chemical reaction becomes faster, the rate of diffusion also increases. Consequently the substrate concentration is closer to its bulk value than it would be under comparable conditions with external mass-transfer control. It may well be that all that is required is a higher enzyme activity than that used by Tran-Minh and Broun to achieve linearity above K_M. As will be seen, Mell and Maloy (13) have shown theoretically that this *effect does take place in amperometric probes which are controlled by internal diffusion*. Recent results in our laboratory indicate that Tran-Minh and Broun's calculation are not correct. Please see the appendix at the end of this chapter.

It is evident that an enzyme electrode is really a very small enzyme reactor in which all mass transfer is controlled by diffusion. At very high levels of the bulk substrate concentration, the surface concentration of product becomes independent of the analyte because the reaction enters the zero-order domain and any kinetic factor which influences reaction rate will also alter the surface product concentration. This is not necessarily true in the first-order reaction rate regime. As will be shown, when enough enzyme is present in the electrode, the surface product concentration will become independent of the rate of the enzyme reaction. Under these conditions, diffusional mass transfer is the most important process. Thus changes in pH, the presence of inhibitors, etc., are not nearly so important. It is therefore highly advantageous to use as much enzyme as can be placed on an electrode.

One cannot obtain closed-form solutions for equations (5.18) and (5.19). It is possible to obtain closed-form solutions for the corresponding zero-order and first-order rate-limiting forms. Since most investigators have made their measurements in the steady state, the solutions for this situation will be considered first.

Zero-Order Limit ($[S] \gg K_M$)

When this condition is imposed, equations (5.18) and (5.19) become

$$D_S \frac{d^2[S]}{dx^2} = +k_2[E] \tag{5.20}$$

$$D_P \frac{d^2[P]}{dx^2} = -k_2[E] \tag{5.21}$$

These equations are readily integrated, and after the boundary conditions are imposed, one obtains the results

$$[S] = [S]_0 - \frac{k_2[E]}{2D_S}(L^2 - x^2) \tag{5.22}$$

$$[P] = \frac{k_2[E]}{2D_P}(L^2 - x^2) \tag{5.23}$$

We see in the zero-order limit that the product concentration is independent of the bulk substrate concentration. Since the electrode responds to the surface ($x = 0$), concentration of product its potential will be independent of $[S]_0$:

$$[P]_{x=0} = \frac{k_2[E]L^2}{2D_P} \tag{5.24}$$

At very high bulk substrate concentration this equation implies that:

1. The electrode potential will be independent of $[S]_0$.
2. The potential will increase as the enzyme concentration is increased or as the electrode thickness (L) is increased.

The results obtained with virtually every enzyme electrode reported thus far agree with the first prediction, that is, the potential when $[S]_0$ is greater than K_M is independent of the substrate concentration. To some extent, there is a tendency for certain electrodes to have slight maxima in their calibration curve. This could be due to product inhibition or, less commonly, due to substrate inhibition.

The validity of the second prediction of the above equation is less certain. For example Guilbault and Montalvo, have shown that there is almost no difference in the calibration curve of a urease electrode between 175 and 230 units of enzyme/ml (10); this result is shown in Figure 5.6. Many workers have reported that the electrode potential first increases very rapidly as the enzyme activity in the gel layer increases, then becomes quite independent of the amount of enzyme (4, 15, 21). This is quite easy to understand for low substrate concentration (see below), but

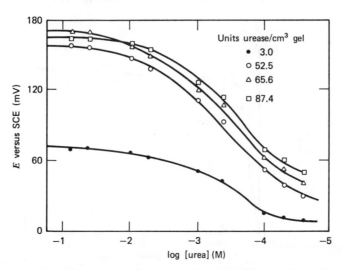

Fig. 5.6. Experimental potentiometric enzyme electrode calibration curve. Results are for a urea electrode based on urease immobilized on a univalent ion-selective electrode for ammonium ions. [Reprinted with permission from G.G. Guilbault and J. G. Montalvo, *J. Amer. Chem. Soc.* **92,** 2533 (1970). Copyright by the American Chemical Society.]

the effect has been demonstrated at high substrate concentration as well (see Figure 5.7). Guilbault and Montalvo have shown that in 83 mM urea, which is well above the K_M of 20 mM in this buffer, only, about 50 enzyme units/ml are needed, and that there is very little increase in potential even after the enzyme activity is quintupled. For a univalent electrode (Nernst slope equal to 0.059 V/decade at 25°C), an increase of 5 in enzyme activity should lead to a 40 mV shift in potential. The data of Figure 5.7 indicate that at most only a 20 mV increase occurs.

To some extent the failure of equation (5.24) may be blamed on the fact that it is after all an approximation, since the substrate concentration inside the trapped layer is the important variable and is less than the bulk substrate concentration ($[S]_0$), which is the basis of the data of Figure 5.7. The digital simulation studies of Tranh-Minh and Broun (20) indicate that in the zero-order region, the product concentration at the surface of the sensor is essentially a linear function of the enzyme concentration; thus the above objection may be invalid. A third possibility is that as more and more enzyme is added, the effective concentration (activity/unit volume) of enzyme does not increase in proportion to the amount added. This is a reasonable explanation for those electrodes in which the enzyme is covalently bound to a solid or occluded in a gel. We believe that this phenomena should be studied in more detail than it has in order to provide a more fundamental basis for the design of new electrodes.

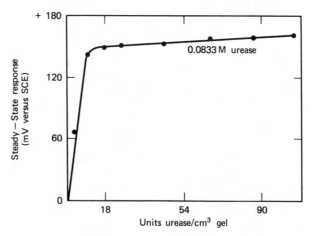

Fig. 5.7. Effect of enzyme concentration in the gel on steady-state Response of a urea electrode. [Reprinted with permission from G. G. Guilbault and J. G. Montralvo, *J. Amer. Chem. Soc.* **92**, 2533 (1970). Copyright by the American Chemical Society.]

First-Order Limit ($[S] \ll K_M$)

According to the digital simulation studies of Tranh-Minh and Broun (20), this limit should be reached when the bulk substrate is less than about 0.1–0.2 K_M over a wide range of enzyme loading factors ($k_2[E]L^2/DK_M \leq 10$). When the first-order limit is imposed on equations (5.18) and (5.19), they become:

$$D_S \frac{d^2[S]}{dx^2} = \frac{k_2[E][S]}{K_M} \tag{5.25}$$

and

$$D_P \frac{d^2[P]}{dx^2} = -\frac{k_2[E][S]}{K_M} \tag{5.26}$$

These linear equations are readily solved subject to the potentiometric boundary conditions:

$$[S] = [S]_0 \cosh \sqrt{\frac{k_2[E]L^2}{K_M D_S}} \frac{x}{L} \left/ \cosh \sqrt{\frac{k_2[E]L^2}{K_M D_S}} \right. \tag{5.27}$$

$$[P] = [S]_0 \frac{D_S}{D_P} \left(1 - \cosh \sqrt{\frac{k_2[E]L^2}{K_M D_S}} \frac{x}{L} \left/ \cosh \sqrt{\frac{k_2[E]L^2}{K_M D_S}} \right. \right) \tag{5.28}$$

Since the first-order region is analytically the most important the substrate and product concentration, profiles are given in Figures 5.8 and 5.9, respectively. As expected, the substrate concentration decreases monotonically as the surface of the sensor is approached and the product concentration increases. This results because the further a molecule of substrate penetrates into the trapped layer, the more time it has to be converted to product. At any point in the solution, an increase in enzyme activity invariably results in a decrease in substrate concentration and an increase in product concentration.

The most important result of the above calculation is that the product concentration is exactly proportional to the *bulk substrate concentration*. The constant of proportionality is a function of the enzyme activity, the thickness of the enzyme layer, the Michaelis constant, and the diffusion coefficients, but for a given electrode and substrate, all of these are constant and therefore can be included in a calibration curve. Potentiometric enzyme electrodes respond to the surface ($x = 0$) concentration of prod-

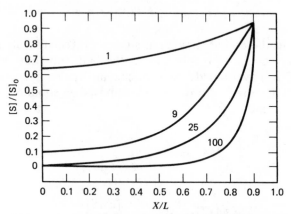

Fig. 5.8. Plot of the steady-state substrate concentration profile under first-order conditions according to the internal diffusion model. All curves are based on equation (5.27). The number of each curve indicates the value of the dimensionless enzyme loading factor $k_2[E]L^2/K_M D_S$.

uct which is given below:

$$[P]_{x=0} = [S]_0 \frac{D_S}{D_P} \left(1 - \text{sech} \sqrt{\frac{k_2[E]L^2}{K_M D_S}} \right) \quad (5.29)$$

We see that the enzyme loading factor $(k_2[E]L^2/K_M D_S)$ is the most important variable in the construction of a probe. The normalized surface

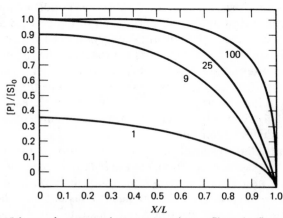

Fig. 5.9. Plot of the steady-state product concentration profile under first-order conditions according to the internal diffusion model. All curves are based on equation (5.28). The numbers on each curve indicate the value of the dimensionless enzyme loading factor $k_2[E]L^2/K_M D_S$.

product concentration as a function of the enzyme loading factor is given in Figure 5.10, where it is assumed that the product and substrate have equal diffusivities. Maximal sensitivity is obtained when the enzyme loading factor is large. The normalized surface product concentration will be greater than 0.99 when $k_2[E]L^2/K_M D_S$ is greater than about 25. The minimum enzyme concentration for a useful electrode can be estimated directly from equation (5.29). Let us assume that the Michaelis constant is about 10^{-3} M which is reasonably typical, and that the substrates diffusion coefficient is about 10^{-5} cm^2/sec, although it may be about 10 times smaller due to the internal porosity and tortuosity of a gel. The results are shown in Table 5.4 for a 100 and 300 μm thick membrane. Guilbault has reported that a useful electrode should contain about 10–20 units/ml in an electrode of thickness of 350 μm (10, 22, 23). His results certainly fall in the neighborhood of the data of Table 5.4.

The above treatment has been simplified by ignoring the effect of slow external mass transfer and phase partitioning on the electrode behavior. These factors have been considered by Blaedel et al. (18, 19). Including such factors, they obtain an equation which still predicts that the *surface product concentration is proportional to the bulk substrate,* but the proportionality constant is considerably more complicated. The results presented above were obtained from their work by taking the appropriate limits.

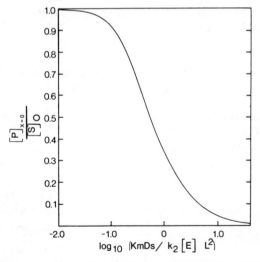

Fig. 5.10. Plot of the normalized surface product concentrations as a function of the enzyme loading factor. The results are limited to a first-order reaction with internal diffusion control. The data were computed from equation (5.29). [Reprinted with permission from P. W. Carr, *Anal. Chem.* **49,** 799 (1977). Copyright by the American Chemical Society.]

Table 5.4   Minimum Enzyme Concentration for
a Potentiometric Enzyme Electrode[a,b]

$\dfrac{[P]_{x=0}}{[S]_o}$	Enzyme concentration (units/ml)	
	100 μm layer	300 μm layer
0.9999	590	66
0.9995	415	46
0.9990	350	39
0.9950	216	24
0.9900	168	18
0.9500	84	9

[a] Computed from equation (5.29) with $D_S = D_P$.
[b] $K_M = 10^{-6}$ mole/ml; $D_S = 10^{-5}$ cm^2/sec.

The vastly different behavior of an electrode under its zero-order and first-order limits points out a very real problem in attempting to compare electrodes prepared and characterized by different groups. It is obvious that under zero-order conditions, ($[S]_0 \gg K_M$) the electrode will be sensitive to all those factors which influence the enzyme activity regardless of the level of activity on the electrode. In contradistinction, at low substrate concentration ($[S]_0 \ll K_M$), the electrode may or may not be sensitive to the same factors. If the enzyme loading factor is high, there will be almost no effect of pH changes, activators and inhibitors, or loss of activity. However, if the enzyme loading factor is low, the electrode will respond to these factors.

5.1.3   Limit of Detection and Potentiometric Enzyme Electrode Calibration Curves

Either the internal or external diffusion models can be used to generate theoretical potentiometric calibration curves [equation (5.9) or (5.28)]. As stated before, an ion-selective electrode will respond to the *surface product concentration*. If an ideal univalent Nernstian electrode is used as the basis of the sensor, then one will ideally obtain a logarithmic response of 0.059 V/decade of substrate concentration at 25°C. The ion-selective electrode can deviate from ideal behavior in two ways: a sub-Nernst slope, and imperfect selectivity. As stated in a preceeding chapter, the effect of nonselectivity can be written as:

$$E = \text{const} + \frac{RT}{zF} \ln([P]_{x=0} + \sum K_I{}^P[I]) \tag{5.30}$$

reiterating that $K_I{}^P$ is the selectivity coefficient of the electrode for a particular interferant I with respect to the primary electroactive species, that is, P the product of the enzyme reaction.

The limit of detection of an ion-selective electrode will depend upon the level of interferent present, or upon an intrinsic property of the electrode such as the solubility of the sensor electrode [e.g., of lanthanum fluoride for an F^- electrode or of $Ag(CN)_2$ for the cyanide electrode]. An exact statistical definition of the limit of detection of potentiometric sensor has appeared (24). It is evident that when the product concentration falls to a level no greater than about 10 times $\Sigma\, K_I{}^P[I]$, the calibration curve will no longer be linear.

This is true because, as shown in Figure 5.10, the surface product concentration will be much smaller than the bulk substrate concentration at low enzyme loading. Consequently, the surface product concentration will become smaller than the sum of interferences at a higher bulk substrate concentration, thereby restricting the lower limit of detection.

Because the internal-diffusion model [equations (5.18) and (5.19)] cannot be solved in closed form, it is difficult to generate a entire calibration curve on this basis alone. We have sketched curves which are consistent with the zero- and first-order limits, and with the numerical result that the surface product concentration will be proportional to the bulk substrate concentration up to $0.2\, K_M$ (20). These results are shown in Figures 5.11 and 5.12, which are plots of $\log_{10}([P]_{x=0} + \Sigma\, K_I{}^P[I])$ versus the $\log_{10}$ of the bulk substrate concentration.

When the enzyme activity of the electrode is quite high, both models [equations (5.9) and (5.28)] predict that the product concentration will

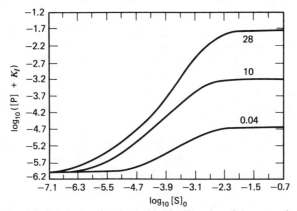

Fig. 5.11. Effect of enzyme loading factor on the linearity of the potentiometric steady-state response curve. In each case the K_M is taken as $10^{-3}\, M$. The number on each curve is the enzyme loading factor $(k_2[E]L^2/K_M D)$. The curves were computed from equations (5.29) and (5.30) with the total interference equal to $10^{-6}\, M$.

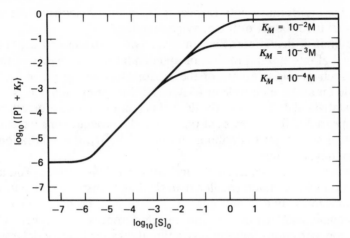

Fig. 5.12. Effect of the Michaelis–Menten constant on the linearity of potentiometric enzyme electrode steady-state response curves. In each case the enzyme loading factor (k_2[E] $L^2/K_M D$) was taken as 28. The number on each curve is the Michaelis–Menten constant for the enzyme system. All results were computed from equations (5.29) and (5.30) with the total interferenced equal to 10^{-6} M.

become equal to the bulk substrate concentration subject to the restrictions that D_S is equal to D_P and the reaction stoichiometry is as shown in reaction (5.1) (i.e., 1:1). Under these conditions, a plot of electrode potential versus concentration for the substrate and for the product should be identical. If two moles of product are generated per mole of substrate, then the product calibration curve should be shifted by $(RT/zF)\ln 2$. This result has been verified experimentally by Blaedel and coworkers (18). We conclude that in the absence of any specific interaction of the reaction product with the trapped enzyme, the *limit of detection of an enzyme electrode can be no better than the detection limit of the base sensor for the reaction product.* When the level of enzyme activity on the electrode is low (k_2[E]$L^2/K_M D_S < 25$), the detection limit may be much worse. This effect is shown in Figure 5.11 which is a theoretical curve computed using equations 28 and 30 with ΣK_I^P[I] set equal to 10^{-6} M.

This figure and Figure 5.12 indicate the effect of the enzyme loading factor and K_M on the electrode's linear dynamic range. The widest dynamic range occurs when both K_M and the enzyme activity are high. As K_M is decreased, all else being constant, the upper limit of linearity decreases toward lower substrate concentrations. As the amount of enzyme is decreased (below about k_2[E]$L^2/K_M D_S \cong 1$), the lower limit of linearity moves toward higher concentration. It is important to note that in the linear region, the slope of the calibration curve will be Nernstian even at very low enzyme loading factors, but the whole line will be shifted to lower potentials. This is in accord with the digital simulation studies of Tran-Minh and Broun (20). Once again, we can generalize and say that at best the *linear dynamic range of an enzyme electrode will be no better than the ion-selective electrode used to build it and can be much worse if low enzyme activities are employed.*

A major exception to the Nernstian behavior anticipated in the above discussion has been found by several groups and studied theoretically. Super Nernstian behavior has been observed. Papastathopoulos and Rechnitz (25) found that their urea electrode had a slope of 90 mV per decade. This device was based on an ammonia gas potentiometric electrode. Even though two moles of ammonia are produced per mole of urea, this should not result in a change in slope of a plot of potential versus logarithm of concentration; only an offset should result. Anfalt, Graneli, and Jagner (26) have shown that this effect will occur when a *gas sensor* is employed. It is due to a shift in solution pH immediately at the sensor (not in bulk solution or in the enzyme). A high sample urea concentration produces so much ammonia that the buffer capacity of the internal electrolyte is changed as the gas enters the membrane. In 0.1 M Tris buffer ($pK_a = 8.1$), theoretical calculations indicate that the slope of a plot of electrode potential versus $\log_{10}$[urea] should vary from 72 to 60 mV per decade as the pH of the buffer is changed from 7.0 to 9.0. Obviously the effect would be decreased if a stronger buffer were employed. It would seem possible to observe such effects with other enzyme electrodes when solutions of very low buffer capacity are tested. Obviously a very complicated situation might result if a major pH gradient existed between the enzyme layer and the analyte solution. A high substrate concentration might be sufficient to turn off the enzyme reaction. This is a distinct danger when pH glass electrodes are used as the base sensor in unbuffered solutions, for example, in the electrodes developed by Papariello et al. (15, 16) for the analysis of penicillin.

5.1.4 Response Time and Recovery Time of Immobilized-Enzyme Electrodes

The rate of attainment of steady state of an enzyme electrode depends upon a great many factors which are summarized in Table 5.5. It is evident that the electrochemical sensor may dictate the response if it is sufficiently slow; *the electrode certainly cannot be any faster than its base sensor.* Montalvo and Guilbault (14) have shown that a polyacrylamide gel coating does have some influence on the response time of a univalent cation electrode when it is immersed in a solution of ammonium ion. They found

Table 5.5   Factors Which Influence Electrode Response
Time

A. Physical factors
   1. Stirring rate
   2. Protective membrane thickness and permeability
   3. Thickness of enzyme layer
B. Chemical factors
   1. Enzyme concentration
   2. Substrate concentration
   3. pH, temperature
C. Instrumental factors
   1. Sensor response time

that the 98% steady response time increased from 23 sec for a 60 μm layer to 42 sec for a 350 μm layer. Mascini and Liberti (27) have studied the effect of external stirring rate, amount of enzyme, substrate concentration, pH, and thickness of the trapping membrane on the steady-state response time of an amygdalin electrode which is based on a cyanide solid-state electrode coupled with the enzyme β-glucosidase. Their electrodes were prepared by spreading a paste of enzyme on the electrode and covering it with dialysis paper. Thus as the amount of enzyme was increased, the electrode became thicker. Their results indicated that the *steady-state was obtained more rapidly at high substrate concentration and when the pH was close to the pH optimum of the enzyme.* Stirring had a very significant effect on both the steady-state potential and the rate of attainment of the final potential. In general the steady-state potential decreased and the electrode responded more rapidly as stirring was increased. Both of these observations are consistent with increased mass transfer to and from the electrode. Their results also demonstrate that the protective dialysis paper thickness is quite important—the *membrane should be no thicker than 20–25 μm for fastest response.*

One should realize that there is a trade-off involved in minimizing the membrane thickness. According to equation (5.9), a decrease in membrane permeability, which is equivalent to an increase in thickness, corresponds to an increase in the rate of the enzyme reaction relative to mass transfer, thereby making the electrode *more immune to chemical perturbations and increasing the linear dynamic range.* It is also evident that the thicker or less permeable is the membrane, the *less dependent the electrode potential will be on external stirring.*

We have considered the transient response characteristics of an idealized enzyme electrode in which the rate-determining factors were assumed to be confined to the trapped enzyme layer (28). The model employed in the study was given previously in equations (5.13) and (5.14) along with the boundary and initial conditions specified by equations (5.15)–(5.17). Fourier's method was used to solve the boundary value problem and due to the nonlinearity of the Michaelis–Menten scheme, the equations were linearized by considering only the first-order ($[S] \ll K_M$) and zero-order ($[S] \gg K_M$) limiting behavior of the electrode. The most important analytical variable is the product concentration at the electrode surface ($x = 0$), which is given by the following equations:

$$\frac{[P]_{x=0}}{[S]_0} = \left(1 - \text{sech} \sqrt{\frac{k_2[E]L^2}{K_M D}} \right) - 2 \sum_{n=1}^{\infty} (-1)^{n+1} \exp \left(-\frac{\lambda_n^2 Dt}{L^2} \right)$$

$$\times \left\{ 1 + \frac{K_M D}{2 k_2[E]L^2} \lambda_n^2 \left[1 - \exp \left(-\frac{k_2[E]t}{K_M} \right) \right] \right\}$$

$$\times \left[\lambda_n \left(1 + \frac{K_M D}{k_2[E]} \lambda_n^2 \right) \right]^{-1} ; [S] \ll K_M \qquad (5.31)$$

$$\frac{[P]_{x=0}}{[S]_0} = \frac{k_2[E]L^2}{D[S]_0} \left[\frac{1}{2} - 2 \sum_{n=1}^{\infty} \frac{(-1)^{n+1}}{\lambda_n^3} \exp \left(-\frac{\lambda_n^2 Dt}{L^2} \right) \right] ; [S] \gg K_M \quad (5.32)$$

where

$$\lambda_n \equiv \frac{2n - 1}{2} \pi, \, n = 1, 2, 3, \ldots \qquad (5.33)$$

These equations can be used to predict the time required to achieve any desired fraction of steady state. It is evident that this time depends upon two parameters for the first-order reaction. The first parameter D/L^2 is obviously a characteristic of the electrode construction and the second, $k_2[E]/K_M$, depends upon the amount of enzyme on the electrode and the enzyme's Michaelis constant. Comparison of the first- and zero-order behavior indicates that in general the electrode will be somewhat faster under zero-order than under first-order conditions. This is in agreement with the results of Mascini and Liberti, whose amygdalin electrode came to steady state in about 1 min at 10^{-1} M but required about 10 min at 10^{-4} M (27). Equation (5.31) indicates that when the amount of enzyme on the electrode is very high ($k_2[E]L^2/DK_M \gg 1$) then the first-order and zero-order time constant becomes similar and are controlled solely by the substrates diffusion coefficient and the electrode thickness. The time required for the electrode potential to settle to within 0.1 and 1.0 mV under a variety of conditions is shown in Table 5.6. An important conclusion is that if enough enzyme is present on the electrode to convert a substantial fraction of the substrate to product at the electrode surface, for example,

Table 5.6 Steady-State Response Time for Potentiometric Enzyme Electrode[a]

Enzyme loading factor[b]	Time to establish steady state[c]	
	1.0 mV	0.1 mV
∞^d	1.43	2.36
100	1.43	2.36
31.6	1.43	2.36
10	1.43	2.39
3.16	1.58	2.52
1.00	1.79	2.76
0.316	1.96	2.98
0.100	2.04	3.09
0.0316	2.07	3.13
0.0100	2.08	3.14
—[e]	1.06	1.98

[a] For the first-order domain only ($[S]_o \ll K_M$).
[b] Defined as $k_2[E]L^2/DK_M$.
[c] Dimensionless time (Dt/L^2) required to arrive at 1 mV or 0.1 mV of the final potential. All computed from equations (5.31) and (5.33).
[d] This is the pure diffusion-controlled result.
[e] This result is for a zero-order response. As indicated by equation (5.32), it is independent of the enzyme load.

when $[P]/[S]_0 > 0.5$ (see Figure 5.10), then the electrode transient response will be nearly diffusion controlled. Plots of the logarithm of the surface product concentration versus Dt/L^2 are given in Figure 5.13. It is evident that an electrode with $k_2[E]L^2/DK_M$ equal to 10 is not significantly slower than a purely diffusion-controlled electrode. An electrode which has enough enzyme to produce a significant steady-state response (for present purposes this can be defined as an electrode whose calibration curve for the substrate is not significantly different from the product) will inherently contain enough enzyme to give a diffusion-controlled transient. These results also indicate that when the amount of active enzyme is extremely low, for example, when the sample pH is much too high or too low relative to the enzyme pH optimum, then the electrode response will be somewhat slower than when diffusion controls the transient response.

It is very easy to diagnose when steps should be taken to increase the enzyme activity. This should be done when:

1. The response curves under first-order conditions are much slower than under zero-order conditions.

2. When the substrate and product calibration curves differ grossly, for example, by more than 20 mV for a univalent species (H^+, NH_4^+).

The total enzyme activity *should not* be increased by increasing the electrode thickness since this will decrease D/L^2. A purer enzyme should be used, its concentration in the gel increased, an activator added, or the pH adjusted closer to the optimum. As pointed out above, Mascini and Liberti (27) observed that at pH 10 versus pH 7 their electrode response time was much longer. Similarly, Meyerhoff and Rechnitz (29) exploited the polyphosphate activation of creatinase to improve the characteristics of an enzyme electrode. One should also consider increasing the temperature of the system since this can have a significant effect on activity.

The effect of temperature on response time is likely to be quite complicated due to the presence of both diffusional (D/L^2) and chemical kinetic terms ($k_2[E]/K_M$) in the rate equations. At high substrate concentration [equation (5.32)], where the only factor is diffusion, one should expect an apparent activation energy of about 5 kcal/mole, which is typical for diffusional processes. In contrast an Arrhenius plot at low substrate concentrations could show two straight line segments representative of both the chemical and diffusional processes. Activation energies of about 10–12 kcal/mole are typical for enzyme reactions.

Sample throughput with an enzyme electrode will depend upon both the rise time when sample is added and the time required to reachieve

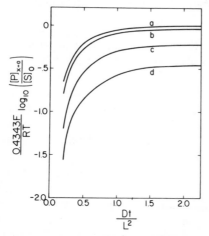

Fig. 5.13. Calculated potentiometric response-time curves for an enzyme electrode. All calculations are based on equations (5.31) and (5.32). The enzyme loading factor ($k_2[E]L^2/K_MD$), for curve a is ∞, b is 10, c is 2.5, and d is 1.0. [Reprinted with permission from P. W. Carr, *Anal. Chem.* **49**, 799 (1977). Copyright by the American Chemical Society.]

the baseline when the sample is removed or the electrode is washed. The principal transient will be diffusion of product out of the electrode. Montalvo and Guilbault (14) have presented wash out response curves for NH_4^+ on a univalent-cation electrode coated with a 350 μm layer of polyacrylamide gel and a thin membrane. Rapid return to the baseline is promoted by the use of a thin enzyme layer and a highly permeable membrane. An automatic electrode rinser has been described (14). Since most enzyme electrodes reach steady state in 3–5 min and the electrodes must be washed between measurements to eliminate sample carry over, *one should expect a throughput of about 15–20 samples per hr with a manual system.*

5.2   PRINCIPLES OF OPERATION OF AMPEROMETRIC ENZYME ELECTRODES

5.2.1   External Mass-Transfer Model

Most of the preceeding discussion pertains exactly only to the operation of potentiometric enzyme electrodes. When an amperometric electrode is employed as the fundamental sensor, the system is necessarily somewhat more complicated because one of the species involved in the overall enzyme reaction must be *electrochemically* consumed. This is best illustrated with the glucose–glucose oxidase electrode:

$$\beta\text{-D-Glucose} + O_2 \xrightarrow{\text{glucose oxidase}} \text{gluconolactone} + H_2O_2 \quad (5.34)$$

One could easily measure the decrease in available oxygen, for example, by reduction at a bare platinum electrode or with a classical Clark-type oxygen sensor. It is also possible to measure hydrogen peroxide by oxidation at a platinum electrode. Unlike the use of a potentiometric probe which does not, of necessity, interact with the chemical reaction in the trapped layer, an amperometric system must do so. For the sake of simplicity in the subsequent discussion, we will assume that a product is electroactive and is measured. Very similar results should apply if a coreactant is measured if one realizes that it is the decrease in coreactant concentration that is measured. Obviously neither measurement approach will give a linear response when the sample concentration exceeds that of any essential coreactant.

Assuming that the product of the reaction is measured, we can write a boundary condition analogous to equation (5.16) which applies to potentiometric electrode. Obviously the flux of substrate at the surface is still zero, but the product concentration will be determined by the electrode potential. Maximal sensitivity will be obtained when the electrode potential is set to completely oxidize or reduce the product, thus:

$$[P] = 0 \quad \text{at } x = 0 \tag{5.35}$$

The measured variable will be the current, which will be related through Faraday's and Fick's First Laws to the flux of product at the electrode surface; thus

$$i = nFAD_P \left. \frac{\partial[P]}{\partial x} \right|_{x=0} \tag{5.36}$$

all terms have their usual meaning; n is the number of electrons involved in the electrochemical oxidation or reduction of the product.

The flux of product to the electrode might be limited by either external mass transfer as was studied by Racine and Mindt (17), or by diffusion in the immobilized-enzyme layer as studied by Mell and Maloy (13, 30). The simpler case in which external mass transfer is limiting will be considered first because it permits a closed-form answer. Racine and Mindt assumed that the enzyme layer was very thin so that all of the reaction product could reach the electrode surface and be measured. Under this condition the current will be limited by the rate of product formation; thus

$$i = -nF \left(\frac{dN_S}{dt} \right) = nF \left(\frac{dN_P}{dt} \right) \tag{5.37}$$

In the steady state, the rate of chemical reaction will be equal to the rate of supply by mass transfer through the outer membrane [equations (5.6) and (5.7)]; thus:

$$\left(\frac{k_2[E][S]}{K_M + [S]} \right) V^* = \phi_S([S]_0 - [S]) \tag{5.38}$$

where [S] is the substrate concentration inside the membrane layer. One can easily obtain an equation for this concentration. The current will be:

$$i = nF\phi_S([S]_0 - [S]) = \frac{nF\phi_S}{2} \left([S]_0 + K_M + \frac{k_2[E]V^*}{\phi_S} \right.$$

$$\left. + \sqrt{\left(K_M + \frac{k_2[E]V^*}{\phi_S} - [S]_0 \right)^2 + 4K_M[S]_0} \right) \tag{5.39}$$

This is obviously a nonlinear result. The zero-order and first-order limits are easily obtained:

$$i = nFk_2[E]V^* \qquad \text{when } [S] \gg K_M; \quad \text{(zero order)} \qquad (5.40)$$

$$i = \frac{nFk_2[E]V^*[S]_0}{K_M + k_2[E]V^*/\phi_S} \quad \text{when } [S] \ll K_M; \quad \text{(first order)} \qquad (5.41)$$

The upper limit of linearity is governed by precisely the same factors as in the case of a potentiometric probe, thus equation (5.12) applies. *Obviously a high upper limit of linearity is promoted by a low permeability, a high K_M, and a high activity of enzyme on the electrode.* We can consider that equation (5.41) has two limiting cases: when the mass transfer is rate-limiting, that is, when ϕ_S is small, or when the rate of chemical reaction is low.

$$i = nF\phi_S[S]_0; \qquad \text{mass-transfer-limited rate} \qquad (5.42)$$

$$i = \frac{nFV^*k_2[E][S]_0}{K_M}; \quad \text{enzyme-limited rate} \qquad (5.43)$$

It is seen that in either case, the measured current is proportional to the substrate concentration in the bulk solution. For a given value of ϕ_S, the slope of a plot of i versus $[S]_0$ increases with the amount of enzyme until the slope becomes independent of amount of enzyme and depends solely upon the membrane's permeability. This is shown clearly in Figure 5.14,

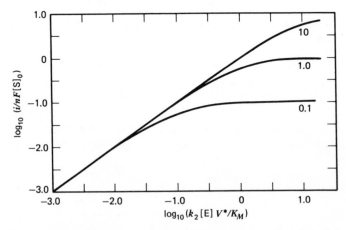

Fig. 5.14. Sensitivity of amperometric enzyme electrodes as a function of the enzyme loading factor and the membrane permeability. The number on each curve is the dimensionless Michaelis constant. Data are based on the external diffusion model and were computed from equation (5.41). These results pertain only when the bulk substrate concentration is much less than the $1K_M$.

which is a plot of $i/nF[S]_0$ versus $k_2[E]V^*/K_M$ at various values of the membrane permeability as computed from equation (5.41). The greatest electrode sensitivity occurs when both the permeability and amount of enzyme on the electrode are as large as possible. *Ultimately the sensitivity will be limited by the rate of mass transfer from the bulk fluid to the outer surface of the electrode membrane.* This will depend upon the intensity of stirring in the test solution.

5.2.2   Internal Mass-Transfer Model (Amperometric Probe)

When the enzyme layer is thick compared to the membrane coating, and diffusion in it is slow relative to the rate of mass transfer from the bulk solution, one must consider the simultaneous occurrence of chemical reaction and diffusion, that is, the analogy to the situation considered by Blaedel et al. for a potentiometric probe (18) as summarized in equations (5.13) and (5.14). The boundary condition at $x = L$ (see Figure 5.3) is evidently the same as with a potentiometric probe but, as discussed above, the electrochemical consumption of the product requires the use of equation (5.35) at $x = 0$. Mell and Maloy (13, 30) have carried out an extensive experimental and theoretical investigation of amperometric enzyme electrodes, which included calculation of both the transient and steady-state current as a function of enzyme activity, Michaelis constant, and substrate concentration, by means of the implicit finite difference digital simulation method of Feldberg (31). Mell and Maloy demonstrated that when the enzyme electrode is covered by a nylon net of unspecified thickness, the charge obtained by integration of the experimental current–time curve followed a square-root-of-time relationship indicating diffusion control. Furthermore the current at an enzyme electrode was 50 times smaller than at a bare platinum electrode of the same area (0.03 cm²).

The simulated current–time curves obtained by Mell and Maloy are shown in Figure 5.15. These are normalized plots of current $(i/nFV^*[E])$ versus time (Dt/L^2). Their results indicate that a steady-state current will be obtained under all conditions, that is, regardless of the enzyme loading factor $(k_2[E]L^2/K_M D)$ or substrate concentration (expressed as $[S]_0/K_M$). The maximum time required to achieve this steady state is given below:

$$t_{ss} \le 1.5L^2/D \qquad (5.44)$$

This predicts that a 100 μm electrode should achieve steady state for a small substrate $(D \cong 10^{-5}$ cm²/sec) in about 15 sec. At high substrate concentration $([S]_0 \gg K_M)$, the current should reach half of its steady-

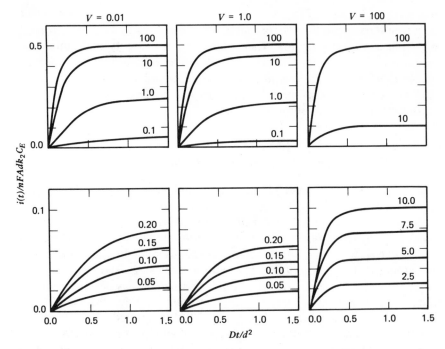

Fig. 5.15. Calculated dimensionless current–time curves obtained by digital simulation for an amperometric immobilized-enzyme electrode. The number above each curve is the enzyme loading factor $k_2[E]L^2/K_MD$. The number on each curve is the ratio of bulk substrate concentrations to the Michaelis constant (K_M). [Reprinted from L. D. Mell and J. T. Maloy, *Anal. Chem.* **47**, 299 (1975). Copyright by the American Chemical Society.]

state value when

$$t_{1/2} \cong 0.1 \frac{L^2}{D} \qquad (5.45)$$

These curves qualitatively indicate that at low substrate concentrations relative to the K_M, the current is linearly related to the substrate concentration but it becomes nonlinear at higher concentration. Furthermore, the maximum possible current ($[S]_0 \gg K_M$) is equivalent to electrolysis of exactly one-half of the product generated, regardless of the enzyme loading factor. This is due to the symmetry of the boundary condition for the product, that is, the concentration of product at both $x = 0$ and L is zero.

A set of simulated calibration curves are shown in Figure 5.16. They indicate that the current is linearly related to the bulk substrate at low concentration but ultimately it is limited by the Michaelis constant. As the amount of enzyme on the electrode is increased, the linearity persists at concentrations in excess of K_M. An *increase in the loading factor from 1 to about 100 results in a shift in the upper limit of linearity by almost a factor of 100.* This is quite analogous to the results obtained for external mass-transfer control in either potentiometric or amperometric probes, but is at variance with the results of Tran-Minh and Broun for internal mass-transfer in a potentiometric probe.

The other interesting feature of the data of Figure 5.16 is that the current at low concentrations is independent of the enzyme loading factor, but the limiting value at high substrate depends upon the loading factor. The data contained in the calibration curve may be analyzed via "electro-chemical Lineweaver–Burke" plots in which the reciprocal of the current is plotted against the reciprocal of the bulk substrate concentration. These curves were linear only at low enzyme loading, that is, when the current was controlled by the enzyme kinetics. Mell and Maloy found that at high enzyme activity $(k_2[E]L^2/DK_M > 10)$, the simulated calibration curve would fit the equation

$$i = \frac{nFV^*k_2[E][S]_0}{k_2[E]L^2/D + 2[S]_0} \tag{5.46}$$

At low enzyme loading factors (≤ 1) the simulated data fit the equation

$$i = \frac{nFV^*k_2[E]}{2}\frac{[S]_0}{K_M + [S]_0} \tag{5.47}$$

The fact that the current is controlled by the amount of enzyme on the electrode is clearly demonstrated by the depth of the analogy between

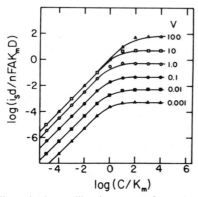

Fig. 5.16. Calculated dimensionless calibration curves for an amperometric immobilized-enzyme electrode. The number on each curve is proportional to the amount of enzyme on the electrode. [Reprinted from L. D. Mell and J. T. Maloy, *Anal. Chem.* **47**, 299 (1975). Copyright by the American Chemical Society.]

equation (5.46) and the conventional chemical Michaelis–Menten rate law. When the substrate concentration is very high, both equations (5.45) and (5.46) approach a common limiting current:

$$i = \frac{nFV^*k_2[E]}{2} \equiv i_{max} \qquad (5.48)$$

It is interesting to compare equation (5.48) to (5.40), which is identical except for a factor of 2.

The most vital analytical feature of the above equations is the result that the current is proportional to the substrate concentration. At low enzyme loading factors, equation (5.46) becomes

$$i = \frac{i_{max}[S]_0}{K_M}; \quad \text{when } [S]_0 \ll K_M \qquad (5.49)$$

This relationship, and especially the conditions under which it is valid, is in exact agreement with the expected results of a kinetic analysis. In contrast, when mass transfer is the principal rate-controlling factor, namely, that equation (5.46) obtains, one finds that

$$i = \frac{nFAD[S]_0}{L} \text{ when } [S]_0 \ll \frac{k_2[E]L^2}{2D} \qquad (5.50)$$

This last result is quite interesting since it is identical to the behavior of a simple (nonenzymatic) membrane amperometric electrode which is directly sensitive to the bulk substrate. Both of the above limits at low substrate concentration have exact analogs in the external mass-transfer-limited electrode as described by Racine and Mindt [see equations (5.42) and (5.43)].

Analytically the major results of the calculations of Mell and Maloy are:

1. The prediction that the current at high enzyme loading factors will be a linear function of the substrate concentration at concentrations greater than K_M.
2. The prediction that the current at low substrate concentrations will increase as the membrane thickness decreases if mass transfer is the rate-limiting factor.
3. The estimation of the maximum steady-state time as given in equation (5.44).

5.2.3   Limit of Detection and Linearity of Amperometric Enzyme Electrodes

By and large all electroanalytical methods which are based on current measurement are limited by the magnitude of the background current, that is, in the absence of the sample species, some current will be detected even with an electrode whose area and potential are both held constant. Quite generally, increasing the electrode area merely serves to increase the background current in the same proportion as the slope of the calibration curve (current versus sample concentration) and therefore the detection limit does not improve.

The limit of detection is fixed then by three factors: the magnitude of the background signal, the sensitivity of the technique (di/d[sample]), and the reproducibility of both of the above. In general, a limit of detection is defined as that sample concentration which produces a signal that is above the background (blank) noise level at some stated confidence interval. A specific definition of these factors depends upon the type of noise encountered, for example, whether the noise is normally (i.e., Gaussianly) distributed random noise, and its frequency dependence. General discussions of this topic can be found in the literature (32) and are beyond the intent of this work. It is clear that the best limit of detection will be obtained by minimizing the background current and maximizing the slope of the calibration curve.

The preceeding discussion of the work of Mell and Maloy indicates that the two major factors which influence the slope, that is, sensitivity of the current–concentration curve, are the amount of enzyme on the electrode and the thickness of the electrode. Let us consider an electrode of a given thickness. The slope of the calibration curve will increase with the amount of enzyme added until it becomes equal to the value dictated by mass transfer. Assuming that both the substrate and its product, which is the species being measured at the electrode, have the same diffusion coefficient then the slope of the calibration curve will be the same regardless of whether the substrate or the product is assayed with the electrode. Thus it is clear that the limit of detection of an amperometric enzyme electrode can be no better than is the limit of the base sensor for the product. Since a bare electrode can be used to measure the product, whereas a trapped layer of enzyme is needed to convert the substrate to product, the *detection limit will generally be poorer for an enzyme electrode as compared to the analogous simple, that is, uncoated, product sensor.*

A particularly severe limitation occurs when the amperometric electrode is designed so that the decrease in coreagent concentration is measured. This occurs most frequently when an oxygen-consuming enzyme (amino acid oxidase, glucose oxidase) is used in conjunction with an oxygen-detecting electrode. When this type of sensor is used, the decrease in current from some initially high value is measured. Thus the sample concentration is computed as the difference between two large quantities. This is a situation fraught with statistical embarassments. To overcome the problem, Updike and Hicks (2) designed their original probe as a differential device containing two sensors, one covered with active enzyme, the other coated with denatured material. The difference in current increases in proportion to the sample concentration and is independent of changes in the oxygen concentration which are not mediated by the immobilized enzyme.

Practical limits of detection of amperometric enzyme electrodes vary quite considerably. A ballpark figure of 10^{-5}–10^{-4} M can be deduced from a survey of the calibration curves shown in the literature. Estimation of the limit of detection is complicated by the fact that solid (nonmercury) electrodes must be pretreated before use in order to decrease the background signal due to presence of oxides and adsorbed materials on the electrode surface. Pretreatment most often consists of cycling the electrode potential and then holding it at the value used for the measurement until the current becomes constant. For example Guilbault and Lubrano (33) pretreated their platinum sensors by holding the electrode at -0.2 V versus a calomel reference electrode until the current decayed to a small level then switched to $+0.05$ V until almost zero current was obtained. Finally the electrode was poised at $+0.6$ V, where peroxide is oxidized, until the current decayed to a suitably low value. This entire pretreatment is not required for each sample, just for the final stage.

Clark (3) has reported an extremely sensitive electrode whose limit of detection for peroxide is only a few nanomolar (10^{-9}–10^{-8} M). This electrode which is shown in Figure 5.1(E) is unique in that *oxygen is supplied internally*. The background current for a 1 mm diameter cylindrical tube of platinum was about 0.03 μA and the noise level was such that a 0.006 μA change could be detected.

In a recent publication, Mell and Maloy (30) describe a technique for enhancing the response of an amperometric enzyme electrode. They have shown that if the electrode is held at a potential where the product is not electroactive and is allowed to accrue, and then, after a suitable dwell period, the potential is jumped to a value where the product is electroactive, the transient current will be increased significantly and sensitivity improved by as much as a factor of 10. They were able to measure glucose at concentrations where the steady-state current was 10 times smaller than the background signal; their results are shown in Figure 5.17. The dashed line indicates the measured background current level. It should be noted that the current indicated in the figure is calculated from the number of coulombs accumulated and the length of time the electrode is active.

The linearity and upper limit of application of amperometric enzyme electrodes are clearly indicated in the work of Mell and Maloy (30). It

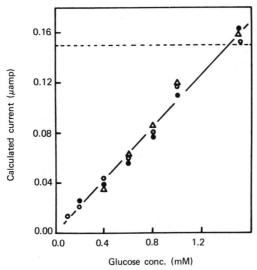

Fig. 5.17. Calibration curve for an amperometric enzyme electrode with electrochemical enhancement. The dashed line represents the background current. The three curves points represent the average current obtained with the indicated glucose concentration at different ratios of on (t) to off (τ) times of the electrode potential. $\triangle$, ●, and ○ represent three of these different ratios. [Reprinted with permission from L. D. Mell and J. T. Maloy, *Anal. Chem.* **48**, 1597 (1976). Copyright by the American Chemical Society.]

should be quite clear that the electrode will be linear above the K_M provided that there is a high specific enzyme activity on the electrode. The upper limit of linearity as a function of the amount of enzyme on the electrode can be readily deduced from equations (5.46) and (5.47). When the amount of enzyme is low ($k_2[E]L^2/K_M D < 1$), equation (5.47) pertains and the upper limit is dependent upon the K_M. We can expect linearity to within 10% as long as $[S]_0$ is less than $0.1K_M$. At high enzyme loading factors, the upper limit of linearity will depend upon the enzyme loading factor. The results of several typical calculations based on equation (5.46) are shown in Table 5.7. This shows that a very high enzyme specific activity is needed to improve the upper limit of linearity if the K_M is as large as a few millimolar.

5.3 PRACTICAL ASPECTS IN THE USE OF ENZYME ELECTRODES

One of the most critical features affecting the usefulness of immobilized-enzyme probes is their long-term stability. For our purposes, this can be

Table 5.7   Upper Limit of Linearity of Amperometric Enzyme Electrodes[a,b]

Enzyme activity (μmole/min·ml)	Limit of linearity (μM)
0.10	0.83
0.50	4.2
1.00	8.3
5.00	4.2×10^1
10.0	8.3×10^1
50.0	4.2×10^2
100	8.3×10^2
500	4.2×10^3
1000	8.3×10^3

[a] Computed from equation (5.46) with $D = 10^{-5}$ cm^2/sec, $L = 0.01$ cm, and the assumption that 10% deviation from linearity is acceptable.
[b] These results apply only if $k_2[E]L^2/K_M D \geq 10$.

defined as resistance to change in the slope of the calibration curve when an amperometric sensor is used, and as the change in absolute potential of a potentiometric electrode.

5.3.1   Stability of Immobilized-Enzyme Electrodes

Those factors which can influence the stability of an enzyme electrode are listed in Table 5.8. It is very difficult to predict the useful lifetime of an electrode, although some rules of thumb will be given later. Some reported electrodes can be used only a few times. For example, the amperometric peroxide electrode of Aizawa, Karube, and Suzuki (34) died abruptly and reproducibly upon its tenth immersion in 1.5 mM substrate [Figure 5.18(F)]. It should be noted however, that this electrode used an extremely thin film of adsorbed enzyme fibrils. Other electrodes such as the glucose oxidase electrode of Guilbault and Lubrano (33) can be used for several months and several thousand determinations [Figure 5.18(E)].

One might expect that the sensitivity of all electrodes would decay monotonically from the instant they are made. The data of Figure 5.18 indicate that this need not be true and that sensitivity can actually improve for a while then fall off. Complicated changes in electrode response are not uncommon; in some cases, there is an induction period of several days after construction of the electrode before it can be used (35). Such problems can generally be attributed to the enzyme and will be discussed later.

Table 5.8   Factors Which Determine Enzyme Electrode
Stability

I.   The stability of the base sensor
II. Physical factors
   a. The thickness of the enzyme layer
   b. The mechanical stability of the trapped layer
III. Chemical factors
   a. Immobilization method
     1. Soluble trapped layer
     2. Physically occluded
     3. Covalent attachment
   b. The total enzyme activity in the trapped layer
   c. The stability of the trapped enzyme
   d. Chemical conditions of use
     1. pH
     2. Temperature
     3. Accumulation of irreversible inhibitors
   e. Storage conditions

A final factor which must be considered before a general discussion of electrode stability is undertaken is the fact that the apparent stability will depend very much on whether the electrode is tested at low or high substrate concentration with respect to the enzyme's Michaelis constant, and whether or not the probe is tested in the linear or nonlinear segment of the calibration curve. Guilbault has shown repeatedly that the stability is best in the first-order, that is, low, substrate-concentration domain. For example, his original urea electrode exhibited a drift of only 0.05 mV per day at low substrate concentrations but was 0.2 mV per day at high substrate concentrations (10). The reasons for this are straightforward and are discussed below in terms of the effect of the enzyme loading factor on stability. It is clear that when electrodes are compared in terms of their stability, one must be aware of this complication.

Let us turn to Table 5.8 and consider those factors which can influence stability. The first serious aspect of the electrode in this regard is the base sensor. It is evident that if this changes, the response must be altered. Fortunately this has not been a limiting factor due to the stability of most ion-selective and amperometric electrodes. Probably the two worst cases are cyanide electrodes and heterogeneous membrane ammonium electrodes. The cyanide electrode which is based on the moderately soluble $Ag(CN)_2$ crystal is not stable in solutions which contain more than 1 mM CN^-.

Even in this case, Llenado and Rechnitz report that their amygdalin electrodes stability ($\cong$3 days under use) is limited by loss of enzyme and

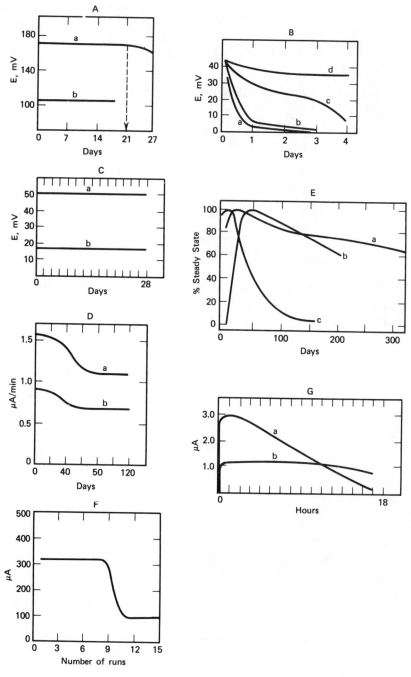

240

not by the base sensor (4). Similarly Guilbault and Nagy (36) report that their nonactin-impregnated silicone rubber membrane has a total drift of only 2 mV over 5 weeks.

Two physical factors can influence the electrode stability: the thickness of the trapped layer (but this is really directly related to the amount of enzyme on the electrode and will be discussed below), and the mechanical properties of the electrode. Ideally the electrode should be constructed so that no enzyme can leak out of it and so that the enzyme layer does not become physically detached from the sensing surface of the electrode. Since electrodes are often, of necessity, stored cold, the repeated expansion and constriction of the fluid, gel, etc., can ruin the mechanical integrity so that "O" ring seals start to leak. Obviously soluble enzymes held in place by a membrane are most susceptible to this problem, but so are gel-occluded enzymes.

Guilbault has discussed the optimum condition for preparing polycrylamide gel entrapped enzymes (10, 37). He recommends that the gel-forming solution be stored in the dark and, of course, the photopolymerization (riboflavin + persulfate) catalyst should not be added until immediately before polymerization. Optimum electrode stability was obtained when the prescription below was followed:

1. 1.15 g of N,N'-methylenebisacrylamide is added to 40 ml of hot (60°C) 0.1 M phosphate buffer (pH 6.8) and dissolved completely.
2. The solution is cooled to 35°C and 6.06 g of acrylamide (monomer) are added.

Fig. 5.18. Stability of enzyme electrodes. (A) Stability of a urea electrode in Tris buffer at 25°C, (a) response to 0.0833 M; (b) response to 1 mM. (B) Stability of a D-amino acid electrode as a function of the FAD solution used for storing the electrode: (a) 0 μM FAD; (b) 2 μM FAD; (c) 50 μM FAD; (d) 0.25 mM FAD. (C) Stability of an asparagine electrode at pH 8.0, (a) 2 mM substrate, (b) 0.2 mM substrate. [(A), (B), and (C) are all reprinted from G. G. Guilbault and E. Hrabankova, *Anal. Chim. Acta.* **56**, 285 (1971), with permission of Elsevier Scientific Publishing Co.] (D) Stability of a uric acid electrode using a rate method. Temperature = 30°C, 0.1 M glycine buffer, pH 9.2. (a) 8 mg % substrate; (b) 4.2 mg % substrate. [Reprinted from M. Nanjo and G. G. Guilbault, *Anal. Chem.* **46**, 1769 (1974). Copyright by the American Chemical society.] (E) Stability of glucose electrodes. Conditions: 5 mM, pH −6.6 stored at 25°C, (a) covalently bound, (b) occluded in gel, (c) soluble enzyme. [Reprinted from G. G. Guilbault and G. J. Lubrano, Anal. Chim. Acta. **64**, 439 (1973), with permission from Elsevier Scientific Publishing Co.] (F) Stability of a peroxide electrode. [Reprinted from M. Aizawa, I. Karube, and S. Suzuki, *Anal. Chim. Acta.* **69**, 431 (1974). with permission of Elsevier Scientific Publishing Co.] (G) Stability of a lactate electrode. Curve a with a thin outer membrane; Curve b with a thicker membrane. [Reprinted from P. Racine and W. Mindt, *Experientia Suppl.* **18**, 525 (1971), with permission of Birkhäuser Verlag.]

3. The solution is well mixed and then filtered into a 50 ml volumetric flask.

4. If the solution is to be used immediately (2 days) add 5.5 mg of both riboflavin and potassium persulfate. Then make up to volume. Otherwise store the solution in dark and add catalyst when needed.

5. Polymerize the required volume of solution by irradiation with a No. 1 photoflood lamp (150 W) for 1 hr at 28°C.

The method of enzyme entrapment can have a major effect on the enzyme stability. As a general rule of thumb, electrodes based on soluble enzymes will be less stable than those based on gel-occluded enzymes, which in turn are less stable than covalently bound or crosslinked enzymes (37). Guilbault states that soluble-enzyme electrodes are stable for 1 week (25–50 assays), physically immobilized enzymes are stable for 3–4 weeks (50–100 assays) and chemically immobilized enzymes can be used for 4–14 months (200–10,000 assays).

Guilbault and Lubrano (33) studied the long-term stability of several types of glucose oxidase electrodes. Their results for a soluble (Type c), physically occluded (Type b), and chemically crosslinked polymer (Type a) are shown in Figure 5.18E. The soluble-enzyme electrode sensitivity falls off immediately after preparation in a roughly exponential fashion. Both the chemically bound and physically occluded electrodes show an increase in sensitivity over the first 25–50 days and then a decay. Guilbault postulates that the rise is due to opening up of channels in the gel which permit increased mass transfer followed by an eventual loss of enzyme activity by denaturation or irreversible inhibition.

A very complicated stability problem arises in the case of an amperometric alcohol electrode based on alcohol oxidase (35, 3). The electrode is based on the reaction sequence:

$$RCH_2OH + O_2 \xrightarrow{\text{alcohol oxidase}} RCHO + H_2O_2 \qquad (5.51)$$

Peroxide can be measured on a platinum electrode by oxidation at $+0.6$ V versus a saturated calomel electrode:

$$H_2O_2 \rightarrow O_2 + 2H^+ + 2e^- \qquad (5.52)$$

Fresh alcohol oxidase does indeed generate peroxide but after three days almost no peroxide was detected. At a potential of -0.6 V, one can reduce both oxygen and peroxide:

$$H_2O_2 + 2e^- \rightarrow 2OH^- \qquad (5.53)$$

$$O_2 + 2H_2O + 4e^- \rightarrow 4OH^- \qquad (5.54)$$

With fresh enzyme at an electrode poised at -0.6 V, Guilbault observed reduction of both peroxide and oxygen. After the enzyme was aged for three days, the alcohol oxidase lost its ability to generate peroxide, but did catalyze a reaction [other than that shown in equation (5.51)] between the substrate and oxygen. The net change in current was actually larger when the enzyme did not generate peroxide. Thus Guilbault and Lubrano recommended that the electrode should be prepared and then used only after 2 weeks. The electrode retained its sensitivity for approximately 4 months.

Blaedel and Jenkins (38) developed a reagentless lactate electrode which employed immobilized lactate dehydrogenase and co-immobilized NAD. Their results indicated that the electrodes stability was dicated by the weakness of the covalent bond between NAD^+ and the matrix.

The major factor to be considered is the total enzyme activity on the electrode. In reality the enzyme loading factor $(k_2[E]L^2/K_MD)$ is the important parameter. We have stated repeatedly that at high enzyme loading the electrode potential or current will be limited by the rate of mass transfer. Suppose that an electrode contains a tenfold excess of enzyme over that which is required to obtain a mass-transfer-limited system. Obviously if 50% of the enzyme is lost, for example, due to leakage, denaturation, or inhibition, the resultant *change* in electrode response *will be trivial* since the response is being controlled by *mass transfer*. Eventually, enough enzyme will be lost so that the chemical reaction rate in the trapped layer becomes rate-controlling; the response will then rapidly decay with further loss of enzyme.

The effect of enzyme loading factor on the stability of a potentiometric probe which is limited by internal diffusion and reaction can be viewed quantitatively from the data presented in Figure 5.10 and Table 5.4. Consider an electrode with a loading factor of 100 which, due to denaturation, gradually decays to a loading factor of 25. Figure 5.10 indicates that less than a 1% change in surface product concentration will occur for this fourfold decrease in enzyme activity. Virtually all reports indicate that the electrode will be stabilized by adding more enzyme. Guilbault has noted that useful electrodes require a minimum of 10–20 units per cm^3 for an electrode of about 350 μm thickness (10).

The difference in stability of an electrode when tested under zero-order and first-order conditions, that is, high and low substrate concentrations, respectively, is due to the difference in the mathematical dependence of the signal on the enzyme activity. As an example, consider a potentiometric probe. Figure 5.10 indicates that at high enzyme activity, the surface product concentration will be independent of the enzyme activity.

In contrast, equation (5.24), which applies at high substrate concentration, indicates that the response is always dependent on the amount of enzyme.

We can conclude that the stability of an electrode under *zero-order conditions* mirrors the enzyme activity but under *first-order conditions* the stability is a function of the electrode geometry, that is, thickness and the amount of enzyme.   Fortunately the electrode will be more stable under the analytically interesting conditions (first-order) than it is under the less stable zero-order conditions. It is obviously very misleading to report that immobiliation per se stabilizes an enzyme if one studies only the first-order stability. The same is true of any studies of the effect of pH, temperature, ionic strength, etc., on the electrode. **We recommend that all types of studies be conducted under both zero- and first-order conditions so that different characteristics of the systems can be validly compared.**

There is no intent here to question the idea that immobilization can increase the stability of an enzyme. Earlier chapters indicate that this is a fact with many enzymes. It should be clear that chemical factors can have a profound effect on the intrinsic stability of an enzyme and an enzyme electrode. Chemical factors also came into play in choosing storage conditions as well as operational ones so that we will consider these simultaneously.

Guilbault and Hrabankova discovered a very interesting effect in the development of a D-amino acid electrode (39). This enzyme has a loosely bound FAD moiety which essentially dialyzes out of the enzyme when it is occluded in a gel. The data shown in Figure 5.18(B), indicate the considerable effect of storage conditions on stability. Obviously FAD leaches out of the electrode. This can be prevented by storing the electrode in an appropriate buffered solution which contains this material.

Meyerhoff and Rechnitz (29) reported that their electrode for creatinine required pretreatment and storage in an activating solution of sodium tripolyphosphate ($Na_5P_3O_{10}$). Without activation and storage in this solution, the electrode response became essentially useless within a few days but when stored in the activator solution, the decrease in slope of the calibration curve was barely measurable over the same period of time.

Solution pH can have a marked effect on electrode stability. This is most evident in the various amygdalin electrodes which have been reported (4,27,40). These electrodes are based on a cyanide ion-selective electrode which is obviously very pH-sensitive. The base sensor is best used at high pH where all of the cyanide is deprotonated. Rechnitz and Llenado reported that the amydalin electrode is stable for about 3 days at pH 10. In contrast Mascini and Liberti (27) used the electrode at pH 7 where it was stable for about 1 week. Due to the difference in immobilization methods and amounts of enzyme employed, it is not certain that pH is the sole factor involved here.

A final very interesting and provocative example of the effect of conditions of storage on electrode stability is evident in the report from Rechnitz on his whole-cell electrodes (5). In this work a colony of bacteria (*Streptococcus faecium*) which convert arginine to ammonia was trapped at the tip of an ammonia gas sensor electrode. The electrode is useful for 3 weeks and maybe regenerated after the response deteriorates by returning it to the original nutrient broth for storage! These devices will be described in more detail in a later section.

5.3.2   Effect of pH and Temperature on Enzyme Electrodes

A variety of pH effects on the potential or current obtained with immobilized-enzyme electrodes have been observed. The general feature of all these curves is that there is an optimum of pH or pH range for their operation. At extreme acidity or alkalinity, the electrode fails to respond to a change in substrate concentration, that is, the measured variable approaches the background blank value. In some cases, for example, the peroxide electrode of Aizawa, Karube, and Suzuki (34), the optimum range is very broad, but in most instances the electrode can be used only over a narrow range of perhaps 2 pH units. The effect of pH on response is due to two factors: the dependence of the enzyme's activity on this important chemical variable, and the effect that pH has on the ability of the electrode to detect the measured species, for example, the product of the enzymatic reaction.

The net effect of pH on the extent of conversion of the substrate to product will inevitably depend upon the enzyme loading factor ($k_2[E]L^2/DK_M$) and whether the electrode is operated in the low or high substrate-concentration region. When the enzyme load is high so that there is an excess of enzyme, a large change in pH, which is equivalent to a change in activity, should have very little effect on the product concentration in the first-order kinetic regime (see Figure 5.10). Thus one should expect pH curves to be much broader and flatter than those shown by the soluble enzyme. In contrast a low enzyme activity or at high substrate concentration, the electrode potential (or current) will be very dependent on the enzyme activity and therefore should be a strong function of pH. At this time there has been no published data which reflect on this point. Most workers are content to assess the effect of pH at only one substrate concentration if they do so at all.

In view of the high charge on a protein, and the complex electrostatic effects which can occur, one should not expect any exact coincidence between the position of the pH optimum of soluble enzyme and the immobilized system, except perhaps at high ionic strength. Guilbault's work (39,43) indicates that when the *same* electrode is used in the steady-state mode or in a kinetic mode, the optima may be at different pH values. When a multiple enzyme system such as the phosphate electrode is tested, the effect of pH can be extremely complicated (44). The potential complexity of pH effects in media of low buffer capacity is well illustrated in a paper by Thomas (45). In such media, he was able to induce *oscillating pH variations* related to the hysteresis caused by slow membrane diffusion.

When potentiometric sensors are employed, it may be very difficult to find a pH which is compatible with the electrode and the enzyme. There are at least two complications in this regard. First, the potentiometric electrode can detect only one form of the product species. For example the cyanide electrode senses CN^-, not HCN, and an ammonium ion electrode detects only NH_4^+ but not NH_3. Obviously, the pH will play a major role in establishing the fraction of the product in a given form. Second, the electrode may be directly sensitive to hydrogen ion. This will be the case when a pH glass electrode is coupled to the trapped enzyme (15,47) and certainly is a problem when a univalent glass electrode is used to detect NH_4^+ (10).

As an example of this effect, the amygdalin electrode described by Llenado and Rechnitz (4) was used at pH 10 in order to ionize most of the cyanide generated by the enzyme. Unfortunately this is well above the optimum pH and is sufficiently high to denature the enzyme. In later work with the same enzyme and electrode, Mascini and Liberti (27) decreased the pH to 7. The net effect of the decrease in pH on the calibration curve was much less than what one would expect based on purely thermodynamic consideration, that is, a displacement of the curve by about 180 mV (3 × 0.059) because the enzyme is very much more active at low pH. Mascini and Liberti also found that the steady-state response was obtained much more promptly and was independent of the amount of enzyme on the electrode over a wide range of pH 7 compared to pH 10. In addition their electrode was active for a longer time when used at pH 7. Since neither group (4, 27) carried out a detailed pH study between 7 and 10, it may well be that an optimum pH representing a compromise between the enzymatic and electrode requirements does exist. It does seem clear, however, that the *electrodes are more tolerant of suboptimal conditions than are the enzymes.*

Sample temperature is a very important variable in enzyme electrode work. There are a number of important questions to be faced. First, is there an optimum temperature for response? Second, is this temperature compatible with the stability of the sensor and enzyme? Third, how closely should the sample temperature be controlled?

At low temperature (0°C), most enzymes are less active than at room temperature. We therefore expect a gradual increase in activity and response as the temperature is raised. Ultimately all enzymes are thermally denatured so an optimum response is expected. This is verified by the results shown in Figure 5.19 for a potentiometric arginase sensor (46).

Very few temperature-response curves have been reported. Guilbault and Lubrano (48) did characterize an amperometric amino acid electrode. Since amperometric electrodes can be used in both steady-state and kinetic (rate) modes they measured the steady-state current and initial rate of change of current. As shown in Figure 5.19 the Arrhenius plot shows two regions. At high temperature, a lower activation energy is observed than at low temperature. This may be due to two different conformations of the enzyme, but it could also be due to a shift in rate control by the enzyme reaction to rate control by diffusion, as has been observed by Buckholz and Ruth (49). It is also interesting to note the difference in the activation energy, that is, slope of the Arrhenius plots for the steady-state and kinetic measurements (43).

In general, enzymes should not be used for analytical purposes at their temperature optimum because of slow thermal denaturation at these temperatures (see Chapter 1). To provide a long lifetime, the electrode should be used at a temperature *at least 10°C below the temperature optimum.*

It is always advisable to control the sample temperature when making electrochemical measurements, whether or not enzymes are employed. In the case of potentiometric systems, the indicator electrode will be temperature-dependent. For example, the asymmetry potential of a glass pH electrode has a temperature coefficient of about 1 mV/°C (50) which is in addition to the temperature coefficient of the buffer. For example, Tris buffer has a temperature dependence of -0.025 pH/°C (51). The saturated calomel electrode which is used as a reference electrode in many potentiometric systems has a temperature coefficient of 0.7 mV/°C (52). Amperometric measurements are sensitive to temperature since they rely upon the diffusion coefficient of the measured species which may be as much as 2–3%/°C. An enzyme electrode will be even more temperature-sensitive; therefore it is advisable to thermostat to within 0.1°C, most particularly when the electrode is used in a rate mode. One last point is that the coefficient which appears in the Nernst equation (0.4343 RT/F) varies from 0.05915 at 25°C to 0.06153 at 37°C.

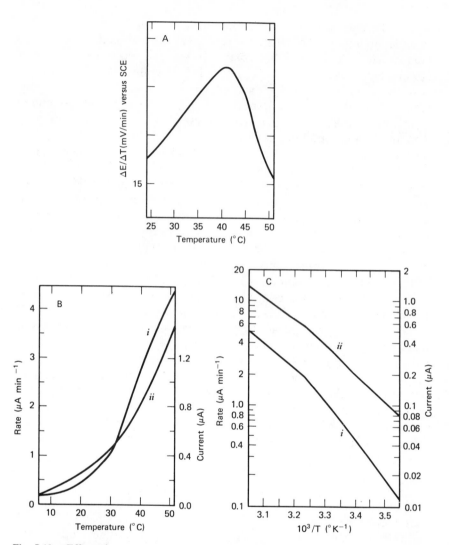

Fig. 5.19. Effect of temperature on the response of enzyme electrode. (A) Potentiometric arginase electrode. [Reprinted with permission from H. E. Booker and J. L. Haslam, *Anal. Chem.* **46,** 1054 (1974). Copyright by the American Chemical Society.] (B) and (C) Amperometric L-amino acid electrode. Curve i reaction-rate method, curve ii indicate steady state (B) observed effect; (C) Arrhenius and von't Hoff plot, [Reprinted from G. G. Guilbault and G. J. Lubrano, *Anal. Chim. Acta.* **69,** 183 (1974), with permission of the Elsevier Scientific Publishing Co.]

5.3.3   Interferences and Problems in Using Enzyme Electrodes

An interference may be considered as any chemical species which alters (either by decreasing or increasing) the response of a sensor to the sample. One can generalize and divide interferences into three general categories, that is, ions or molecules which effect the base sensor, those which alter the enzyme, and those which react with the measured species, that is, the product. These problems are summarized in Table 5.9.

Electrochemical interference is more common with ion-selective electrode sensors in comparison to amperometric sensors. When a glass-type ammonium-ion sensor is used in conjuction with response or the amino acid oxidases, a great many cations including Li^+, K^+, Na^+, and H^+ can strongly interfere. These electrodes cannot be used at all below pH 6 due to the H^+ problem and certainly cannot be used in common buffers (phosphate, bicarbonate, veronal) with alkali metal ions (Tris buffer is the usual buffer). At first glance, choice of the buffer appears to be a trivial problem. It really is not since it is well known (53) that in Tris and citrate buffers, for example, the urease reaction produces ammonia and ammonium carbamate,

$$H_2O + NH_2\!-\!\overset{\displaystyle \|}{\underset{\displaystyle O}{C}}\!-\!NH_2 \rightarrow NH_4^+ + H_2NCOO^- \qquad (5.55)$$

but in phosphate and maleate media the reaction goes all the way to bicarbonate and two ammonium ions. The biochemical literature (also see chapter 2) clearly indicates that the chemical nature of the buffer species can markedly effect behavior of an enzyme (54). Thus one cannot always have complete freedom in circumventing electroactive interference problems.

In the particular case of ammonium/ammonia detectors, one has considerable flexibility. One example is the nonactin membrane, which has

Table 5.9   Types of Interferences in Analysis with Enzyme Electrodes

1. Electrochemical interferences
    a. Species which are detected by potentiometric electrode ($K_I^p \neq 0$).
    b. Species which are reduced or oxidized at same potential as the enzyme product.
2. Enzymatic interferences
    a. Enzyme inhibitors, especially irreversible inhibitors.
    b. Enzyme activators.
    c. Nonspecific and impure enzymes.
3. Chemical interferences
    a. Species which deplete coreagents.
    b. Species which consume the measured material.

better selectivity for NH_4^+ over Na^+ than the glass-type sensor and has been used for work in blood, plasma and serum (36), although its potassium selectivity is still not acceptable for accurate analysis. Gas-sensing electrodes are very much more specific for ammonia relative to other species than are the glass and nonactin sensors. Unfortunately the sensor requires a pH which is high relative to an enzyme's pH optimum.

Direct electrochemical interference with amperometric electrodes is not as serious a problem as in the case of potentiometric sensors. The Clark O_2 electrode is free of all interferences except for reducible gases; it is somewhat sensitive to proteins adsorbing on the gas-permeable membrane. Amperometric detection of peroxide and NADH is relatively free of electrochemical interferences in serum or plasma, but one must carefully choose the electrode to obtain well-defined voltammograms (38). At this point there has not been enough applications work to indicate many electrochemical problems with amperometric electrodes. Evidently any electrode which *detects product generation* must be set up so that *one corrects for the presence of endogenous product in the sample*. Similarly any electrode which detects the decrease in coreactant concentration must be employed only under conditions where the initial concentration of the coreactant is rigorously controlled. Ideally, a differential measuring system, for example, one similar to that of Hicks and Updike (2), should be used.

The enzyme itself is probably the major source of problems in terms of chemical interferences with enzyme probes. In the most fundamental sense, an analysis with an enzyme probe is a kinetic analysis unless the enzyme is present in such excess that the product generation is controlled by mass transfer or the reaction is complete. Thus an enzyme probe may be quite sensitive to the presence of inhibitors or activators in a given sample and may produce erroneous results in such samples. Guilbault has used the inhibition action of anticholinesterase materials to detect organophosphate insecticides (55) and has developed a phosphate detector based on the inhibition of the hydrolysis of glucose-6-phosphate by alkaline phosphatase (44). The phosphate detector was relatively nonselective and also responded to tungstate, arsenate, molybdate, borate and other species. Working curves, even with amperometric detectors, were extremely nonlinear. We have already mentioned the use of tripolyphosphates to activate a creatine electrode (29). Other inhibitimetric electrodes have been reported (56, 57).

Perhaps the most troublesome problem in using enzyme electrodes is that only a few of the enzymes thus far employed are absolutely specific. An extreme example is the case of the amino acid oxidase electrodes (22, 39, 48, 58, 59) and the alcohol oxidase electrode, which is sensitive to a variety of alcohols, carbonyls, and carboxylic acids (35). There are of course many very specific enzyme electrodes including the urea and AMP (21) electrodes which are insensitive to a number of closely related substrates. Even when a highly selective enzyme is used, one must *ensure that it is not contaminated with an impurity which could induce sensitivity to other species*. This is in fact an often-overlooked advantage of immobilized enzymes, that is, since the enzyme is to be reused, one can economically take the pains to check each lot of enzyme for contaminating proteins and purify it if necessary. This might not be feasible in an assay with a soluble enzyme that is discarded upon use.

A very real problem which can occur is the direct chemical interference by reaction of some material either with a coreagent or product of the enzyme reaction. Whether or not this will cause an error in the analysis will depend upon what is measured. For example ascorbic acid consumes O_2; an amperometric glucose electrode which detects the decrease in O_2 concentration will show an error. The above reaction does not produce peroxide, thus a peroxide sensor would give the correct result unless the interfering reaction consumes all of the O_2 or enough of it to cause a change in the rate of production of peroxide. In a related sense, uric acid is known to consume peroxide but does not cause a decrease in O_2 concentration in a glucose analysis with glucose oxidase (60).

Williams, Doig, and Korosi circumvented many of the above types of difficulties by using substitute coreagents in their amperometric enzyme electrodes for glucose in which benzoquinone was used instead of O_2 in a glucose-oxidase-based electrode, and ferricyanide was substituted for NAD^+ in an assay of lactic acid (61). It was essential in their work to use a high concentration of the alternate electron acceptor to overwhelm the endogeneous reactants.

5.4 NEWER CONCEPTS IN ENZYME PROBES

One of the most attractive features of enzyme electrodes is their "probe" type nature, that is, they can be taken to the sample and placed in it. This is an extremely convenient feature for industrial process control, environmental monitoring, and is essential in bio-implantation systems. It is becoming clear that electrochemical sensors are limited in two regards. In one sense most of the detectors are too specific. The enzyme per se can provide this feature. Secondly, very few of the present electrodes are useful below about 10–100 μM, particularly in real samples due to background signals and interferences. These limitations of electrochemical probes have engendered investigations of alternative types of devices as the basis for enzyme probes, most particularly the enzyme-coated thermistor as a general detector, and the enzyme-coated fiber-optic chemiluminescent probe as a high-sensitivity device.

5.41   Thermometric Probes

Several papers by Cooney and Weaver have appeared in which a thin layer of enzyme is fixed on an encapsulated thermistor probe (62, 63). A differential measurement principle in which two thermistors, one coated with enzyme and the other inactive, are placed in a well-stirred sample. The local heat production due to substrate consumption is registered as a disbalance of a differential bridge. These authors obtained linear plots of the temperature change versus glucose concentration for both immobilized hexokinase and glucose oxidase probes. The upper limit of linearity is about $10K_M$ in the case of the phosphorylation reaction and about $3.5\ K_M$ for O_2 with the glucose oxidase reaction. Thermochemical probes are inherently universal since they are based on a very general property of all chemical reactions, that is, generation of heat. Even when the catalyzed reaction per se is thermally neutral, one can obtain heat by protonation or deprotonization of a buffer or by use of ancillary reactions. For example, if peroxide is produced, the addition of catalase will cause breakdown of peroxide to water and oxygen. This is an extremely exothermic process ($\Delta H = -24$ kcal/mole).

The development of thermistor enzyme probes is limited primarily by an unfortunate set of circumstances. The rate of heat production at the surface of the probe is limited by the diffusion of matter into the probe, but this heat is dissipated into the bulk fluid by thermal diffusion which is a much faster process ($D_{matter}/D_{heat} \cong 10^{-2}$ in water at 25°C). A crude model (63) indicates that the expected temperature change will be

$$\Delta T = \frac{\delta k_2[E]\Delta H}{\mathcal{H}} \frac{[S]_0}{(1 + k_2[E]V^*\delta)/AK_M D_S} \tag{5.56}$$

where D_S is the mass diffusivity, ΔH the reaction enthalpy, $[S]_0$ the bulk substrate concentration, $[E]$ the enzyme concentration in the trapped layer, δ the thickness of the diffusion layer and $\mathcal{H}$ is the thermal conductivity of the enzyme layer. Quick calculations indicate that temperature changes on the order of about 1 mC° are anticipated when $[S]_0$ is 1 mM, provided that conversion is complete. A more detailed theoretical model of a thermistor probe has been developed by Bowers (65). It also indicates that very small temperature changes will be observed.

In order to measure such small temperature changes, a differential thermistor pair is needed and it must be placed in a low thermal noise environment. Even under essentially adiabatic conditions, the temperature resolution of thermistors is no better than a few micro degrees centigrade (66).

Work by Jespersen (67, 175), in which a thermistor was placed in a droplet of mercury metal and the enzyme immobilized by adsorption on the surface of the mercury, indicates that it may be possible to overcome the inherently unfavorable ratio of mass to thermal diffusivity by placing the thermistor in contact with a material whose thermal conductance directs the heat toward the sensor. Such methods in conjunction with more sensitive temperature sensors may make thermal enzyme probes useful analytical devices.

Most recently, Tran-Minh and Vallin (68) immobilized an enzyme directly on a thermistor and placed this assembly in a miniature cell. Unfortunately, so little experimental detail is given that it is impossible to discern the mechanism of signal development. Since in some cases, the authors state that a glass jacket must be placed around the thermistor, we suspect that the sensor is detecting net heat production in the cell and not simply that generated by diffusion at the surface, and that this is not a probe device but a reactor. Response to H_2O_2, glucose, and urease was obtained using the enzymes catalase, glucose oxidase/catalase, and urease, respectively. The device is not a practical sensor since very high concentrations (>1 mM) are needed to produce substantial responses.

5.4.2 Chemiluminescent Probes

Most recently Seitz and Freeman (69) have investigated the possibility of immobilizing an enzyme on the tip of a fiber optic device (see Figure 5.20). A light-emitting reaction such as the firefly reaction, which requires the enzyme luciferase, could be used to monitor the concentration of ATP in a test solution. Alternatively the reagents required for chemiluminescent detection of peroxide, for example, luminol in base, could be mixed with the sample solution and an enzyme which generates peroxide in base, (e.g., peroxidase) immobilized on the probe. Because the detected entities are photons, this type of probe has the advantage that the rate of achievement of steady state will not be limited by mass diffusion in the gel. Response times of only a few seconds have been obtained. Since only a small amount of light will be generated in a thin layer of gel, it is essential to correct for ambient light levels. At present this is done by placing a light shield over the sample container. It might be possible to set up a pair of probes, one active and one serving as a reference, to avoid this complication. With cooled photomultiplier tubes as the ultimate light de-

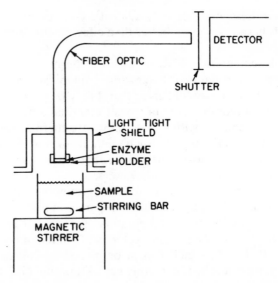

Fig. 5.20. Diagram of chemiluminescent immobilized-enzyme probe. [Reprinted from T. W. Freeman and W. R. Seitz, *Anal. Chem.* **50**, 1242 (1978). Copyright by the American Chemical Society.]

tector, the limit of detection has been estimated to be in the range 10^{-6} to 10^{-7} M (69).

5.5   APPLICATIONS OF IMMOBILIZED-ENZYME PROBES

A number of major reviews on the principles and the applications of immobilized-enzyme electrodes have appeared in the past five years (12, 37, 70–92). The intent of this section is to review some of the various types of enzyme electrodes, particularly those which represent an *important design or application.* In addition, this section contains what we believe to be a comprehensive tabular review of all original papers through December 1978 that are devoted to immobilized-enzyme electrodes. Citation of papers published since that date are found at the end of the literature references of this chapter. No attempt was made to enumerate all review papers which have appeared in symposium proceedings. We encountered several difficulties in compiling these lists. First, there seems to be some confusion as to what constitutes an enzyme electrode. Clearly, if some type of electrode is used to measure either a rate or the equilibrium position of an enzymatically catalyzed reaction which takes place in solution, this is not a true enzyme electrode (93). Second, several papers have appeared in which an electrode is used in conjunction with a flow-through column of immobilized enzyme or immobilized on a stirrer. These are not enzyme electrodes, but will be reviewed in this book in Chapter 8 (94, 95). Third, several papers have appeared whose intent was to study an enzyme electrode as a fuel cell. We believe that these electrodes could be used to detect substrates and even though no analytical data was present in the paper, they have been included in our tabulation.

A final problem in compiling this table is that many of the papers which have appeared are of quite uneven quality in terms of the description of the experimental conditions. In many cases there was no indication of how much enzyme was placed on the electrode, in others the linear dynamic range was not specified. There may have been no tests of precision or accuracy, specification of pH, ionic strength, or buffer composition. We have attempted to calculate or estimate linear dynamic ranges and other details where possible from the figures in such papers. We opted to include as much experimental information as possible for those papers which presented such data. This has the unfortunate consequence that there are a large number of blank entries in these tables. Note that all entries under pH in parenthesis indicate an optimum pH. One last word of caution to the reader: we have not critically reviewed this literature due to the large number of papers. Thus any erroneous estimates made in the original papers may not have been filtered out and may appear in this table.

Before presenting a compilation, we will review a number of the major types of electrodes. First we will discuss the urea and glucose sensitive probes. These are the archetypal, perhaps classic, cases of substrate-sensitive-electrodes. Second, we will describe a few of the enzyme-sensitive electrodes which are based on the use of an insoluble or trapped substrate as well as the use of a simple enzyme electrode which detects a second enzyme. Third, enzyme electrodes used to detect inhibitors are described. Finally, one of the most interesting new developments, that is, immobilized coenzyme and completely reagentless recycling electrodes will be reviewed.

5.5.1 Glucose Electrodes

It is only just that we consider the glucose electrode before all others since it was the first device to be suggested and constructed and has been brought to a higher state of refinement than any other electrode. As pointed out previously, the basic concept of an enzyme electrode was first proposed by Clark and Lyons (1) in 1962. A sketch of their device is shown in Figure 5.1(A). Updike and Hicks developed the first operational glucose electrode in 1967 and it was based on the amperometric measurement of O_2 depletion in an immobilized glucose oxidase gel (2). It is shown in Figure 5.1(B). A dual sensor was used so that only the O_2 consumption due to glucose was measured. The dual-electrode system cancels out background current due to species other than oxygen, compensates for changes in the initial O_2 concentration, and corrects for O_2 utilization by other reactions which are not moderated by the immobilized enzyme, for example, reaction of uric and ascorbic acid with O_2.

Notin and coworkers (96) developed a glucose electrode in which the enzyme was immobilized on a cellulose acetate membrane directly on the platinum tip of an O_2 electrode. A linear range of 0 to 200 mg/l of glucose was obtained; results with the electrode correlated (correlation coefficient = 0.990) with colorimetry and were precise to several percent. The electrode was stable for several days.

Several glucose electrodes have been formulated which measure peroxide production either by direct electrochemical oxidation or by coupling peroxide production to a second electrochemical oxidation or by coupling peroxide production to a second electrochemically active species such as I^- (97). In the Clark peroxide-sensitive glucose electrode, the H_2O_2 is measured amperometrically by oxidation at + 0.7 V on a platinum electrode. The enzyme can be trapped over the anode or on a polycarbonate membrane. A commercial instrument available from Yellow Spring Instrument Co. (Yellow Springs, Ohio) is linear up to 40 mg of glucose/100 ml. It is operated at a temperature of 37°C and is insensitive to normal serum levels of ascorbic acid, uric acid, formic acid, glutathione, and cysteine (98–100). It requires only 25 µl of serum for operation and maintains a precision of about 1%.

The glucose sensor of Guilbault and Lubrano (33, 43) is used to measure H_2O_2 at a platinum anode at a potential of + 0.6 V versus the saturated calomel electrode (SCE). As mentioned before, the electrode must be pretreated to obtain a low limit of detection. Calibration curves were linear down to about 10^{-5} M in peroxide and could be used to quantitate glucose by measurement of either the steady-state current or the rate of change of current. Rate methods only require about 12 sec for the actual measurement, not including washout time. The major feature of these electrodes is that they are stable for up to 10 months and exhibit a drift of only 0.1% per day. The technique was compared to the o-toluidine colorimetric method on a series of quality control sera. It had a correlation coefficient of 0.998 and a coefficient of variation at 1.7% at a glucose concentration of 200 mg/100 ml. In a more recent paper (58), Nanjo and Guilbault showed that detection of O_2 consumption on a bare platinum cathode was considerable more sensitive than measurement of H_2O_2 production, that is, about 5 μM glucose could be detected. Either a rate or equilibrium assay can be employed.

Several papers have appeared in which the detection of glucose is based on measurement of iodide via the coupled reaction:

$$2H^+ + H_2O_2 + 2I^- \rightarrow 2H_2O + I_2 \tag{5.57}$$

the change in iodide concentration can be detected with the highly selective iodide ion-selective electrode (97), or the amount of iodine produced may be monitored amperometrically (13, 30, 101, 102). The above reaction requires a catalyst such as peroxidase or simply Mo(VI) to be useful. Potentiometric detectors for glucose are linear over only one decade in concentration and are difficult to optimize since the lower limit of detection is promoted by a low iodide concentration and the width of the dynamic range is greatest at a high iodide level. Nonetheless, one can strike a compromise which is based on the expected range of glucose concentrations (97). This electrode has several types of interferences. First thiocyanate, sulfide, cyanide, and silver (I) are sensed by the iodide electrode. Second, some species interfere with the glucose–glucose oxidase, hydrogen peroxide–peroxidase reactions, for example, uric acid, ascorbic acid, and iron(II) all competed with iodide for peroxide. Third, several sugars also interfered including: cellubiose, maltose, and 2-deoxyglucose.

Amperometric detection of iodine generated via reaction (5.57) has a considerably wider dynamic range. The 0.5 mm thick electrodes of Mell and Maloy required about 200 sec to achieve steady state, and these authors estimated that the diffusion coefficient of glucose in a polyacrylamide gel was about 2.4×10^{-6} cm^2/sec (13). Furthermore, they estimated that only 1% of the total enzyme was active after immobilization and very low currents were obtained due to blockage of the electrode surface by the gel. In an attempt to improve detection limits, a new approach termed "amperometric enhancement," in which the electrode potential is cycled on and off, improved the detection limits so that less than 0.1 mM glucose could be detected (33).

The work of Shu and Wilson (101, 102) should be mentioned because they used a new electrode material for enzyme immobilization, that is, carbon paste in a rotating-disc configuration. The enzyme was immobilized by addition of octylamine to the paste. Glucose oxidase was bound to the paste via glutaraldehyde. They found that steady state could be obtained in less than 50 sec. Very little in the way of analytical implications can be drawn from this work because it was mainly concerned with the theoretical analysis of the catalytic ring–disc electrode. It is nonetheless an important paper because it laid a firm foundation for future developments.

Williams, Doig, and Korosi (61) developed a unique electrode for glucose based on replacement of oxygen with benzoquinone as the oxidant. The hydroquinone produced was measured amperometrically by oxidation on platinum:

glucose + benzoquinone

$$+ \text{ H}_2\text{O} \xrightarrow{\text{glucose oxidase}} \text{gluconic acid + hydroquinone} \quad (5.58)$$

Linear calibration curves were obtained from 0.5 mM to about 5 mM, but it was necessary to use a dual electrode to cancel out a high blank due to endogeneous compounds in serum.

A recent entry in the glucose-electrode area is the multienzyme system of Coulet and Bertrand, which can detect glucose via glucose oxidase and a peroxide electrode, as well as maltose by the presence of glucoamylase (178).

5.5.2 Urea Electrodes

The urea electrode developed by Montalvo and Guilbault (10, 103, 104) was the first potentiometric enzyme electrode [see Figure 5(A)] and has gone through several generations of development. Their first electrode was based on the combination of urease trapped in polyacrylamide and held in intimate contact with a univalent cation electrode which was sensitive to ammonium ions but not at all selective toward them in comparison to many other common cations, including H^+, K^+, and Na^+. This series of papers clearly demonstrated that enzyme electrodes could be developed into a sensitive, fast, convenient, and selective analytical device. For this reason among others, Professor Guilbault was awarded the American Chemical Society's Fisher Award in Analytical Chemistry in 1977.

The major difficulty of the potentiometric urea electrode has been poor selectivity of the base sensor. To improve this situation and minimize the errors involved in handling real samples, the following attempted improvements have been implemented:

1. To determine urea in blood and urine, 2 g of cation-exchange resin is added to the sample, which is diluted by adding a 1 ml aliquot of 50 ml of 0.5 M Tris buffer at pH 8. The potential of a glass-based urea electrode is then measured after 1–2 min against the potential of an identical univalent glass electrode which was not coated with gel or enzyme (105). Guilbault and Hrabankova found that addition of the resin can cause an

effect on the electrode potential, but calibration curves versus urea down to 10^{-5} M could be obtained without interference in the presence of as much as 5 mM and 0.1 mM sodium and potassium, respectively. Unfortunately the potassium content of serum is too high and variable to permit direct routine measurement in this fashion, although results comparable to the Fearon colorimetric technique were achieved within a few percent.

2.   The electrode selectivity problem has been attacked directly by using either an NH_3 sensor (26) or a CO_2 sensor (106) with the enzyme directly coupled to or in close contact with the gas-permeable membrane. Anfalt, Graneli, and Jagner obtained rapid response (1 min) at 1 mM urea when urease was attached to an Orion ammonia membrane via glutaraldehyde linkage. Optimum response times were obtained at pH 9 in 0.1 M Tris buffer. Super-Nernstian slopes were obtained at pH 7 and the response was slower. The slope of the calibration curve varied only from 57.5 mV/decade to 56 mV/decade over 20 days but the entire calibration curve shifted 13 mV during this time. A precision of 2–6% was claimed and due to the nature of the electrode, no interference from ionic species should be observed other than changes in liquid-junction potentials. In a related study, Guilbault and Shu trapped an enzyme solution on a nylon net with dialysis paper (106). Although very fast response was obtained at high concentration, for example, less than 1 min at 0.1 M, in excess of 4 min were needed at 1 mM urea. Linear calibration curves with Nernstian slope were obtained at pH 6.2 in 0.1 M phosphate buffer from 10^{-2} to 10^{-4} M urea. The only interferant was acetic acid. Although the calibration curve was stable for 3 days, the response time increased significantly.

3.   The electrode selectivity could also be improved by use of a nonactin membrane as the base sensor (36), for example, the NH_4^+/K^+ and NH_4^+/Na^+ selectivity ratios are 6.5 and 750, respectively. Such electrodes respond to changes in ion activity in as little as 10 sec and are rather stable (2 mV/month drift). Calibration curves versus urea were linear from 10^{-2} to 10^{-4} M in 0.1 M Tris buffer at pH 7. The electrodes were not insensitive to normal serum levels of potassium ions but calibration curves could be obtained down to 10^{-4} M urea in the presence of 0.7 mM K^+. The electrode came to 98% of steady state in 180 sec at low urea concentration. In order to use the electrode in serum, an elaborate standardization procedure was essential (107); the electrode was calibrated in a solution containing a fixed level of K^+ and the sample was diluted with buffer until it read at the same potential as the standard.

4.   It is clear that although the NH_3 gas-sensor electrode is by far the most selective electrochemical device for use with an urea electrode, it suffers from very slow response and washout characteristics. Mascini and

Guilbault attempted to solve this problem by casting a glutaraldehyde crosslinked albumin-urease film directly on the surface of the Teflon membrane of an NH_3 sensor. They were able to measure urea in blood. Analysis time was under 5 min in all cases (108).

5. The most recently reported enzyme electrode for urea is an ultra micro device, which has the potential for use for intracellular studies, developed by the Rechnitz group (109) (see below).

Other methods have been developed for urea assays with electrodes, but these fall into the category of enzyme reactor systems and include the use of the air-gap electrode (41) and the immobilized-enzyme stirrer (95).

5.5.3 Electrodes for Measurement of Enzymes

Several groups have investigated the possibility of using the concepts of an immobilized enzyme to determine the activity of enzymes in solution. Thus far three approaches to this idea have appeared:

1. Montalvo has designed an ultra-micro flow cell in which urea is passed between an ammonium ion electrode surface and a dialysis membrane. When urease activity is present in the bulk solution surrounding the electrode, it hydrolyzes some of the urea which diffuses out into solution. The ammonium ions generated can diffuse back into the dialysis paper compartment and be sensed by the ammonium-ion electrode. Collection of ammonium ions is obviously very inefficient and the method is not very sensitive (110, 111). A related method for measurement of cholinesterase activity via a pH-sensitive glass electrode has been developed and is based on the use of a high molecular weight polyethyleneimine buffer around the electrode. The sample is flowed through an adjacent membrane compartment. A precision and accuracy of 2–4% can be obtained in 1.5 min of assay time (112).

2. Guilbault and Iwase (55) reported on an immobilized-substrate electrode in which acetylcholine was adsorbed on an ion-exchange resin and held with a thin nylon mesh around a flat glass electrode [see Figure 5.2(D)]. The rate of change of pH was a linear function of cholinesterase activity from about 0.1 to 1.0 units/100 ml of sample. The basic problem with this device is that a wide dynamic range requires a high buffer capacity but this obviously diminishes the net pH change.

3. Booker and Haslam (46) have described an electrode which measures the activity of arginase. Since this enzyme liberates urea from L-arginine, one can employ a urease–urea electrode based on a cation-se-

lective electrode to measure the rate of generation of urea:

$$\text{L-arginine} + H_2O \xrightarrow{\text{arginase}} \text{L-ornithine} + \text{urea} \qquad (5.59)$$

Their urea electrode was quite linear from about 5×10^{-3} to 5×10^{-5} M urea when used in 0.1 M Tris buffer at pH 7, and had a slope of 50 mV/decade. Three methods related to the initial slope, fixed, and variable time techniques of kinetic analysis were tested. The rate of change of potential was linear with respect to arginase activity from about 1 to 80 units. The characteristics of the analysis system such as pH and temperature effects were dependent upon both enzymes. Arginase has an optimum pH in the range 9–9.5 for the metal-ion [Co(II), Ni(II), Mn(II)] activated form, but at these high pH values, most of the ultimate product of the double enzyme reaction is ammonia. The system was used to measure the K_M of arginine (1.2×10^{-3} M) and was found to be in good agreement with the literature value. Booker and Haslam concluded that spectrophotometric measurement of the enzyme activity was more sensitive, but the potentiometric method was more convenient, faster, cheaper, and did not require sample blanks.

5.5.4 Electrodes for Inhibitors

Several immobilized-enzyme systems have been developed for the analysis of enzyme inhibitors. Perhaps the best known of these is the anti-cholinesterase monitor for organophosphate pesticides and nerve gases (56), but these will be treated under reactor systems since they are generally used in a flow-through reactor configuration.

Guilbault and Nanjo (44) developed an electrode sensitive to inorganic phosphate based on the following reaction sequence:

$$\text{glucose-6-phosphate} + H_2O \xrightarrow{\text{alkaline phosphatase}} \text{glucose} + HPO_4^{2-} \qquad (5.60)$$

$$\text{glucose} + O_2 \xrightarrow{\text{glucose oxidase}} \text{gluconolactone} + H_2O_2 \qquad (5.61)$$

Inorganic phosphate is an inhibitor of reaction (5.60); thus it will limit the rate of glucose production and therefore the current due to O_2 reduction. An actual electrode was made by co-immobilization of both enzymes and glutaraldehyde to form a spongy polymer. The insolubilized proteins were then trapped over a platinum electrode and used at 30°C in 0.1 M buffers (pH 7.4–8.4). 10 mM magnesium is needed to activate the alkaline phosphatase. The effect of glucose, glucose-6-phosphate, and magnesium concentrations as well as pH were studied. The electrode was inhibited by high concentrations of zinc(II) (1 mM) and calcium (40 mM). Analytically, the device can be used either by adding phosphate and glucose-6-phosphate simultaneously or by addition of phosphate subsequent to the attainment of steady state after glucose-6-phosphate is added. In the former case, the rate of decrease in O_2 reduction current is followed; in the latter the rate of increase toward a higher steady state is measured. Regardless of what variable is measured, the electrode response functions are quite nonlinear but very sensitive to the phosphate level in the sample. The other major drawback of the electrode is that it is sensitive to many oxyanions.

Guilbault and Cserfalvi (113) studied several systems based on production of NADH or NADPH as potential enzyme electrodes for phosphate. Due to the complex nature of the processes, this does not seem to be a fruitful approach.

One of the major reasons for the intense interest in immobilized enzymes is of course the economic advantage involved relative to soluble enzymes. In many instances, the cost of analysis is not limited by that of the enzyme, particularly if it is already immobilized, but by the cost of coenzymes. A detailed cost analysis is presented elsewhere in this book (see Table 8.5) and should be consulted. It would be a major advantage if coenzymes such as NADH, NADPH, FADH, and ATP could be immobilized and recycled. Furthermore, immobilizations might serve to stabilize these often labile materials. Finally if both the enzyme and coenzyme could be immobilized, then a major step toward truly reagentless analysis will have been made.

5.5.5 Immobilized Coenzyme Electrodes

The work of Davies and Mosbach (114) was the first step in the direction of an immobilized-coenzyme system. Methods for both glutamate and pyruvate based on the generation of ammonium ions and their detection by an ammonium-ion-sensitive electrode were developed. Dextran-bound NAD$^+$ was made by the linkage of NAD-N^6-(N-(6-aminohexyl)acetamide) through the terminal amino group to cyanogen-bromide-activated dextran. The coenzyme concentration was approximately 30 μmoles per gm of dry gel. The essential enzymes glutamate dehydrogenase and lactate dehydrogenase were held along with the immobilized coenzyme on top of the electrode by a dialysis membrane.

The reaction cycle shown below can be carried out in either direction depending upon the initial ratio of glutamate to pyruvate in the sample:

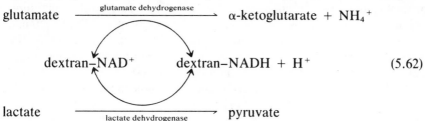

$$\text{glutamate} \xrightarrow{\text{glutamate dehydrogenase}} \alpha\text{-ketoglutarate} + NH_4^+$$

$$\text{dextran--NAD}^+ \qquad \text{dextran--NADH} + H^+ \qquad (5.62)$$

$$\text{lactate} \xrightarrow{\text{lactate dehydrogenase}} \text{pyruvate}$$

All measurements were carried out in Tris buffer (pH 8) at 25°C and the electrodes were immersed immediately before use in 50 mM Tris (pH 8) containing 10 μM EDTA and 100 μM ADP. For the determination of glutamate, the sample was made 2 mM in pyruvate and for the measurement of pyruvate the sample was adjusted to 10 mM glutamate. Linear potentiometric calibration curves were obtained for glutamate from 10^{-3} to 10^{-4} M (slope = 15–20 mV/decade). The response time was 3–4 min. The electrode system is not at all stable, and upon storage at 4°C the response falls off by 50% in about 4 days. Pyruvate concentrations in the range of 10^{-5}–10^{-4} M could be measured, but the calibration curve was nonlinear. Although no work was carried out, it was postulated that the electrode could be used to measure both lactate and α-ketoglutarate under appropriate conditions.

A very novel approach to a reagentless system in which both an enzyme and its coenzyme were both insolubilized by covalently bonding has been described by Blaedel and Jenkins (38). Although their approach to lactate determination is not presently workable due to the nonlinearity of the calibration curve and the short half-life of the immobilized coenzyme, it did lay the ground work for future applications in this area. A diagram of their flow through electrode is given in Figure 5.1(D). Basically it consists of a combination of lactic acid dehydrogenase and NAD$^+$ attached to a membrane via glutaraldehyde linkages or a membrane holding an Agarose-bound NAD$^+$ gel in a solution of LDH. The membrane traps the component in close contact with a glassy carbon anode which is poised at a potential of +0.75 V versus a silver/silver chloride reference electrode. The glassy carbon electrode measures the amount of NAD$^+$ formed when lactic acid is oxidized by NAD$^+$ to pyruvic acid. The reduced coenzyme (NADH) is inherently recycled electrochemically by reoxidation to NAD$^+$. Coenzyme concentrations of 4 mM in the Agarose were achieved by binding probably via the N^6 position on the adenine ring.

The electrode was sensitive to lactate concentrations from about 80 μM, at which concentration the signal was equal to the background noise level, up to about 4 mM, where the response became independent of further increases in substrate concentration. Due to diffusional limitations, the current with immobilized NAD^+ was much smaller ($\cong \frac{1}{10}$) than when the coenzyme was provided in the bulk sample. In general, the response decreased by 15–20% over the first several hours of operation. This was attributed to a loss of immobilized NAD^+ rather than the enzyme. It should be noted that NADH is rather stable in the pH 8.0, 0.05 M phosphate buffers used in the work. Detailed stability and pH studies of the glutaraldehyde membrane-bound LDH/NAD, and trapped LDH, Agarose-bound NAD in both the reagentless and reagent-provided mode were carried out.

5.6   RECENT DEVELOPMENTS IN BIOCHEMICAL SENSORS

Tables 5.10 and 5.11 contain several entries as late as mid-1979. In order to do some justice to several new areas of biochemical sensors which are closely related to the immobilized-enzyme probe, we felt it necessary to add this brief section. Three of the most significant recent developments are: the introduction of the immobilized-enzyme microprobe, the use of whole cells and tissue sections as the biochemically active component of an electrode, and the development of immunologically based devices. Rechnitz and his coworkers, particularly Kobos and Meyerhoff, must be considered to be in the forefront of the activity in all of these recent developments.

The earliest ultra microelectrode was reported by Silver who immobilized glucose oxidase on the tip of a sub-micron Pt wire (164). His report was a very unspecific brief communication. As part of their work on the development of a micro air-gap sensor for cellular ammonia, the Rechnitz team attached urease to a 10 μm NH_3 sensor (109). They found that this device responded to urea in the range 10^{-5}–10^{-2} M, although the Nernstian range was very limited and response time was long. The miniature sensor had many of the characteristics of commercial macro NH_3 sensors.

One of the most fascinating analytical concepts in the area of enzyme technology is the use of whole cells and tissue sections for analytical purposes. The use of live bacteria is not at all unusual in the fermentation industry, but the use of bacterial electrodes was only suggested by Divies as late as 1975 (129). Rechnitz has developed four bacterial electrodes [arginine (130), aspartic acid (131), cysteine (132), and glutamic acid (133)] and mention has been made of an electrode based on liver tissue sections (134). Kobos has developed a nitrate sensor based on the combination of *Azobacter vinelandii* and an ammonia sensor (176).

Table 5.10 Survey of Immobilized Enzyme Potentiometric Electrode Studies

Item number	Substrate	Enzyme(s) used	Source or type	Amount of enzyme used	Buffer composition	pH	Thickness of enzyme layer
1	Urea	Urease	—	175 mg/100 ml	N.A.	N.A.	60–350 μm
2	Urea	Urease	—	5400 units/g	0.5 M, Tris	8.0	stocking-thickness nylon
3	Urea	Urease	Cal-Bio	5400 units/g, 175 mg/ml, 200 μl	0.1 M, phosphate	6.2	N.A.
4	Urea	Urease	Sigma	175 mg/ml of 21 units/mg	0.1 M, Tris	7.0	N.A.
5	Urea	Urease	Sigma	0.7–10 mg/ml 21 units/mg	0.1 M, Tris	7.0	varied in study
6	Urea	Urease		~10 units total	0.5 M, Tris	8.5	N.A.
7	Urea	Urease	Sigma	3700 units/g, 150 mg/ml, 10 μl on electrode	0.1 M, Tris	7.0	~80 μm
8	Urea	Urease	Sigma	4980 units/g, 52 mg/ml, 0.1 ml or 0.2 ml	0.1 M, Tris	7–9	"very thin"
9	Urea	Urease	Worthington	300 units/ml	0.1 M, Tris	7.0	thin film, ~30 μm
10	Urea	Urease	N.A.	3500 units/g, 100 mg/ml	0.5, 1 and 10 mM Tris	7.0	200 units, 200 μl
11	Urea	Urease	N.A.	—	Tris	7.0	400 μm
12	Urea	Urease	Worthington	60 units/mg, 5–150 mg/ml	0.1 M, Tris	7.0	N.A.
13	Urea	Urease	Worthington	73 units/mg, 2–5 mg/ml	0.05 M, Tris	8.0	75–300 units
14	Urea	Urease	Sigma Type VII	27 units/electrode	0.2 M, Tris	8.5	50–100 μm
15	Urea	Urease	—	—	—	—	50–350 μm

265

Table 5.10 (*Continued*)

Item number	Substrate	Enzyme(s) used	Source or type	Amount of enzyme used	Buffer composition	pH	Thickness of enzyme layer
16	Urea	Urease	Merck	1000 units/ml of enzyme on the electrode	0.1 M, Tris	7.0	—
17	L-Amino acids	L-Amino acid oxidase	Snake venom, Sigma	N.A.	N.A.	7.2	350 μm
18	L-Amino acids	L-Amino acid oxidase	Snake venom, Sigma	20–200 mg/dl	N.G.	7.2	N.A.
19	L-Amino acids	L-Amino acid oxidase	Sigma, Snake venom	0.46 units/mg	0.1 M, Tris	6.5–8 (7.5)	N.A.
20	L-Amino acids	L-Amino acid oxidase	Sigma, Snake venom	100 mg/dl soaked on nylon net	0.1 M, Tris	6.5–8 (7.5)	N.A.
21	L-Amino acids	L-Amino acid oxidase	N.G.	N.A.	0.1 M, Tris	7.0	~30 μm
22	L-Tyrosine	Tyrosine decarboxylase	Sigma	90 units/g, 50 mg/ ml, 200 μl	0.1 M, Citrate	5.5	200 μl of solution
23	L-Phenylalanine	L-Amino acid oxidase	Sigma, Snake venom	0.46 units/mg, 10 mg enzyme/150 mg polymer	N.G.	5.0	N.G.
24	L-Phenylalanine	L-Amino acid oxidase + peroxidase	Horseradish peroxidase + L-aminoacid-oxidase	—	0.1 M, acetate	(5.0)	N.G.
25	L-Asparagine	L-Asparaginase	Galard-Schlesinger	580 units/mg. About 80 units/electrode, only 3.1 units are active	0.1 M, Tris	7–9 (8.0)	50 μm
26	Asparagine	Asparaginase	Sigma from *E. coli*	250 units/ml of gel	0.1 M, Tris	7.5–8.5 (8.2)	~0.2 ml

266

#	Substrate	Enzyme	Source	Amount	Buffer	pH	Thickness
27	Phenylalanine, tyrosine, lysine	Amino acid decarboxylase	Sigma	0.2 units + 3.5 mg, Albumin/electrode	0.02 M, phosphate	5.8	"very thin"
28	Glutamine	Glutaminase	Nutritional Biochemical Co.	150 mg/ml	0.3 M, Tris, acetate, or phosphate	5–7	N.G.
29	Arginine	Intact *Streptococcus faecium* cells	—	N.A.	N.A.	7.4	N.G.
30	Lysine	Lysine decarboxylase	*E. Coli* B and *B. cadaveris*	3–5 mg of purified enzyme per electrode	0.5 M, acetate	5.8	N.G.
31	Glutamic, Glutamate or Pyruvic acids	Glutamate dehydrogenase; lactate dehydrogenase	N.A.	360 units and 410 units respectively	0.05 M, Tris	8.0	N.G.
32	Glutamic acid	Glutamate decarboxylase	—	Varied	0.1 M, phosphate	5.0	—
33	D-Amino acids	D-Amino acids oxidase	Sigma	0.02 units/mg and 25 units/mg. Used 50–1000 units/ml of gel	0.1 M, Tris	7.5–8.5 (8.2)	~200 µl
34	D-Amino acids	D-Amino acid oxidase	Sigma	N.G.	0.1 M, Tris	7.5–8.5 (8.2)	40 µl
35	Amygdalin	β-Glucosidase	Sigma/almonds	50 mg/gm gel	Beckman 22326 Buffer	12.7	N.G.
36	Amygdalin	β-Glucosidase	Sigma	3.7 units/mg; 100 mg per ml of polymer; 10 mg of enzyme/membrane	Borate, 0.1 M	10.4	300 µm

Table 5.10 (*Continued*)

Item number	Substrate	Enzyme(s) used	Source or type	Amount of enzyme used	Buffer composition	pH	Thickness of enzyme layer
37	Amygdalin	β-Glucosidase	Sigma	1 mg of enzyme paste	phosphate borate	7.0 10.0	N.G.
38	Glucose	Glucose oxidase and peroxidase	Sigma—from *Aspergilus niger*	10 mg oxidase/150 mg of polymer; 18 units/mg + 100 units/mg peroxidase	1.0 M, phosphate 0.1 M, phosphate	5.0 6.0	N.G.
39	Glucose	Glucose oxidase		15,000 units/g, 200 mg/ml	1 mM, phosphate	6.9	~200 μm
40	Penicillins	Penicillinase	Riker, *B. cereus*, β-lactamase	6300 units/mg, 125 mg/ml	must be very dilute	6–7 (6.4)	8–10 drops
41	Penicillins	Penicillinase	Riker, *B. cereus*, β-lactamase	40 mg of 5000 units/mg in 1 ml adsorbed on electrode	must be very dilute	(6.9)	1.3 mm disc
42	Penicillins	Penicillinase	N.G.	40,000 units/mg, 50 mg/ml	5 mM, phosphate	6.8	~200μm
43	Penicillins	β-Lactamase			dilute phosphate	—	varied
44	Penicillins	β-Lactamase	Calbiochem	1.0 units/electrode	0.005–0.1 M, phosphate	7.0	—
45	Creatinine	Creatininase	*Clostridium*	15 μl of 2.9 units/ml were trapped	0.15 M, tripolyphosphate	8.5	15 μl of solution on electrode
46	Uric acid	Uric acid oxidase (Uricase)	*Candida utilis*	3.5 units/mg, 3 mg on 20 mg of resin	0.1 M, phosphate	6.5	N.G. total ~20 mg of solids
47	5'-Adenosine monophosphate	5'-AMP Deaminase	Sigma-Type IV	90 units/ml, 10 μl on electrode	0.05 M, Tris	7.5	10 μl of solution on electrode

Item no.	Substance	Enzyme (E.C.)	Source	Immobilization	Buffer	pH	Linearity
48	AMP	AMP Deaminase E.C.3.5.4.5	Sigma	—	0.05 M, Tris	7.5	—
49	Acetylcholine	Cholinesterase	Sigma, electric eel	235 units/electrode	0.01 M, phosphate	7.0	—
50	Acetylcholine	Cholinesterase	Sigma, electric eel	—	0.01–0.1 M, phosphate	6.0–8.0	50 μm
51	Lactic acid	Lactate dehydrogenase + catalase	Sigma	—	0.1 M, phosphate + 0.1 mM, EDTA + ferricyanide	6.0	10 μl on electrode
52	Nitrite	Nitrite reductase	Spinach leaves	160 mg and 100 mg BSA, 4 units/sample	~0.3 M, Tris/ HCO_3^-	7.1–7.8	N.G.
53	Fluoride–urea	Urease	Sigma-VI	0.05–1 unit	0.025 M, phosphate	7.0	60–120 μm
54	Cholinesterase	Acetylcholine-chloride	—	—	2 mM, Tris	7.9–8.3	~0.2 g
55	Arginase	Urease	—	200 and 400 mg/ ml of gel, 21 u/ mg	0.1 M, Tris	7–9.5 (7.5)	N.G.

Item no.	Temperature (°C)	Protection method	Immobilization method	Sensor	Response time	Linearity
1	N.A.		Polyacrylamide occlusion	Beckman Univalent Cation Electrode	25–60 sec for 98%	0.2–80 mg/dl Nernst
2	25	Nylon net support and outer dialysis paper			60–90 sec	10^{-4}–10^{-2} M Nernst
3	25	Nylon net and dialysis paper	Not immobilized, 200 μl of enzyme slurry soaked on nylon net	Instrumentation Laboratory CO_2 electrode	1–6 min	10^{-4}–10^{-2} M Nernst

Table 5.10 (*Continued*)

Item no.	Temperature (°C)	Protection method	Immobilization method	Sensor	Response time	Linearity
4	25	N.A.	Acrylic acid polymerized and covalent binding to protein via diazonium salt	Nonactin in silicone rubber	60–180 sec depends on thickness of enzyme layer	3×10^{-3}–$3 \times 10^{-5}M$ Nernst
5	25	Dialysis paper	Commercial Enzygel	Nonactin in silicone rubber	60–180 sec	10^{-2}–$10^{-4}M$ Nernst
6		100 μm net over enzyme layer	Trapped paste layer	Air-gap NH_3 electrode	1 min at >10 mM; 4 min at < 1 mM	2×10^{-2}–$10^{-4}M$ not Nernst
7	25	Dialysis membrane	Polymerized by gluteraldehyde on sensor membrane	NH_3 gas sensor	~5 min at ~1 mM	5×10^{-4}–$7 \times 10^{-2}M$ 90.7 mV/decade
8	25	None used	Albumin–enzyme glutaraldehyde crosslinked membrane	NH_3 gas sensor	1.5–2 min at 10 mM, pH 9	5×10^{-4}–$5 \times 10^{-2}M$
9	25	None	Trapped liquid layer	Beckman Univalent Electrode	36 sec	10^{-4}–10^{-2} M
10	25	Dialysis membrane	Polyacrylamide occlusion	pH–glass	N.A.	5×10^{-5}–5×10^{-3} M
11	25	Dialysis membrane	Polyvinylalcohol occlusion	pH–glass	~10 min	N.A.
12	25	None	Albumin–urease copolymer membrane crosslinked glutaraldehyde	Beckman Univalent Electrode	2 min	10^{-4}–10^{-2} M, 55 mV/decade
13	25	None	Glutaraldehyde crosslink to BSA on Teflon membrane	Beckman Univalent Electrode	2–5 min	10^{-5}–10^{-2} M depends upon type of anion-exchange resin used
14	Room	None	—	NH_3 gas	1–5 min	10^{-4}–10^{-2} M
15	—	—		NH_4^+ glass	—	—

No.	Temp (°C)	Location	Immobilization method	Electrode	Response time	Concentration range
16	25	Dialysis membrane	Polyacrylamide and solution trapped by membrane	Nonactic silicone rubber NH_4^+	60–80 sec	10^{-2}–10^{-4} M ~45 mV/decade
17	25	Outer membrane	Polyacrylamide gel	Beckman Cation Univalent 39137	60–120 sec 98%	2×10^{-3}–2×10^{-4} M
18	25	Outer membrane	Trapped liquid	Beckman Cation Univalent, 39137	30–80 sec 98%	2×10^{-3}–2×10^{-4} M
19	25	Outer membrane	Polyacrylamide gel	Beckman Cation Univalent, 39137	1–2 min	N.A.
20	25	Outer membrane	Trapped liquid	Beckman Cation Univalent, 39137	1–2 min	10^{-3}–10^{-4} M
21	25	None	Albumin–enzyme glutaraldehyde crosslinked membrane	Beckman Univalent Electrode	1 min 99%	10^{-4}–10^{-2} M
22	25	Not immobilized, 200 units of enzyme on net and dialysis paper outer membrane	Not immobilized	Instrumentation laboratory CO_2 electrode	1–5 min	10^{-4}–5×10^{-3} M Nernst
23	25	Outer membrane	Polyacrylic acid covalent binding via diazonium salt	Nonactin silicone rubber	60–80 sec depends on thickness and enzyme concentration	2×10^{-3}–2×10^{-4} M Almost Nernst
24	25	Outer membrane	Polyacrylic acid covalent binding via diazonium salt	Ag_2S–AgI iodide electrode	30 sec for slope 60–180 for steady state	10^{-4}–10^{-3} M from slope $\Delta E/\Delta t$ versus conc.
25	30	None	Glutaraldehyde binding to nylon net	NH_3 gas sensor	~1.3 min	8×10^{-3}–8×10^{-5} M at pH 8.5
26	25	Outer membrane	Polyacrylamide occlusion	Beckman Univalent Cation Electrode	1–2 min	10^{-3}–10^{-4} M Non-Nernst

Table 5.10 (*Continued*)

Item no.	Temperature (°C)	Protection method	Immobilization method	Sensor	Response time	Linearity
27	20	None	Albumin–glutaraldehyde–enzyme cross linked membrane	CO_2 gas sensor	~8 min	2.5×10^{-3}–1.5×10^{-2} M for phenylalanine
28	25	Outer membrane	Liquid trapping	Beckman 39137 or Thomas 4923-Q10	>2 min	10^{-1}–10^{-3} M Non-Nernst
29	N.G.	Dialysis	Liquid trapping	NH_3 gas sensor	N.G.	10^{-5}–10^{-2} M
30	Room	None	Cast glutaraldehyde crosslinked BSA–enzyme membrane on a gas-permeable membrane	CO_2 gas	5–10 min	10^{-4}–0.03 M
31	25	Dialysis	Coenzyme covalently bound; enzyme trapped in dialysis paper	Beckman Univalent Electrode	3–4 min	10^{-4}–10^{-3} M glutamate, 20 mV/decade, 10^{-4}–10^{-3} M pyruvate, 15 mV/decade
32	37	Dialysis membrane	Mainly glutaraldehyde polymerization and outer-membrane entrapment	CO_2 gas-type electrode	5–30 min	0.4–10 mg%
33	25	Outer membrane	Polyacrylamide occlusion	Beckman Univalent Cation Electrode	1–2 min	10^{-3}–10^{-4} M
34	25	Outer membrane	Liquid trapping	Beckman Univalent Cation Electrode	1–2 min	N.G.
35	25	Dialysis membrane	Polyacrylamide occlusion	CN^- –solid-state electrode	<1 min	5×10^{-3}–5×10^{-7} M Non-Nernst 20 mV/decade
36	25	None	Polyacrylamide occlusion	CN^- –solid-state electrode	10–30 min	10^{-3}–10^{-5} M, 48 mV/decade

No.	Ref.	Membrane	Immobilization	Electrode type	Response time	Range
37	N.G.	Dialysis membrane 20 µl thick	Trapped liquid layer	Heterogeneous AgL type	5 min, depends upon amount of enzyme	10^{-4}–10^{-1} M, 47 mV/decade at 10 pH 53 mV/decade at 7 pH
38	26	Dialysis membrane	Liquid, polyacrylamide occluded, covalent bound	Ag$_2$S–AgI Iodide solid-state electrode	2–8 min depends on thickness, concentration and temperature	10^{-3}–3×10^{-4} M
39	25	Dialysis membrane	Trapped liquid layer	pH–glass	N.G.	10^{-3}–10^{-1} M Non-Nernst
40	25	Dialysis membrane	Polyacrylamide occlusion	pH–glass	15–30 sec	10^{-1}–10^{-2} M 38–52 mV/decade
41	N.G.	None	Adsorption on glass frit	pH–glass	~1 min	3.5–1100 µg/ml, 10^{-5}–3×10^{-3} M Nernst
42	25	Dialysis membrane	Trapped liquid layer	pH–glass	>2 min	10^{-3}–10^{-2} M
43	—	Dialysis membrane	Attached to Sepharose and trapped by dialysis membrane	pH–glass	<2 min	up to 25 mM
44	—	None	Soaked into dialysis membrane/trapped liquid layer	pH–glass Glass H$^+$ electrode	2 min	1–10 mM
45	27	Dialysis membrane	Trapped liquid layer	NH$_3$ gas sensor	6–10 min < 5 mM; 2–5 min > 5 mM	7×10^{-5}–10^{-2} M, 47 mV/decade
46	37	Dialysis membrane	Enzyme adsorbed on hydroxyethylcellulose. Slurry placed on electrode	CO$_2$ gas sensor	5–15 min	2×10^{-4}–3×10^{-3} M 57 mV/decade
47	27	Dialysis membrane	Trapped liquid layer	NH$_3$ gas sensor	6 min at 10^{-3} M; 2 min at 10^{-2} M	8×10^{-5}–1.5×10^{-2} M 46 mV/decade
48	28	Cellophane membrane	Trapped liquid layer	NH$_3$ gas sensor	—	—

Table 5.10 (*Continued*)

Item no.	Temperature (°C)	Protection method	Immobilization method	Sensor	Response time	Linearity
49	25	—	Glutaraldehyde crosslinked enzyme to BSA to form a membrane	pH glass	<2 min	2×10^{-4}–10^{-2} M
50	25	None	Glutaraldehyde crosslinked gelatin–enzyme coated on the electrode	pH glass	—	0.01–1 mM
51	25	Dialysis membrane	Enzyme occluded in gelatin	Redox membrane electrode	20–30 min	Very poor
52	30	None	Copolymer of albumin and enzyme via glutaraldehyde crosslinking	Air gap for NH_3	~1 min at 1 mM	5×10^{-2}–10^{-4} M Nernstian
53	25	None	Urease–human serum albumin crosslinked with glutaraldehyde on silicone membrane	CO_2 sensor	5–6 min	0.3–10 mM
54	23	Nylon cloth	Adsorption on cation-exchange resin	pH glass	~2 min	0.01–1 units/ml of enzyme
55	25	Membrane	Polyacrylamide occlusion	Beckman Univalent Cation	96% in 15 sec, 98% in 30 sec	1–50 units of enzyme/ 24 ml of solution

Item no.	Stability	Comments
1	>14 days without cellophane membrane	This device is the basis of the potentiometric urea electrode and has been described in the text (10, 14, 103, 104).
2	>3 weeks	In this work the univalent-cation-based urea electrode was applied to the first time for measurement in serum. Cation-exchange resin removal of Na^+ and K^+ was essential. The electrode potential was measured versus an uncoated cation electrode (105).

274

#	Lifetime	Description
3	>3 days but becomes sluggish	This electrode was the first potentiometric gas sensor used with an enzyme electrode; it is based on a CO_2 gas-permeable membrane. The membrane electrode is slow to respond and the potential time curve is determined not by the enzyme reaction but by the electrode response time. The response was fastest at pH 7.0 (106).
4	>1 week	A nonactin impregnated silicon rubber membrane was used to minimize Na^+ interference with the urea electrode. Potassium ions still present some difficulty (36).
5	>1 month, less than 2 mV shift at 0.5 mM urea	The above electrode was tested in serum. Potassium ions at the normal serum level interfered to some extent so that a very nonlinear calibration curve in a matrix containing the normal serum level was obtained. A standard dilution technique was devised to permit measurement in serum (107).
6	≫3 weeks and 300 runs	This is not a true enzyme electrode since the enzyme was not in direct contact with the sensor. An air-gap configuration was employed in which a sample was injected into a cell containing the immobilized enzyme. The pH was raised by addition of sodium hydroxide and the ammonia measured. There are no interferences in this technique (41).
7	>3 weeks	A trapped liquid layer of urease was held next to an NH_3 gas membrane. The response time was measured in blood and serum relative to that in water. Low enzyme concentrations (25–75 mg/ml) generally showed both decreased calibration curve slopes and increased response time. A precision of 3% was obtained and the method correlated quite well with a colorimetric method for ammonium based on the glutamate de-hydrogenase enzyme technique (25).
8	~20 days, only 1 mV decrease in slope	This unusual electrode was prepared by evaporating an enzyme directly on the gas-permeable membrane of an ammonia electrode. The enzyme was then crosslinked by glutaraldehyde. Relative standard deviation of 2–6% were obtained (26).
9	N.A.	This electrode was a part of a wider study of several electrodes. The most interesting feature is that 70% of the enzyme activity was retained (20, 127).
10	>2 weeks	A pH-sensitive glass electrode was coupled to immobilized urease. The sensitivity of the device is a strong function of the sample pH and buffer capacity. Even though promising results were obtained in-water solutions, it is doubtful that this approach will be useful in real samples (46).
11	N.A.	This was a modification of item 10 in which the enzyme was occluded in polyacrylamide gel (46).
12	N.A.	This paper is an important theoretical paper on the basis of steady-state potentiometric enzyme electrodes. It shows that with a properly immobilized enzyme, the calibration curves for urea and ammonium ions are nearly identical (18).
13	N.A.	This device is really a micro reactor. The enzyme was immobilized as an anion-exchange resin to minimize diffusion of ammonium ions out of the electrode area to maximize the sensitivity. It is a unique approach in that both a reactor and a separator were employed (115).

Table 5.10 (*Continued*)

Item no.	Stability	Comments
14	200–1000 assays; >2 months	Electrode was a special Teflon membrane to achieve fast electrochemical response to NH_3. Response time varied greatly with pH. Fastest at pH 8.5. They point out the importance of the buffer capacity of the internal electrolyte. The amount of enzyme influenced the slope, response, and recovery time. The electrode potential was constant to within ± 5 mV for 200 days (108).
15	—	Theoretical and experimental study of the effect of product inhibition on an enzyme electrode (152).
16	~12 days	This is an application in which the rate of dissolution of animal feedstuffs containing urea was measured. Response time depends upon thickness of enzyme layer. For thin films 90% of steady-state was obtained in as little as 10 sec (149).
17	>1 week	This was a preliminary report, see item 24 (59).
18	>1 week	This was a preliminary report, see item 24 (59).
19	N.A.	The enzyme was immobilized in a polyacrylamide gel supported by a nylon net. A variety of L-amino acids were detected including phenylalanine, leucine, methionine, alanine, and proline. Sensitivity decreased in the order given. The electrode showed a drift of 0.05–0.35 mV/day at 0.2 mM phenylalanine and 1.5–4.2 mV/day at 2 mM phenlalanine (22).
20	>2 weeks	This type of electrode was more stable than the one cited in item 16. It was prepared by soaking a nylon net in a concentrated enzyme solution (100 mg/ml). The relative improvement in stability was attributed to the fact that riboflavin which is used to polymerize the polyacrylamide inhibits the enzyme. A series of experiments in which the enzyme concentration (2–200 mg/ml), pH (6.0–8.0), and temperature (25–37°C) were systematically varied and this effect on a phenylalanine calibration curve determined was reported. The addition of catalase to the original enzyme solution in amounts up to 50 mg/ml improved the electrode response and stability (22).
21	N.A.	This was part of a theoretical study in which the diffusion equations were solved numerically (127).
22	<10 hr, slope decreased to 47 mV/decade overnight	This sensor is similar to the CO_2–urea electrode described in item 3. The electrode response became independent of the amount of enzyme at about 4 mg/ml but a large excess, 50 mg/ml, is needed for stability. The electrode is stable only for about 1 day, but this can be improved by using polyacrylamide to immobilize the enzyme (106).
23	N.G.	The enzyme was bound to polyacrylic acid via a diazonium linkage. The selectivity of a nonactin silicone rubber membrane permits its use in presence of potassium ion in serum (116).

24	>3 weeks	The electrode described in item 17 was improved by addition of peroxidase and use of an iodide electrode via the coupled reaction $2H^+ + H_2O_2 + 2I^- \rightarrow 2H_2O + I_2$. Large amounts of H_2O_2 inhibit the electrode. The rate of change of electrode potential was found to be a better indicator of the amount of phenylalanine than the steady-state potential. Cysteine interfered electrochemically due to its interaction with silver sulfide. Other amino acids also give a response (116).
25	>15 days still 95% active	The electrode response is very fast due to open area of the nylon net support since no dialysis membrane was used. The electrode is insensitive to urea, L-alanine, L-glutamate and creatinine (117).
26	>28 days	An electrode based on a trapped liquid layer was only stable for 1 day. Polyacrylamide-occluded enzyme was more stable. Bubbling the solution with O_2 as opposed to air saturation did not improve the response. The enzyme is specific for asparagine (39).
27	>10 days at 4°C	Specific electrodes were developed for each substrate based on the use of a *chromatographically* purified commercial decarboxylase. In general amino acid decarboxylases are more specific when purified than are the deaminases. A relative standard deviation of 2% was obtained (128).
28	≅1 day	The hydrolytic bacterial enzyme was used. At low pH the calibration curve was quite nonlinear due to the interference of H^+ with the cation electrode. At pH greater than 5.5, the electrode response was very slow because this is above the pH for optimum activity. The electrode requires at least 40 mg/dl of enzyme for good response. Polyacrylamide occlusion failed to stabilize the response (23).
29	~3 weeks	This is a preliminary report of a new principle in which bacteria containing a high activity for a given substrate are trapped on the electrode. It was stated that when the electrode response decreased, it could be improved by returning the electrode to a nutrient broth in which the bacteria are regenerated. Additional reports have now appeared (5, 129–135).
30	50 days	Electrode does not respond to any other amino acid. Electrode preparation was optimized by varying the amount of enzyme, BSA, and glutaraldehyde. Nearly Nernstian slope is obtained. Effect of pH on slope was studied. The sensitivity varied with time over a period of 40 days. It is stored in pyridoxyl-5-phosphate (158).
31	response decreases after 1 day	This device was described in the text in detail (114, 166).
32	—	Enzyme was trapped, attached by adhesive as well as by glutaraldehyde to the membrane. The enzyme was crosslinked to itself with glutaraldehyde. The response time varied with both enzyme activity and substrate concentration. Response slope was non-Nernstian and exceeded 60 mV/decade (154).
33	>21 days	This electrode was prepared by placing a mixture of equal parts of enzyme solution 125 units/ml and polyacrylamide reagents on a nylon net and polymerizing on the electrode. Approximately 40 µl of enzyme solution was used for each electrode. The electrodes give excellent calibration curves for phenylalanine,

Table 5.10 (*Continued*)

Item no.	Stability	Comments
34	—	alanine, valine, methionine, leucine, norleucine, and isoleucine; proline did not react at all. The electrode must be stored in a buffer containing FAD to prevent its loss by dialysis (39). The above studies were also carried out with immobilization by soaking a nylon net in various concentrations of the enzyme. These electrodes give good initial responses but stability was quite poor. Studies were carried out from pH 6–9.5. Bubbling with O_2 did not improve the response (39).
35	>2 weeks	This is a preliminary report on an amygdalin electrode. See item 36 for a more detailed study (40).
36	~4 days	The enzyme β-glucosidase was immobilized by polyacrylamide occlusion on a cyanide electrode. The gel is mounted vertically and a small sample placed on it; the reference electrode is dipped into the sample. It is possible that a large fraction of the small sample was reacted. Response was slow due to the high pH used to get good electrode response and also due to poor immobilization technique. At pH 11 the enzyme is irreversibly denatured. Copper, cadmium, and mercury inhibit the system (4).
37	~1 week	A paste of enzyme was coated on the electrode and trapped. The main purpose of the work was to elucidate the effect of pH, stirring, and membrane thickness on the sensitivity and response time of the electrode. The response was much faster at pH 7 than at pH 10 and was independent of the amount of enzyme used at low pH but not at high pH. The steady-state potentials were more negative at pH 10 than 7 because more product is present as the anion (CN^-) at the higher pH (27).
38	>30 day for 10% change at 1 mM	The enzyme was immobilized by simple trapping, polyacrylamide occlusion, and covalent binding. Covalently bound enzyme was the most stable. Since the electrode is based on the consumption of iodide by peroxide oxidation, the linearity is limited but it can be optimized in the expected range. Several ions including SCN^-, HS^-, CN^-, and Ag^+ interfere electrochemically. Cellubiose and maltose interfere only when present at greater than 10^{-4} M (97).
39	>2 weeks	This electrode is based on the measurement of the pH change generated at the electrode by the enzyme reaction. The buffer capacity of the solution affects the pH charge very strongly and the electrode must be pre-equilibrated in a standard buffer prior to each measurement and presaturated with O_2. The major advantage of such devices is their insensitivity to cations (46).
40	~2 weeks, some decrease in speed of response	Several electrode configurations including polyacrylamide-occluded and trapped liquid layers were tested. The response time was shown to depend upon the porosity of a glass frit when it was used to immobilize the enzyme around the pH electrode. Slope of the calibration curve was shown to depend upon the amount of enzyme on the electrode. Response time increased at high substrate concentration. Also noted that the response time increased as the sample volume increased. The very large local pH changes (>2 units) caused a decrease in the rate of the enzyme reaction. No interference from penicilloic acid (15).

278

Item	Lifetime	Description
42	>1 week	See discussion of item 10 and 39 (46).
43	2 days	Brief communication on the use of a penicillin electrode in fermentation control (153).
44	6 days	Buffer capacity of solution has great effect on sensitivity. Store in 0.025 M sodium phosphate, pH 7.0 for 1 hr prior to use. A pH-stat was used to hold the pH of the test solution at pH 7.0 during the run. The system was used to monitor a penicillin fermentation broth. Average deviation from standard additions assay is 10% (148).
45	~4 days, lost 2 mV in slope	This electrode is based on a trapped liquid layer of enzyme placed in contact with an ammonia membrane. The enzyme is activated by tripolyphosphate. No interference was observed for creatine, urea or arginine (29).
46	~10 days, 2 mV shift in curve	A uricase–albumin copolymer did not work as well as when the enzyme was simply adsorbed on hydroxymethyl cellulose. Phosphate activated the enzyme but borate inhibited it (118).
47	~4 days	Device is similar to that in item 45. The calibration curve reflects both enzyme and sensor properties. Enzyme is active in succinate, citrate, and Tris buffers. Enzyme activity increased by a factor of 2 did not improve its stability nor did immobilization on calcium phosphate or use of an albumin–glutaraldehyde copolymer. No interference was observed with 5'-ADP, 5'-ATP, 3',5'-cyclic AMP, adenine, or adenosine. It was used subsequently to study AMP binding to enzymes (21, 135).
48	—	This work reports the use of a previously developed AMP electrode for the study of the binding of AMP to D-fructose-1-6-diphosphatase. See reference 21 for details on electrode (135).
49	—	Slope of the logarithmic calibration curve depends on solution pH in the range 7.0–8.5. The effect of pH and temperature on response was studied in detail. From 10–40°C, temperatures have no effect (155).
50	~40 days	Optimum conditions of buffer concentration, ionic strength, linearity, stability, sensitivity, pH of the bulk, and response time were studied. The linear range depends strongly on buffer conditions and whether a rate method or steady state method was used (168).
51	several days	Based on $Fe(CN)_6^{3-}$ oxidation of lactate. Potential changes were detected with a membrane redox electrode comprised of ferrocene in polyvinylchloride. Catalase tends to stabilize the enzyme (173).
52	N.G.	This is not a true enzyme electrode because the enzyme is not attached to the electrode surface. It is similar to the device in item 6. Sulfite, sulfate, nitrate and perchlorate do not interfere (94).
53	—	Fluoride reversibly inhibits the urea–urease reaction and therefore alters the steady-state electrode potential. The limit of detection decreases as the amount of enzyme used decreases. The percent inhibition with a given level of fluoride *increases* with the urea concentration (171, 172).
54	10 runs	Acetylcholine was immobilized by adsorption on a cation-exchange resin and used to assay for cholinesterase activity. The *rate* of change of pH was used to determine the enzyme concentration. The rate was independent of pH from 7.9–8.3 but depends strongly on ionic strength (55).
55	N.G.	This electrode was described in detail in the text (46).

Table 5.11 Survey of Enzyme Electrodes Based on Amperometric Detectors

Item numbers	Substrate	Enzyme(s) used	Source or type	Amount of enzyme used	Buffer composition	pH	Thickness of enzyme layer
1	Glucose	Glucose oxidase	Sigma II	N.G.	0.1 M, phosphate	6.0	N.G.
2	Glucose	Glucose oxidase	Sigma II	18 units/mg	0.1 M, phosphate	6.0	N.G.
3	Glucose	Glucose oxidase	Sigma II	—	0.1 M, phosphate	6.0	N.G.
4	Glucose	Glucose oxidase	Sigma II	0.1 g/ml of gel	0.1 M, phosphate	6.0	N.G.
5	Glucose	Glucose oxidase	—	13–18 units/mg	0.1 M, phosphate	7.3	~0.2 g total protein
6	Glucose	Glucose oxidase	—	13 units/mg	0.04 M, phosphate + 0.026 M NaCl, 0.004 M KCl	7.4	150 units
7	Glucose	Glucose oxidase	Boehringer	1.5 mg/100 ml	N.G.	N.G.	N.G.
8	Glucose	Glucose oxidase	—	130 units/mg; 10–100 mg/100 ml	N.G.	N.G.	25–50 μm
9	Glucose	Glucose oxidase	Nutritional Biochemical Co.	0.2 g/ml, 30 units/mg	0.2 M, KI	6.1	0.5 mm
10	Glucose	Glucose oxidase	Nutritional Biochemical Co.	N.G.	See item 9	6.1	500 μm
11	Glucose	Glucose oxidase	Sigma	15 units/mg	0.2 M, KI	6.1	1–50 μm
12	Glucose	Glucose oxidase	N.G.	N.G.		4.0–7.5	N.G.
13	Glucose	Glucose oxidase	N.G.	N.G.	0.1 M, phosphate	7.35	N.G.
14	Glucose	Glucose oxidase	N.G.	N.G.	phosphate	7.3	N.G.
15	Glucose	Glucose oxidase	Aspergillus niger	700 units/g	1 M, phosphate	7.2	N.G.
16	Glucose	Glucose oxidase	Sigma II, Aspergillus niger	0.001 to 1 unit per electrode	0.1 M, phosphate	6.6	varied

No.	Substrate	Enzyme	Source	Amount	Buffer	pH	
17	Glucose	Glucose oxidase	Aspergillus niger, Nutritional Biochemistry	—	0.05 M, phosphate	7.4	0.008 in.
18	Glucose	Glucose oxidase	—	—	—	—	—
19	Glucose	Glucose oxidase	Asperigillus niger, Sigma Type II	0.01–1 units/electrode	0.1 M, phosphate	6.6	varied
20	Glucose	Glucose oxidase	Asperigillus niger, P. L. Biochemicals	N.G.	0.01 M, phosphate + 0.154 M, NaCl	7.35	—
21	Glucose	Glucose oxidase	—	—	isotonic citrate	7.4	—
22	Glucose	Glucose oxidase	Yellow Springs Instrument	—	phosphate	7.3	—
23	Glucose	Glucose oxidase	—	—	—	6.0	—
24	Glucose	Glucose oxidase	Boehringer	—	various	5.6 (optimum)	—
25	L-Amino acids	L-Amino acid oxidase	N.A.	0.3 units/mg, 100 mg enzyme + 100 mg albumin	0.1 M, phosphate	7.3	~0.2 g total protein
26	L-Amino acids	L-Amino acid oxidase	Sigma I	80 mg enzyme/200 mg carrier	0.1 M, acetate, phosphate, Tris	6–9 (7.8)	N.G.
27	L-Amino Acids	L-Amino acid oxidase	Calbiochem	200 mg/ml of gel	0.1 M, phosphate	6–8 (7.0)	0.1 ml
28	Amino acids	Amino acid oxidase	Crotalus adamenteus Sigma	5 units/mg	0.05 M, phosphate	7.4	0.008 in.
29	L-Amino acids	L-Amino acid oxidase	Crotalus adamenteus	0.002 to 0.09 units/electrode	0.1 M, phosphate	7.8	varied
30	Amino acids	Amino acid oxidase	Crotalus adamenteus	0.02 to 0.1 units/electrode	0.1 M, phosphate	7.8	varied
31	Lysine	Lysine decarboxylase + diamine oxidase	Bacterium cadaveris Pea	8.3 n katal/mg, 550 katal/mg	0.1 M, phosphate	6.5 (optimum)	—

Table 5.11 *(Continued)*

Item numbers	Substrate	Enzyme(s) used	Source or type	Amount of enzyme used	Buffer composition	pH	Thickness of enzyme layer
32	Arginine	Arginine decarboxylase + diamine oxidase	E. coli Pea	3.3 n katal/mg; 580 katal/mg	0.1 M, phosphate	6.2 (optimum)	—
33	Glutamate alcohol lactate	Appropriate dehydrogenases	Ferak, Berlin	—	0.01 M, phosphate	8.0	—
34	Uric acid	Uric acid oxidase	Sigma IV and V	100 mg enzyme (2.5 units/mg) + 100 mg albumin	0.1 M, glycine	7.5–11 (9.2)	200 mg total protein
35	Lactic acid	Lactate dehydrogenase	N.A.	3–10 units/mg	0.04 M, phosphate + 0.020 M NaCl, + 0.004 M KCl	7.4	150 units
36	Lactic acid	Lactic acid dehydrogenase	Oriental yeast	N.G.	0.1 M, phosphate	7.7	—
37	Lactic acid	Lactic acid dehydrogenase	—	0.6–4 mg/ml	0.2 M, phosphate	7.2	—
38	Lactic acid	Lactic acid dehydrogenase	Worthington	8 mg NAD + 6 mg LDH/ml, 200 µl/electrode	0.05 M, phosphate	7–9.4 (8.0)	100–200 µm
39	Lactic acid	Cytochrome b_2	Sigma	6 units/100 mg	phosphate	7.3	—
40	Lactic acid	Lactate dehydrogenase	Anaerobic bacteria	20–5000 units/ml	0.2 M, phosphate	7.2	40–50 µm
41	Lactic acid	Cytochrome c	Baker's yeast	2 units/ml of solution in electrode	0.1 M, phosphate 2 mM Fe(CN)$_3^{3-}$	7.3	—

282

42	Lactic acid	Lactate dehydrogenase	Sigma	13–32 units/ electrode	0.1 M pyrophosphate; 0.02 M ferrocyanide; 1.5 mM, NAD$^+$	9.0	—
43	Alcohols	Alcohol oxidase	Basidiomycete (Wyeth)	0.02 units/mg	0.1 M, phosphate	6–10 (8.2)	~200 mg copolymer
44	Alcohols	Alcohol oxidase	Basidiomycetes Oriental yeast	1 mg/2 µl	phosphate—Tris	8.5	very thin
45	Ethanol	Alcohol dehydrogenase		0.3 units/ml solution + collagen	0.1 M, phosphate	7.7	N.G.
46	Aldehydes	Alcohol oxidase	Basidiomycete (Wyeth)	0.02 units/mg based on H_2O_2 generation	0.1 M, phosphate	6–10 (8.2)	~200 mg copolymer
47	Carboxylic acids	Alcohol oxidase	Basidiomycete (Wyeth)	0.02 units/mg based on H_2O_2 generation	0.1 M, phosphate	6–10 (8.2)	~200 mg copolymer
48	Diamines	Diamine oxidase	Pisum sativuum	15 units/mg 0.2 units total	0.1 M, phosphate	7.0	very thin
49	Phosphate				0.25 M, Tris, and 0.1 M, KCl, 0.02 M, MgCl$_2$	7.0–8.6	
50	Phosphate	Alkaline phosphatase and glucose oxidase		13 units/per electrode and 45 units/per electrode	0.1 M, glycine + 0.01 M, Mg^{2+}	4–10 (8.4)	5 mg
51	Sulfate	Aryl sulfatase	Helix pomatia	0.28–2.5 units/ electrode	0.5 M, acetate	3.7–6.7 (3.7–4.1)	very thin
52	H$_2$O$_2$	Catalase	N.G.	N.G.	N.G.	4–10 (6.2)	1 µm
53	Galactose	Galactose oxidase	N.G.	N.G.	N.G.	N.G.	N.G.

Table 5.11 (*Continued*)

Item numbers	Substrate	Enzyme(s) used	Source or type	Amount of enzyme used	Buffer composition	pH	Thickness of enzyme layer
54	Galactose	Galactose oxidase	Worthington	30–130 units per electrode	0.07 M, phosphate	7.3	—
55	Sucrose	Invertase + glucose oxidase	N.G.	N.G.	Citrate–succinate	5.4	N.G.
56	Sucrose	Invertase + Mutarotase + Glucose oxidase	E.C.3.2.1.26; E.C.5.1.3.3; E.C.1.1.3.4	5000 units/g; 6000 units/g; 45000 units/g	0.1 M, acetate	4–8	50 μm
57	Phenols	Polyphenol oxidase	Potatoes, Mushrooms	~60 units	0.1 M, phosphate	5.0–8.5	—
58	Phenol	Tyrosinase	Mushrooms, Worthington Biochemicals	300 units/electrode	0.25 M phosphate; 0.02 M, ferrocyanide	6.5	—
59	Cholesterol	Cholesterol oxidase	Wako	6.4 units/mg	0.5 M, phosphate	7.0	60 μm

Item no.	Temperature (°C)	Protection method	Immobilization method	Sensor	Response time	Linearity
1	25	N.G.	Chemically bound to polyacrylamide	Pt; +0.6 V versus SCE	1 min	0.5–15 mM
2	25	Membrane 25 μm	Enzyme bound to polyacrylamide or polyacrylic acid. Net spacer	Pt; +0.6 V versus SCE	1 min	0.5–15 mM
3	25	Membrane 25 μm	Glucose oxidase slurry trapped by membrane	Pt; +0.6 V versus SCE	1 min	0.5–15 mM
4	25	Membrane 25μm	Polyacrylamide occlusion, 90 μm nylon net spacer	Pt; +0.6 V versus SCE	1 min	0.5–15 mM

No.		Membrane	Immobilization method	Electrode	Response time	Range
5	30	Membrane 25 μm	Albumin–glutaraldehyde copolymer of enzyme	Pt; -0.6 V versus SCE	1–2 min	5×10^{-6}–10^{-4} M
6	N.G.	Dialysis membrane	Trapped layer	Pt; $+0.4$ V versus SCE	3–10 min	1–20 mM
7	N.G.	N.G.	Cellulose acetate impregnated with polyacrylamide-occluded enzyme	Clark O$_2$ electrode	N.G.	0–200 mg/dl
8	25–40	Nylon net impregnated with silicon plastic cement	Polyacrylamide gel occlusion	Dual Clark O$_2$ electrode	30 sec–3 min for 98%	0–100 mg/100 ml nonlinear
9	N.G.	Nylon net	Polyacrylamide gel occlusion	Pt; -0.2 V versus SCE	2–3 min at 0.5 M	N.G.
10	N.G.	Nylon net	Polyacrylamide gel occlusion	Pt; -0.2 V versus SCE	2–3 min	0.1–1.5 mM
11	N.G.	None	Albumin–enzyme glutaraldehyde copolymer membrane	Rotating ring disc carbon paste, -0.2 V	<50 sec at 400 rpm	10–75 mg/dl
12	38	Dialysis membrane	Trapped liquid layer	Pt	1 min	0–200 mg/dl
13	37	N.G.	Polyacrylamide occlusion and glutaraldehyde crosslink	Pt	N.G.	N.G.
14	37	Polycarbonate membrane	Immobilization on "resinous" material	Pt; $+0.7$ V versus SCE	30–40 sec	1–400 mg/dl
15	37	N.G.	Polyacrylamide occlusion	N.G.	N.G.	N.G.
16	—	Cellophane	Immobilized membrane made by crosslinking BSA to the enzyme with glutaraldehyde	Pt; $+0.6$ V versus SCE	1–2 min	5×10^{-4}–10^{-2} M
17	—	Dialysis tubing	Cigarette paper soaked in suspension of the enzyme. Also adsorbed on diatomaceous earth	Pt; H$_2$O$_2$ at $+0.65$ V versus Ag/AgCl	3–8 min	poor

Table 5.11 *(Continued)*

Item numbers	Substrate	Enzyme(s) used	Source or type	Amount of enzyme used	Buffer composition	pH	Thickness of enzyme layer
18	—	—		Pt; at +0.6 V versus Ag/AgCl	—		0 to 250 mg/%
19	room	Cellophane	BSA–enzyme membrane formed by glutaraldehyde crosslinking	Pt; H_2O_2 at 0.6 V	1–2 min		—
20	37	None	Enzyme bound to rayon–acetate cloth with glutaraldehyde	Modified dual Clark O_2 sensor	2 min		50–150 mg/dl
21	37	None	Immobilized with glutaraldehyde on rayon	Dual Clark O_2 electrodes	2 min		0–400 mg/%
22	—	—	Glutaraldehyde crosslinked to cellulose acetate	Pt; H_2O_2	45 sec		0–4.9 g/l
23	25	None	Triazine crosslinking to lens paper	Pt; H_2O_2	50 sec		1–30 mM
24	25	None	Acyl oxide immobilization on porous collagen membrane	Pt; H_2O_2	4 min		10^{-7}–10^{-3} M
25	30	Membrane 25 μm	Albumin–glutaraldehyde copolymer of enzyme	Pt; −0.6 V versus SCE	1–2 min		0.1–1 mM phenylalanine
26	10–50	Membrane	Enzacryl diazonium salt covalent binding	H_2O_2 at +0.6 V on Pt versus SCE	1 min steady-state sec-rate	12	10^{-5}–10^{-3} M
27	N.G.	Dialysis membrane	Polyacrylamide occlusion	H_2O_2 on Pt potential scanned	30–60 sec, 90% response		1–400 μM
28	—	Dialysis tubing	Cigarette paper soaked in suspension of the enzyme. Also adsorbed on diatomaceous earth	Pt; H_2O_2 at +0.65 V versus Ag/AgCl	3–8 min		poor

No.	Temp (°C)	Support	Immobilization method	Electrode/Detection	Response time	Linear range
29	—	Cellophane	Immobilized membrane made by crosslinking BSA to the enzyme with glutaraldehyde	Pt; +0.35 V versus SCE	1–2 min	—
30	room	Cellophane	BSA-enzyme membrane formed by glutaraldehyde crosslinking	Pt; H_2O_2 at 0.35 V	1–2 min	—
31	30	None	Crosslink to nylon net via glutaraldehyde. Both enzymes and bovine albumin were crosslinked	O_2 electrode with polypropylene membrane -0.7 V	1–2 min	0–100 μM
32	30	None	Crosslink to nylon net via glutaraldehyde. Both enzymes and bovine albumin were crosslinked	O_2 electrode with polypropylene membrane -0.7 V	1–2 min	0–100 μM
33	30	Membrane	Cyanuric chloride to Dextran	Pt; +0.75 V	—	Alcohol; 100 mM, lactate; 1.5 mM, glutamate; 0.5 mM
34	30	Nylon cloth	Albumin–glutaraldehyde copolymer of enzyme	O_2 at -0.6 V versus SCE on Pt	2–3 min steady state, rate 1 min	1–10 mg %
35	N.G.	Dialysis membrane	Trapped liquid layer	Pt; +0.4 V versus SCE	<3–10 min	0.1–2 mM
36	25	None	Enzyme coadsorbed with collagen on Pt electrode	Pt electrode oxidation of NADH	~20 min	0–30 mM nonlinear
37	21	Dialysis membrane	Trapped liquid layer	Ferrocyanide determined at +0.25 V on Pt	~1 min	0.1–5 mM
38	25	Dialysis membrane	Glutaraldehyde crosslink and trapped liquid layer types. Also used system trapped on Agarose.	NADH reduced all ~0.7 V on glassy carbon	3–4 min	0.1–4 mM nonlinear

287

Table 5.11 (*Continued*)

Item numbers	Substrate	Enzyme(s) used	Source or type	Amount of enzyme used	Buffer composition	pH	Thickness of enzyme layer
39	25	Cellophane membrane	Trapped liquid layer	Pt; +0.4 V versus SCE	7 min		1–20 mM
40	25	Membrane	Trapped liquid layer	Pt or vitreous carbon +0.25 V versus AgCl	30 sec		0.1–8 mM
41	—	Cellulose	Trapped by membrane	Fe(CN)$_6^{4-}$ oxidized on Pt at +0.25 versus Ag/AgCl	40 sec		0–1.5 mM
42	23	—	Polyacrylamide block occlusion trapped between electrode grid	Pt grid	10 min to 98% response		0.02–50 mM
43	30	Nylon cloth	Glutaraldehyde copolymer with albumin	O$_2$ at −0.6 V on Pt	~2 min		Up to 5 mg/% ethanol
44	10–40 (24)	Dialysis membrane	Enzyme slurry trapped with dialysis membrane and diatomaceous earth	H$_2$O$_2$ at +0.6 V on Pt	~2 min		1–25 nM for methanol
45	25	None	Enzyme coadsorbed with collagen on Pt electrode	Pt electrode oxidation at NADH	N.G.		0–1 M nonlinear
46	30	Nylon cloth	Glutaraldehyde copolymer with albumin	0$_2$ at −0.6 V on Pt	2 min		N.G.
47	30	Nylon cloth	Glutaraldehyde copolymer with albumin	O$_2$ at −0.6 V on Pt	2 min		N.G.
48	10–40 (30)	None	Glutaraldehyde crosslink to polyamine	O$_2$ measured on Pt	<15 sec		0.02–0.4 mM
49	25	N.G.	N.G.	NADH, NADPH on Pt, +.65 V	N.G.		N.G.

	Temp (°C)	Membrane/support	Immobilization method	Electrode/detection	Response time	Range of sensitivity
50	30	Nylon net	Glutaraldehyde copolymer of alkaline phosphatase and glucose oxidase	O_2 measured at -0.6 V on Pt	1–2 min	1–10 mM range of sensitivity
51	25	None	Glutaraldehyde–albumin copolymer method	Oxidation of 4-nitro-catechol $+0.8$ V, Pt	2–3 min	10^{-4}–10^{-2} M
52	20	None	Collagen–enzyme electrochemically adsorbed membrane	Clark O_2 electrode, Teflon membrane	1.5 min	0–1.5 mM
53	37	Polycarbonate	Immobilization on resinous material	Pt; $+0.7$ V versus SCE	N.G.	1–100 mg/dl
54	37	None	Glutaraldehyde crosslinked to 1 μm cellulose acetate membrane	Pt; H_2O_2	40 sec	0–500 mg %
55	35	None	Collagen membrane with both enzymes bound by glutaraldehyde	Pt; $+0.7$ V versus SCE	1–2 min	N.G.
56	25	—	Collagen fibril membrane crosslinked with glutaraldehyde	Pt, O_2	1–5 min	0–10 mM
57	30	Nylon net	Crosslinking to nylon with glutaraldehyde and also crosslink to BSA + trapped polymer with nylon net	Clark O_2 at -0.7 V versus Ag/AgCl	—	0–0.1 mM
58	room	None	Polyacrylamide occlusion	Pt grid electrode	~10 min	0.3–100 μM
59	25	BSA membrane on top of collagen enzyme membrane	Crosslinked collagen fibril membrane to BSA with glutaraldehyde	Teflon–Clark O_2 assembly	2–3 min	up to 0.2 mM

Table 5.11 (*Continued*)

Item number	Stability	Comments
1	N.G.	This electrode is based upon detection of hydrogen peroxide by oxidation. A reproducibility of 1% for steady-state methods was obtained. A rate method based the change in current per unit time had a precision of better than 2% when the assay was completed in less than 12 sec (33).
2	20 days	In this paper (items 2–4) a set of three different types of enzyme immobilization were studied. The enzymes were immobilized by chemical binding to polyacrylamide gels by either diazo linkages or hydrazine linkages.
3	10 months	They was immobilized by physical occlusion in polyacrylamides and by trapping in a membrane. The
4	10 months	general trend in stability was that the chemically bound enzyme was stable much longer than the physically occluded enzyme, which was more stable than the trapped enzyme. The sensitivity of both the chemically bound enzymes and the physically occluded enzyme increased over the first 20–40 days, then decreased monotonically over the next 10 months (43).
5	4 months	A glucose measurement was made by determining the decrease in oxygen at a platinum electrode. The electrode proved to be more sensitive than detection of hydrogen peroxide by oxidation at a platinum electrode (58).
6	N.G.	Glucose was detected by use of differential amperometric electrodes. The oxidizing agent was benzoquinone which reacted with glucose to produce hydroquinone. The hydroquinone was detected by its oxidation back to quinone. Benzoquinone proved to be an excellent electron acceptor, and at a concentration of 1 g/l was able to compete very effectively with oxygen for the oxidation of glucose (61).
7	N.G.	In this study a Clarke oxygen electrode was used to detect the decrease in oxygen which accompanied the glucose/glucose oxidase reaction. The working curve was nonlinear at high concentration (>150 mg/l of glucose) A 2% relative standard deviation was obtained for 20 readings over a period of 24 hr. The technique was correlated with colorimetric methods for glucose in blood and serum (96).
8	N.G.	In this work a differential oxygen measuring amperometric electrode was made. One electrode was covered with glucose oxidase immobilized in an acrylamide gel. The other electrode contained thoroughly inactivated glucose oxidase. The technique therefore only measures changes in oxygen concentration due to reaction at the active electrode (2).
9	N.G.	In this method hydrogen peroxide was detected by its oxidation of iodide to iodine. The iodine was measured electrochemically by reduction back to I^- on a platinum electrode. The work was mainly concerned with theoretical predictions of the behavior of the amperometric glucose electrode therefore very little analytical data was included (13).

10	N.G.	This paper was concerned with the possibility of enhancing the electrode response in the above work by applying a pulse potential to the electrode. Iodine was allowed to buildup while the electrode potential was resting and measured by pulsing the potential to a negative value where reduction of I_2 would occur. Theoretical calculations and experimental work showed that the sensitivity could be improved by a factor of as much as 10 over a continuous measurement of I_2 production. Signal levels were considerably lower than the background blank signal (30).
11	<15 %, decrease after 1 month at 4°C	In this work a rotating ring disc electrode was used to study the catalysis kinetics of glucose oxidase. The enzyme and an inert protein (albumin) were immobilized by glutaraldehyde linkage to amino groups which were placed in the carbon paste of the ring disc. The production of I_3^- was measured. As the rotation speeded up, the electrode sensitivity was increased, but the linear range of the electrode decreased. The sensitivity of the electrode has a peak in its plot versus the rotation rate. The sensitivity of the electrode proved to be approximately the same as that of a stationary electrode (101, 102).
12	>2 months at 38°C	In this work, hydrogen peroxide was measured by differential amperometry at a pair of electrodes only one of which contained enzyme. The electrodes were placed in a flow-through cell of volume 250 μl. It can handle up to approximately 25 μl of whole blood. There was almost no stirring effect. A microelectrode version of the same device was placed on a rat's brain. The effect of injected insulin on the glucose concentration was easily discernible within a response time of about 4 sec. The differential arrangements of electrodes allowed any error due to consumption of oxygen by ascorbic acid and other reducing agents to be subtracted out (7).
13	7–10 days	In this study, immobilized enzyme was sandwiched around a set of platinum electrodes. The intent of this study was not analytical but rather to develop an immobilized fuel-cell device. Nonetheless the analytical implications were made quite clear and the device should be useful for analysis (119). The concept has been applied to determination of phenols (136).
14	N.G.	In this work, hydrogen peroxide was detected by oxidation on a platinum electrode. A double membrane system consisting of a polycarbonate membrane with an exclusion limit of 300–500 Å to block the passage of cells and proteins and an inert cellulose acetate membrane to exclude small molecules such as uric acid, glutathione, ascorbic acid, formic acid, and cystene was employed. Consequently these small molecules did not interfere with the measurement of peroxide. The sample is diluted by approximately a factor of 1–10 and the technique can be employed directly in whole blood. A relative standard deviation of 1% was obtained for glucose measurements (100).

Table 5.11 (Continued)

Item number	Stability	Comments
15	N.G.	This paper was not analytical in intent but rather was a fuel-cell-type application in which constant-current voltammetry was employed. The results showed that the potential of a pair of electrodes was related to the glucose concentration and therefore could be employed analytically if so desired (120).
16	>80 days	Both reaction rates (di/dt) and steady-state signals were measured. The thickness of the protective membrane influenced both the steady-state response and speed of response. The response time was increased by a factor of 5 with a doubling of the thickness, but the steady-state was halved. Membrane thickness varied with concentration of protein. A peak was obtained in a plot of steady-state current versus units of enzyme. The steady-state rate responses varied in a complex way with time (144).
17	—	Very preliminary study. Same work as in item 28 (163).
18	—	First report of an ultra micro glucose electrode; 0.1 μm tip. Device was implanted in rat cerebral cortex. Very little detail given (164).
19	100 days	Various reagent ratios for the immobilization were investigated. Both steady-state and rate methods were developed (159).
20	—	Whole blood was pumped over the sensor. The ultimate objective of this work was to make a sensor for control of an artificial pancreas (160, 161).
21	—	Attempt to develop continuous extra-corporeal glucose measurement via slow withdrawal of blood. Shows continuous response during glucose tolerance test (162).
22	300 assays	Based on Yellow Springs Instruments Glucose Analyzer. This is a clinical evaluation of a commercial instrument. Uses 25 μl of serum; coefficient of variation is 1.2% within a run. Between-runs variation is 5.8%. 100% recovery of added glucose is obtained (151).
23	2 weeks	6–10 milli-units of activity per electrode was obtained. Very brief paper (147).
24	2 years at 0°C	A differential electrode system in which two platinum electrodes both were covered with collagen membranes is employed. A dip type and a flow-through set-up were both investigated. The enzyme-coupling procedure was studied in detail. Both pH and type of buffer are important. The K_M of the immobilized enzyme was 3 mM. Steady-state and rate methods were both studied. Detection limit is 10^{-8} M (174).
25	>4 months	In this work the decrease in O_2 was measured at a Pt electrode. The device was sensitive to many different amino acids including phenylalanine, methionine, leucine, tyrosine, cysteine, and isoleucine. The sensitivities varied with substrate. Both rate methods based on the change in current with time, and steady-state methods were developed. The detection of O_2 decrease was shown to be more sensitive than meas-

292

26   N.G.

urement of peroxide production and had fewer interferences from electro-oxidizable compounds such as ascorbic and uric acid (58).

In this work, hydrogen peroxide was measured by oxidation on a Pt electrode. The device was sensitive to cysteine, leucine, tyrosine, phenylalanine, tryptophan, and methionine, but the slopes of the calibration curves were quite different. The enzyme had an optimum pH at 7.8–8.0 in phosphate and acetate buffers. Both rate and steady-state methods were developed. The effect of temperature on both the rate and the steady-state methods indicated that two rate-controlling processes, possibly enzymatic at low temperature and mass transfer at high temperature, are involved in the measurement (47).

27   10–12 days at 4°C storage

Hydrogen peroxide was measured amperometrically on a Pt surface. A small flow-through cell which contained the enzyme immobilized on the electrode was employed. As a result the signal was in the form of a current peak as the solution was swept through the cell. One could measure either the peak current or the integral under the peak. Linear response was obtained for methionine, tryptophan, histidine, phenylalanine, leucine, and asparaginine over about one order of magnitude. The peaks were swept through the cell in about 4 min (121).

28   —

Leucine, phenylalanine, methionine, tryptophan, and norleucine gave signals but alanine and glycine did not. Calibration curve is very nonlinear (163).

29   —

See comments for Item 14 (146).

30   100 days

Various reagent ratios for the immobilization were investigated. Both steady-state and rate methods were developed (159).

31   ~20 days

Only hydroxylysine interferes. Various ratios of enzymes, BSA, and glutaraldehyde were tested. Membranes were inactive when albumin was omitted. pH was systematically varied. Temperature was varied from 25–45°C. The two enzyme systems have a pH optimum located between those of the individual enzymes. Pyridoxal-5-phosphate reactivates old electrodes (150).

32   ~20 days

Only lysine interferes. All else see item 31 (150).

33   Lactate, 40 hr, Glutamate <10 hr

The enzymes and NAD$^+$ were attached with cyanuric chloride to Dextran and trapped with a membrane over a platinum electrode. Test solutions were pumped into a microcell. Results depend upon presence of added phenazine methosulphate. The lactate electrode was stable for 40 hr (157).

34   ~100 days

The consumption of oxygen by the enzyme-catalyzed reaction was measured by reduction of oxygen on a Pt electrode. The calibration curve which is a plot of rate change in current per unit time was linear up to about 3 mg/100 ml, but showed a deviation of 25% from linearity at 15 mg/100 ml. Cysteine and ascorbic acid interfere with the electrode but a good correlation was obtained versus a chemical method in which the absorbance of allantoin, the reaction product, was measured (42).

Table 5.11 (*Continued*)

Item number	Stability	Comments
35	N.G.	The method is based upon the substitution of ferricyanide for NAD^+ in the oxidation of lactic to pyruvic acid. The ferrocyanide produced is measured via a differential pair of Pt electrodes. Since the K_M was so low (1.2×10^{-3} M) samples of serum had to be diluted to put them on the linear part of the calibration curve (61).
36	~8 runs	A coupled coenzyme system in which NAD bound to the electrode was regenerated by addition of flavin mononucleotide was developed. A plot of steady-state current versus lactic acid concentration was non-linear above about 10 mM and the electrode response time was very slow (20 min) (122).
37	Can be used for many (?) runs	In this work ferricyanide was used to oxidize lactic acid to pyruvic acid; the generated ferrocyanide was measured amperometrically and the current was proportional to the lactic acid concentration. The current–time curve shows a peak at about 1 min and then slowly decreased toward the baseline. The current was independent of the enzyme concentration in the electrode from 0.6 to 4.0 mg/ml. A linear working curve from 10^{-4} to 10^{-2} M lactate was obtained (123).
38	Decreases 20% in first 2 hr	In this work, a study of the reagentless lactate electrode described in the text was carried out (38).
39	14 hr	A three-electrode polarograph was used to measure the amount of ferrocyanide generated by the oxidation of lactic acid by ferricyanide. The enzyme was trapped in a thin layer behind a permeable membrane. Excellent calibration curves were obtained. The effect of membrane permeability on the linearity and stability of the system was experimentally tested and compared with theory (17).
40	N.G.	A small thermostated flow-through cell in which an immobilized enzyme lactate electrode is mounted was described. In a stirred solution, the steady-state was achieved in 30 sec and a linear working curve was obtained between 0.1 and 8 mM. A comparison of the experimental and theoretical steady-state currents was excellent. The technique was compared to a spectrophotometric measurement of the substrate in blood (137–141).
41	~4 weeks	Ferricyanide served as oxidant and, due to reoxidation of ferrocyanide, indicated the amount of reaction. 1 μl of stock enzyme solution (2 units/mg) is placed on the electrode. The sample is pumped past the sensor. An excellent interference study was run including other oxidizable acids and the effect of reducing agent (137).

294

42	~2 weeks	NADH produced by NAD$^+$ oxidation of lactate is oxidized by ferricyanide. The zero current potential of a platinum grid electrode is used to measure lactate concentration (156).
43	~4 months	An alcohol oxidase enzyme was used. Initially, hydrogen peroxide production is quite evident but after a few days it decreases to a negligible level. Since H_2O_2 is not produced, the O_2 consumption measured by a Pt cathode becomes more evident. The electrodes sensitivity reaches a maximum in two weeks and is stable for 4 months thereafter. Both steady-state and rate methods are developed. Methanol gives a strong response is less than that due to ethanol. Butanol also gives a good response. The electrode can be used to measure a variety of alcohols, aldehydes, and carboxylic acids (35).
44	50% decrease in slope in 24 hr	The generation of hydrogen peroxide was measured on a Pt anode. Maximum response was obtained for methanol, ethanol, and least response to propanol. Treatment of the electrode with glutaraldehyde stabilized the response (3, 142).
45	~ 8 runs	The oxidation of NADH on a Pt sensor was measured. The electrode was prepared by electroadsorption of collagen fibrils containing glutaraldehyde crosslinked alcohol dehydrogenase. The NADH was reduced by an excess of FMN to obtain the steady-state signal, which was reproducible to only $\pm 10\%$ (122).
46	~4 months	This electrode is based on the results obtained in reference 35.
47	~4 months	This electrode is based on the results obtained in reference 35.
48	26% decrease at 14 days	The consumption of O_2 was measured at a Pt cathode. The electrode was sensitive to putrescine, cadaverine, hexamethylene diamine, histamine, and spermidine. The enzyme was polymerized on a thin film so as to fill in the pores of a nylon net. A very fast response due to the thin film was obtained with a relative standard deviation of 3% (125).
49	N.G.	A complex set of enzymes were tested to see if response to phosphate could be developed. Although the approach was demonstrated with soluble enzymes, it proved to be impossible to immobilize all of the needed enzymes to make a self-contained electrode (113).
50	~3 months	This electrode is described in detail in text (44).
51	25–50 runs	This electrode is based on the sulfate inhibition of the hydrolysis of 4-nitrocatecholsulfate. A Pt anode is used to measure the amount of hydrolyzed substrate. Two types of electrodes, that is, trapped liquid layer and glutaraldehyde–albumin crosslinked polymer, were developed. The electrode is sensitive to F^-, HPO_4^{2-}, and MoO_4^{2-}, but these interferences can be minimized (126).
52	9–10 runs	Oxygen production was measured on a Pt cathode. Peroxide was converted to oxygen by the use of catalase adsorbed in a thin film of collagen on a Pt gauze electrode (34). Brief mention is made of a uric acid electrode based on the same concept (143).

295

Table 5.11 *(Continued)*

Item number	Stability	Comments
53	N.G.	This electrode is similar to that described in item 14. To measure galactose, the glucose oxidase is replaced with galactose oxidase. Ferricyanide must be added to the buffer to prevent enzyme deactivation (100).
54	>25 days	Dihydroxyacetone was the only interferent of 39 compounds listed. Electrode is stabilized by use of a buffer containing 0.053 M NaCl, 7 mM sodium benzoate, catalase, 10 mg % $K_3Fe(CN)_6$, and 0.2 mg % $CuCl_2 \cdot 2H_2O$. The concentration of glutaraldehyde used to crosslink is very critical to both initial activity and stability. Electrode is used in a flow-through mode. Many common reducing agents had no effect on the sensitivity (146).
55	N.G.	This device is based on the one described in item 14. To measure sucrose, the enzyme invertase is added to the electrode to split the substrate into glucose and fructose. The glucose is then measured after mutarotation with a glucose electrode based on glucose oxidase (100).
56	10 days	The decrease in steady-state O_2 current upon addition of sucrose is measured (165).
57	<1 month	Electrode is sensitive to p-cresol, phenol, pyrocatechol, pyrogallol, and dihydroxyphenylalanine (167).
58	—	Phenol is enzymatically oxidized to $ortho$-benzoquinone by oxygen. Excess ferrocyanide reduces the benzoquinone. The change in zero-current potential of a Pt indicator electrode is related to the phenol concentration. Slope of log phenol versus potential was 30 mV/decade. Catechol p-cresol and resorcinol also interfere. Reproducible results are obtained only when solution is bubbled with 100% O_2 (136).
59	N.G.	Measures decrease in O_2. Used a double-enzyme membrane. After preparation of collagen–cholesterol oxidase membrane, a BSA membrane was formed by glutaraldehyde crosslinking (145).

296

Bacterial electrodes have three important features:

1. In principle they avoid the use of purified enzymes.
2. In principle and in practice the electrode can be regenerated by immersion in nutrient broth.
3. Some means must be found to overcome the poor selectivity of bacteria which contain a myriad of enzymes which convert various substrates to a common product.

The four bacterial sensors reported by Rechnitz's group are based on *Streptococcus faecium* (arginine), *Bacterium cadaveris* (aspartate), *Sarcina flava* (glutamine) and *Proteus morganii* (cysteine). The major objective of Rechnitz's work in this area has been to improve the long-term stability of enzyme electrodes by use of the ability of bacterial colonies to regenerate themselves in appropriate growth media. In the case of the aspartate electrode, the sensor based on bacteria was stable for 10 days while the sensor based on the purified enzyme (aspartate ammonia lyase) from the same bacterium was at best stable for only 3 days. Since the enzyme electrode and the bacterial electrode were both stored in nutrient media, the improved stability of the bacterial probe is due to growth and replenishment of inactive cells. It should be noted that all of this work was carried out with an air-gap ammonia electrode to insure electrochemical selectivity. The bacteria are concentrated into a paste, placed on the electrode and held there with a dialysis membrane. Typically, 1–20 mg (nearly 10^9 cells) are placed on each electrode. A thick bacterial paste causes an increase in response time; thus, even with a total of 1–2 mg of cells per electrode, response times of 20 min were encountered.

The bacteria used for arginine are viable at pH 10; since this alkalinity is ideal for an NH_3 gas sensor, a series of tests were run to determine the optimum pH. It was shown that the slope of the calibration curve is much lower at pH 10 than at pH 7.7. This indicates that the characteristics of the enzyme within the bacteria are still very important. The major practical limitation of bacterial electrodes is likely to be the fact that a great many enzymes coexist in a given bacterial cell; since many of these may be capable of producing a common product, selectivity will be difficult to achieve. The above arginine electrode was shown to be insensitive to histidine, lysine, citrulline, and urea, but the order of response to arginine, glutamine, and asparagine was similar to the order of NH_3 production by the bacteria.

Neither the aspartate nor the arginine electrode has a precisely Nernstian slope. The arginine electrode shows a slope of -52.5 mV/decade between 0.1 and 6.5 mM on the first day whereupon the slope decreases to -42.0 mV/decade and remains approximately constant for at least 40 days. The explanation of this behavior has not been found.

The glutamine electrode, which is based upon the presence of glutamine deaminase in the bacteria used, is *highly selective*. No interference from aspargine, aspartic acid, histidine, alanine, or arginine was observed. As in the case of the previous sensors, the response in the 10^{-2}–10^{-4} M range was sub-Nernstian.

The bacterial cysteine electrode relies upon the production of H_2S and its detection by an H_2S gas sensor. The major limitation of the device is its sensitivity to CO_2. This sensor is not selective in that significant response is obtained in solutions containing homocysteine and cystathionine, although methionine does not interfere.

Finally, a new sensor for arginine based on the use of a thin slice of liver tissue has been reported. The sensor requires the presence of urease to convert the urea generated by the tissue slice to ammonia, which is detected by the gas sensor electrode (134).

Although they are not really immobilized-enzyme electrodes, some mention must be made of the ingenious biological sensors, based on complex cellular processes, which have been developed by Rechnitz and his coworkers. Sheep red cell ghosts can be loaded with an electrochemical marker which is freed from the cell by action of antibodies (169). The marker is then sensed by an ion-selective electrode. In a related approach, living bacterial cells can be loaded with a marker. Cell-wall disruptive enzymes, for example, lysozyme, can then be measured by the rate at which the marker compound appears in the solution (170). The last development which we must mention is a new technique in which one monitors the resistance of a lipid-bilayer membrane. Certain antibiotics can alter the resistance of the membrane, thereby causing a measurable event which can be used to estimate the concentration of antibiotic (177).

Definition of Terms

A	electrode or probe area (cm^2)
D_P, D_S, D	diffusion coefficient of product, substrate and a common value for both in the enzyme layer (cm^2/sec)
E	electrode potential (volts)
$[E]$	enzyme concentration in the active enzyme layer (mole/cc)
F	the Faraday (coul/equivalent)
ΔH	heat of reaction (cal/mole)
i	current (amperes)
k_1, k_{-1}, k_2	rate constants of the Michaelis–Menten reaction

$k_2' \equiv k_2[E]$ enzyme activity per unit volume (mole/ml·sec)

K_d partion coefficient of substrate between bulk and enzyme activity phases

K_I^P interference ratio of species I with respect to P for an ion selective electrode

K_M Michaelis constant (moles/ml)

L   thickness of enzyme layer (cm)

$L' - L$ thickness of semipermeable membrane

n equivalents of electrons per mole of material electrolyzed

N_S moles of substrate

N_P moles of product

[P]   product concentration (moles/cc)

R   the gas constant (volts/coul·°K)

[S]   substrate concentration (moles/ml)

$[S]_0$ initial or bulk substrate concentration (mole/ml)

T   temperature (°K)

t time from start of reaction (sec)

$t_{1/2}$ time needed to get to $\frac{1}{2}$ of steady state (sec)

t_{SS} time needed to get to the steady state (sec)

V*   volume of trapped enzyme layer

x distance from origin of coordinate system (cm)

z charge on an ion $(+1, -1$, etc.)

δ thickness of a diffusion layer (cm)

$\mathcal{H}$ thermal conductivity (cal/°C·sec·cm²)

ϕ_S, ϕ_P, ϕ permeability of a membrane for the substrate, product and a common value for both (ml/sec)

APPENDIX

Due to the apparent conflict between the results of the theoretical simulations of Tran-Minh and Broun on diffusion in potentiometric enzyme probes (20) and the experimental and theoretical results of all other groups on amperometric enzyme electrodes (13, 17), we undertook a second look at the simulation of the nonlinear differential equation (5.13, 5.15, 5.16) that govern these devices (242). Since Tran-Minh and Broun used the finite difference technique, we employed the method of orthogonal collocation, which is demonstrably better behaved in terms of numerical analysis; that is, it is stabler and more accurate. Our simulations clearly

Table 5.12   Minimum Enzyme Concentration Needed to Achieve Linear Response[a]

[S]$_0$ (mM)[b]	$K_M(mM)$		
	1.0	0.1	0.01
100.	1035	653	164
10.	164	103	65.3
1.0	26	16.4	10.3
0.1	4.1	2.6	1.6

[a] These were calculated from the simulations presented in reference 242. Enzyme concentrations are in μmole/min:cc. The substrate diffusion coefficient is assumed to be 10^{-5} cm^2/sec, enzyme thickness (L) is 100μ. Upper limit of linearity is set at 0.05 $\log_{10}$ units deviation from Nernstian response.
[b] Bulk substrate concentration at the upper limit of linearity.

indicate that the amount of enzyme on the electrode can increase the upper limit of linearity well beyond the K_M. The most significant results of our calculation are given in Table 5.12. This table allows one to estimate the concentration of active enzyme (units/cc) needed to obtain linear response at the indicated bulk substrate concentration. The upper limit of linearity was defined as the bulk substrate concentration at which the deviation from Nernstian behavior just exceeds 0.05 $\log_{10}$ units.

References

1. L. C. Clark and C. Lyons, *Ann. N.Y. Acad. Sci.* **102**, 29 (1962).

2. S. J. Updike and G. P. Hicks, *Nature* **214**, 986 (1967).

3. L. C. Clark, Jr., "A Family of Polarographic Enzyme Electrodes and the Measurement of Alcohol", *Biotechnol. Bioeng. Symposium No. 3,* (Wiley, New York, 1972), p. 377.

4. R. A. Llenado and G. A. Rechnitz, *Anal. Chem.* **43**, 1457 (1971).

5. G. A. Rechnitz, *Chem. Eng. News.,* 23 (Oct. 25, 1976).

6. J. D. Czaban and G. A. Rechnitz, *Anal. Chem.* **47**, 1787 (1975).

7. L. C. Clark, Jr., and E. W. Clark, in *Oxygen Transport to Tissue,* H. I. Bicher and D. F. Bruley, Eds. (Plenum Press, New York, 1973).

8. G. Baum, *Anal. Biochem.* **42**, 487 (1971).

9. K. P. Hsiung, S. S. Kuan, and G. G. Guilbault, *Anal. Chim. Acta* **84**, 15 (1976).

10. G. G. Guilbault and J. G. Montalvo, *J. Am. Chem. Soc.* **92**, 2533 (1970).

11. G. G. Guilbault, "Enzyme Electrode Probes," in *Immobilized Enzymes,*

Antigens, Antibodies and Peptides, H. H. Weetall, Ed. (Marcel Dekker, New York, 1975), p. 366.

12. G. G. Guilbault, "Enzyme Electrode Probes," in *Immobilized Enzymes, Antigens, Antibodies and Peptides,* H. H. Weetall, Ed. (Marcel Dekker, New York, 1975), p. 393.

13. L. D. Mell and J. T. Maloy, *Anal. Chem.* **47,** 299 (1975).

14. J. G. Montalvo, Jr. and G. G. Guilbault, *Anal. Chem.* **41,** 1897 (1969).

15. L. F. Cullen, J. F. Rusling, A. Schleifer, and G. J. Papariello, *Anal. Chem.* **46,** 1955 (1974).

16. G. S. Papariello, A. K. Mukherji, and C. M. Schearer, *Anal. Chem.* **45,** 790 (1973).

17. P. Racine and W. Mindt, *Experentia Suppl.* **18,** 525 (1971).

18. W. J. Blaedel, T. R. Kissel, and R. C. Boguslaski, *Anal. Chem.* **44,** 2030 (1972).

19. R. C. Boguslaski, W. J. Blaedel, and T. R. Kissel, "Kinetic Behavior of Enzymes Immobilized in Artificial Membranes," in *Insolubilized Enzymes,* M. Salmona, C. Saronio, and S. Garattinio, Eds. (Raven Press, New York, 1974).

20. C-T. Minh and G. Broun, *Anal. Chem.* **47,** 1359 (1975).

21. D. S. Papastathopoulos and G. A. Rechnitz, *Anal. Chem.* **48,** 862 (1976).

22. G. G. Guilbault and E. Hrabankova, *Anal. Chem.* **42,** 1779 (1970).

23. G. G. Guilbault and F. R. Shu, *Anal. Chim. Acta* **56,** 333 (1971).

24. C. Liteanu, E. Hopirtean, and I. C. Popescu, *Anal. Chem.* **48,** 2013 (1976).

25. D. S. Papastathopoulos and G. A. Rechnitz, *Anal. Chim. Acta* **79,** 17 (1975).

26. T. Anfalt, A. Graneli, and D. Jagner, *Anal. Letters* **6,** 969 (1973).

27. M. Mascini and A. Liberti, *Anal. Chim. Acta* **68,** 177 (1974).

28. P. W. Carr, *Anal. Chem.* **49,** 799 (1977).

29. M. Meyerhoff and G. A. Rechnitz, *Anal. Chim. Acta* **85,** 227 (1976).

30. L. D. Mell and J. T. Maloy, *Anal. Chem.* **48,** 1597 (1976).

31. S. W. Feldberg in *Electroanalytical Chemistry,* A. J. Bard, Ed., Vol. 3, Chapter 4 (Marcel Dekker, New York, 1969).

32. A. Hubaux and G. Vos, *Anal. Chem.* **42,** 849 (1970).

33. G. G. Guilbault and G. J. Lubrano, *Anal. Chim. Acta* **60,** 254 (1972).

34. M. Aizawa, I. Karube, and S. Suzuki, *Anal. Chim. Acta* **69,** 431 (1974).

35. M. Nanjo and G. G. Guilbault, *Anal. Chim. Acta* **75,** 169 (1975).

36. G. G. Guilbault and G. Nagy, *Anal. Chem.* **45,** 417 (1973).

37. G. G. Guilbault, *Handbook of Enzymatic Methods of Analysis* (Marcel Dekker, New York, 1976).

38. W. J. Blaedel and R. A. Jenkins, *Anal. Chem.* **48,** 1240 (1976).

39. G. G. Guilbault and E. Hrabankova, *Anal. Chim. Acta* **56,** 285 (1971).

40. G. A. Rechnitz and R. Llenado, *Anal. Chem.* **43**, 283 (1971).
41. G. G. Guilbault and M. Tarp, *Anal. Chim. Acta* **73**, 355 (1974).
42. M. Nanjo and G. G. Guilbault, *Anal. Chem.* **46**, 1769 (1974).
43. G. G. Guilbault and G. J. Lubrano, *Anal. Chim. Acta* **64**, 439 (1973).
44. G. G. Guilbault and M. Nanjo, *Anal. Chim. Acta* **78**, 69 (1975).
45. D. Thomas, *Physiol. Veg.* **14**, 843 (1976).
46. H. E. Booker and J. L. Haslam, *Anal. Chem.* **46**, 1054 (1974).
47. H. Nilsson, A. Akerlund, and K. Mosbach, *Biochim. Biophys. Acta* **320**, 529 (1973).
48. G. G. Guilbault and G. Lubrano, *Anal. Chim. Acta* **69**, 183 (1974).
49. K. Buckholz and W. Ruth, *Biotechnol. Bioeng.* **18**, 95, (1976).
50. B. Karlberg, *Talanta* **22**, 1023 (1975).
51. R. A. Durst, *Clin. Chem.* **23**, 298 (1977).
52. L. Meites, Ed., *Handbook of Analytical Chemistry*, (McGraw-Hill, New York, 1963), pp. 5–13.
53. N. D. Jespersen, *J. Am. Chem. Soc.* **97**, 1662 (1975).
54. S. N. Buhl, K. Y. Jackson, R. Lubinski, and R. E. Vanderlinde, *Clin. Chem.* **22**, 1872 (1976).
55. G. G. Guilbault and A. Iwase, *Anal. Chim. Acta* **85**, 295 (1976).
56. L. H. Goodson and W. B. Jacobs, *Midwest Research Institute Report*, MRI 1173, (1974).
57. L. H. Goodson and W. B. Jacobs, in *Enzyme Engineering*, E. K. Pye and L. B. Wingard, Jr., Eds., Vol. 2 (Plenum Press, New York, 1974), p. 393.
58. M. Nanjo and G. G. Guilbault, *Anal. Chim. Acta* **73**, 367 (1974).
59. G. G. Guilbault and E. Hrabankova, *Anal. Letters* **3**, 53 (1970).
60. W. J. Blaedel and J. M. Uhl, *Clin. Chem.* **21**, 119 (1975).
61. D. L. Williams, A. R. Doig, and A. Korosi, *Anal. Chem.* **42**, 118 (1970).
62. C. L. Cooney, J. C. Weaver, S. R. Tannenbaum, D. V. Faller, A. Shields, and M. Jahnke, in *Enzyme Engineering*, E. K. Pye and L. B. Wingard, Eds, Vol. 2 (Plenum Press, New York, 1974).
63. J. C. Weaver, C. L. Cooney, S. P. Fulton, P. Schuler, and S. R. Tannenbaum, *Biochim. Biophys. Acta* **452**, 285 (1976).
64. J. C. Weaver, C. L. Cooney, S. R. Tannenbaum, and S. P. Fulton, "Possible Biomedical Applications of the Thermal Enzyme Probe," in *Biomedical Applications of Immobilized Enzymes and Proteins*, T. M. S. Chang, Ed., Vol. 2 (Plenum Press, New York, 1977), p. 191.
65. L. D. Bowers, Ph.D. Dissertation, University of Georgia, 1974.
66. L. D. Bowers and P. W. Carr, *Thermochim. Acta* **10**, 129 (1974).
67. K. S. V. Santhanam, N. D. Jespersen, and A. J. Bard, *J. Am. Chem. Soc.* **99**, 274 (1977).
68. C. T. Minh and D. Vallin, *Anal. Chem.* **50**, 1874 (1978).

69. T. W. Freeman and W. R. Seitz, *Anal. Chem.* **50**, 1242 (1978).
70. G. G. Guilbault, *Crit. Rev. Anal. Chem.* **1**, 391 (1970).
71. G. G. Guilbault, *Pure Appl. Chem.* **25**, 727 (1971).
72. G. G. Guilbault in *Biotechnology and Bioengineering Symposium No. 3* (Wiley, New York, 1972), p. 361.
73. D. A. Gough and J. D. Andrade, *Science* **180**, 380 (1973).
74. H. H. Weetall, *Anal. Chem.* **46**, 602A (1974).
75. G. J. Moody and J. D. R. Thomas, *Analyst* **100**, 609 (1975).
76. G. A. Rechnitz, *Chem. Eng. News,* **53**, 29 (Jan. 27, 1975).
77. G. A. Rechnitz, *Science* **190**, 234 (1975).
78. N. Lakshminanayanaiah, *Membrane Electrodes* (Academic, New York, 1976).
79. L. D. Bowers and P. W. Carr, *Anal. Chem.* **48**, 544A (1976).
80. P. M. Rajczanyi, *Kem. Kozlem* **39**, 223 (1973).
81. G. G. Guilbault, "Analytical Uses of Immobilized Enzymes," in *Enzyme Engineering,* E. K. Pye and L. B. Wingard, Jr., Eds., Vol. 2 (Plenum Press, New York, 1974), p. 377.
82. G. G. Guilbault, "Analytical Uses of Immobilized Enzymes" in *Insolubilized Enzymes,* M. Salmona, C. Saronio, and S. Garattinio, Eds. (Raven Press, New York, 1974), p. 199.
83. G. G. Guilbault, *Bull. Soc. Chim. Belg.* **84**, 679 (1975).
84. G. G. Guilbault and T. J. Rohm, *Int. J. Env. Anal. Chem.* **4**, 51 (1975).
85. V. M. Nelboeck and D. Jaworek, *Chimia* **29**, 109 (1975).
86. I. V. Berezin, *Zh. Anal. Khim.* **31**, 786 (1976).
87. I. V. Berezin, *Usp. Khim.* **45**, 180 (1976).
88. T. M. S. Chang, Ed., *Biomedical Application of Immobilized Enzymes and Proteins,* Vols. 1 and 2 (Plenum Press, New York, 1977).
89. D. A. Gough, S. Arsenberg, C. K. Colton, J. Giner, and S. Soldener, "The Status of Electrochemical Sensors for *in vivo* Glucose Monitoring," Proceedings of a Workshop in Freiburg, March 1976 (George Thieme Publisher, Stuttgart, 1977), p. 10.
90. J. G. Schindler and W. Rieman, "Multi Component Electrochemical Analysis for Physiology and Medicine," Project Report Phillipps–University Marburg/Lahn (1977), p. 66.
91. R. P. Buck, *Anal. Chem.* **50**, 17R (1978).
92. D. N. Gray, M. Keyes, and B. Watson, *Anal. Chem.* **49**, 1067A (1977).
93. N. R. Larsen, E. H. Hansen, and G. G. Guilbault, *Anal. Chim. Acta* **79**, 9 (1975).
94. C. H. Kiang, S. S. Kuan, and G. G. Guilbault, *Anal. Chim. Acta* **80**, 209 (1975).
95. G. G. Guilbault and W. Stokbro, *Anal. Chim. Acta* **76**, 237 (1975).

96. M. Notin, R. Guillien, and P. Nabet, *Ann. Biol. Chim.* **30**, 193 (1972).

97. G. Nagy, H. von Storp, and G. G. Guilbault, *Anal. Chim. Acta* **66**, 443 (1973).

98. L. C. Clark and G. Sachs, *Ann. N.Y. Acad. Sci.* **148**, 133 (1968).

99. L. C. Clark, U.S. Patent 3,529,455 (1970).

100. F. Williams, Jr., A. Brunsman, J. Huntington, J. Johnson, and D. Newnan, "Amperometric-Oxidase Enzyme Probes for Biochemical Analysis," Application Notes, Yellow Springs Instrument Co., Yellow Springs, Ohio (1976).

101. F. R. Shu and G. S. Wilson, *Anal. Chem.* **48**, 1679 (1976).

102. R. A. Kamin, F. R. Shu and G. S. Wilson in "Electrochemical Studies of Biological Systems," American Chemical Society Symposium Series (1976).

103. G. G. Guilbault and J. G. Montalvo, Jr., *Anal. Letters* **2**, 283 (1969).

104. G. G. Guilbault and J. G. Montalvo, *J. Am. Chem. Soc.* **91**, 2164 (1969).

105. G. G. Guilbault and E. Hrabankova, *Anal. Chim. Acta* **52**, 287 (1970).

106. G. G. Guilbault and F. R. Shu, *Anal. Chem.* **44**, 2161 (1972).

107. G. G. Guilbault, G. Nagy, and S. S. Kuan, *Anal. Chim. Acta* **67**, 195 (1973).

108. M. Mascini and G. G. Guilbault, *Anal. Chem.* **49**, 795 (1977).

109. C. P. Pui, G. A. Rechnitz, and R. F. Miller, *Anal. Chem.* **50**, 330 (1978).

110. J. G. Montalvo, Jr., *Anal. Chem.* **41**, 2093 (1969).

111. J. G. Montalvo, Jr., *Anal. Biochem.* **38**, 357 (1970).

112. K. L. Crochet and J. G. Montalvo, Jr., *Anal. Chim. Acta* **66**, 259 (1973).

113. G. G. Guilbault and T. Cserfalvi, *Anal. Letters* **9**, 277 (1976).

114. P. Davies and K. Mosbach, *Biochim. Biophys. Acta* **370**, 329 (1974).

115. W. J. Blaedel and T. R. Kissel, *Anal. Chem.* **47**, 1602 (1975).

116. G. G. Guilbault and G. Nagy, *Anal. Letters* **6**, 301 (1973).

117. R. Wawro and G. A. Rechnitz, *J. Membr. Sci.* **1**, 143 (1976).

118. T. Kawashima and G. A. Rechnitz, *Anal. Chim. Acta* **83**, 9 (1976).

119. E. J. Lahoda, C. C. Liu and L. B. Wingard, *Biotechnol. Bioeng.* **17**, 413 (1975).

120. L. B. Wingard, Jr., C. C. Liu, and N. L. Nagda, *Biotechnol. Bioeng.* **13**, 629 (1971).

121. G. Nagy and E. Pungor, *Hung. Sci. Instr.* **32**, 1 (1975).

122. S. Suzuki, F. Takahashi, I. Satoh and N. Sonobe, *Bull. Chem. Soc. Japan* **48**, 3246 (1975).

123. H. Durliat, M. Comtat, J. Mahenc, and A. Baudras, *J. Electroanal. Chem.* **66**, 73 (1975).

124. H. Durliat, M. Comtat, J. Mahenc, and A. Boudras, *Anal. Chim. Acta* **85**, 31 (1976).

125. Z. Toul and L. Macholan, *Coll. Czech. Chem. Commun.* **40**, 2208 (1975).
126. T. Cserfalvi and G. G. Guilbault, *Anal. Chim. Acta* **84**, 259 (1976).
127. C. T. Minh, E. Selegny, and G. Broun, *Compt. Rend.* **275**, 309 (1972).
128. A. M. Berjonneau and D. Thomas, *Path. Biol.* **22**, 497 (1974).
129. C. Divies, *Ann. Microbiol.* **126A**, 175 (1975).
130. G. A. Rechnitz, R. K, Kobos, T. L. Reichel, and C. R. Gebauer, *Anal. Chim. Acta* **94**, 357 (1977).
131. R. K. Kobos and G. A. Rechnitz, *Anal. Letters* **10**, 751 (1977).
132. M. A. Jensen and G. A. Rechnitz, *Anal. Chim. Acta.* **101**, 125 (1978).
133. G. A. Rechnitz, T. L. Reichel, R. K. Kobos, and M. E. Meyerhoff, *Science* **199**, 440 (1978).
134. G. A. Rechnitz and M. Meyerhoff, *Chem. Eng. News,* 16 (Oct. 9, 1978).
135. T. L. Reichel and G. A. Rechnitz, *Biochem. Biophys. Res. Comm.* **74**, 1377 (1977).
136. J. G. Schiller, A. K. Chen, and C. C. Liu, *Anal. Biochem.* **85**, 25 (1978).
137. Ph. Racine, H. O. Klenk, and K. Kochsiek, *Z. Klin. Chem. Klin. Biochem.* **13**, 533 (1975).
138. Ph. Racine, W. Mindt, and P. Schlapfer, Proc. Electrochemical Bioscience and Bioengineering Electrochemical Society Meeting, Chicago (1973).
139. P. Racine, W. Mindt, C. Rossel and P. Schlapfer, Proc. 10th International Conference on Medical and Biological Engineering, August 1973, Dresden, 1973.
140. P. Racine, R. Englehard, J. C. Higelin and W. Mindt, *Med. Instr.* **9**, 11 (1975).
141. W. Mindt, P. Racine and P. Schlapfer, *Ber. Bunsen Ges.* **77**, 804 (1973).
142. L. C. Clark and G. Sachs, *Ann. N.Y. Acad. Sci.* **148**, 133 (1968).
143. S. Suzuki, M. Aizawa, and I. Karube, "Electrochemical Preparation of Enzyme-Collogen Membrane and Its Applications," in *Immobilized Enzyme Technology,* H. H. Weetall and S. Suzuki, Eds. (Plenum Press, New York, 1974), p. 253.
144. G. J. Lubrano and G. G. Guilbault, *Anal. Chim. Acta* **97**, 229 (1978).
145. I. Satoh, I. Karube, and S. Suzuki, *Biotechnol. Bioeng.* **19**, 1095 (1977).
146. P. J. Taylor, E. Kmetec and J. M. Johnson, *Anal. Chem.* **49**, 789 (1977).
147. S. C. Martiny and O. J. Jensen, "An Enzyme Electrode Based on Immobilized Glucose Oxidase and Ion Enzyme Electrodes" in *Biological Medicine,* M. Kessler, ed., University Park Press, Baltimore, Maryland, 1976, 198–199.
148. H. Nilsson and K. Mosbach, *Biotechnol. Bioeng.* **20**, 527 (1978).
149. I. Fritz, G. Nagy, L. Fodor, and E. Pungor, *Analyst* **101**, 439 (1976).
150. L. Macholan, *Coll. Czech. Chem. Commun.* **43**, 1811 (1978).

151. K. S. Chua and I. K. Tan., *Clin. Chem.* **24**, 150 (1978).

152. D. F. Ollis and R. Carter, Jr., "Kinetic Analysis of a Urease Electrode" in *Enzyme Engineering,* E. K. Pye and L. B. Wingard, Jr., Eds., Vol. 2, (Plenum Press, New York 1974), p. 271.

153. S. O. Enfors, *Proc. Analyt. Div. Chem. Soc.* **14**, 106 (1977).

154. B. K. Ahn, S. K. Wolfson and S. J. Yao, *Bioelectrochem. Bioenergy,* **2**(2), 142 (1975).

155. C. Tran-Minh, R. Guyonnet, and J. Beaux, *Comp. Rend.* **286**, 115 (1978).

156. A. K. Chen and C. C. Liu, *Biotechol. Bioeng.* **19**, 1785 (1977).

157. A. Malinauskos and J. Kulys, *Anal. Chim. Acta* **98**, 31 (1978).

158. W. C. White and G. G. Guilbault, *Anal. Chem.* **50**, 1481 (1978).

159. G. G. Guilbault and G. J. Lubrano, *Anal. Chim. Acta* **97**, 229 (1978).

160. S. P. Bessman and R. D. Schultz, *Trans. Am. Soc. Artif. Intern. Organs,* **19**, 361 (1973).

161. S. P. Bessman and R. D. Schultz, "Progress Toward a Glucose Sensor for the Artificial Pancreas," in *Ion Selective Microelectrodes,* H. J. Berman and N. C. Hebart, Eds. (Plenum Press, New York, 1974).

162. E. C. Layne, R. D. Schultz, L. J. Thomos Jr., G. Slama, D. F. Sayler, and S. P. Bessman, *Diabetes* **25**, 81 (1976).

163. L. C. Clark, Jr., "A Polarographic Enzyme Electrode for the Measurement of Oxidase Substrates," in *Oxygen Supply,* M. Kessler, Ed. (University Park Press, Baltimore, MD, 1973), p. 120.

164. I. A. Silver, "An Ultra Micro Glucose Electrode," in *Ion and Enzyme Electrodes in Biology and Medicine,* M. Kessler, Ed. (University Park Press, Baltimore, MD, 1976), p. 189.

165. I. Satoh, I. Karube, and S. Suzuki, *Biotechol. Bioeng.* **18**, 269 (1976).

166. W. W. C. Chan, P. Davies, and K. Mosbach, "Immobilized Enzymes and Ligands in Biological Research," in *Ion and Enzyme Electrodes in Biology and Medicine,* M. Kessler, Ed. (University Park Press, Baltimore, MD, 1976), p. 182.

167. L. Macholan and L. Schanel, *Coll. Czech. Chem. Commun.* **42**, 3667 (1977).

168. P. Durand, A. David, and D. Thomas, *Biochem. Biophys. Acta* **527**, 277 (1978).

169. D'Orazio and G. A. Rechnitz, *Anal. Chem.* **49**, 2083 (1977).

170. D.'Orazio, M. E. Meyerhoff, and G. A. Rechnitz, *Anal. Chem.* **50**, 1531 (1978).

171. C. Tran-Minh and J. Beaux, *Anal. Chem.* **51**, 91 (1979).

172. C. Tran-Minh and J. Beaux, *Compt. Rend. C* **287**, 191 (1978).

173. T. Shinbo, M. Sugiura, and N. Kamo, *Anal. Chem.* **51**, 100 (1979).

174. D. R. Thevenot, R. Sternberg, P. R. Coulet, J. Laurent, and D. C. Goutheron, *Anal. Chem.* **51**, 96 (1979).

175. S. Rich, R. M. Ianniello, and N. D. Jespersen, *Anal. Chem.* **51**, 204 (1979).
176. R. K. Kobos, D. J. Rice, and D. S. Flournoy, *Anal. Chem.* **51**, 1122 (1979).
177. M. Thompson, P. J. Worsford, J. M. Holuk, and E. A. Stubley, *Anal. Chim. Acta* **104**, 195 (1979).
178. P. R. Coulet and C. Bertrand, *Anal. Letters* **12**, 581 (1979).

The following additional references from late 1978 to date have appeared to late to review in the text:

179. C. J. Olliff, R. T. Williams, and J. M. Wright, "A Novel Penicillin Enzyme Electrode," *J. Pharm. Pharmacol.* **30**, 45 (1978).
180. S. O. Enfors and N. Molin, "Response Characteristics of pH-Based Enzyme Electrodes for Analysis of Penicillin and Glucose in Fermentation Broth," *Prepr.-Eur. Congr. Biotechnol.* **1**, 35–38 (1978).
181. L. C. Clark, Jr., C. Emory, C. J. Glueck, and M. Campbell, "The Cholesterol Electrode: Use of the Polarographic Oxidase Anode with Multiple Enzymes," *Enzyme Eng.* **3**, 409 (1978).
182. T. L. Riechel and G. A. Rechnitz, "Hybrid Bacterial and Enzyme Membrane Electrode with Nicotinamide Adenine Dinucleotide Response," *J. Membr. Sci.* **4**, 243 (1978).
183. J. Kulys and K. Kadziauskiene, "Bioelectrocatalysis. Lactate-Oxidizing Electrode," *Dokl. Akad. Nauk SSSR* **239**, 636 (1978).
184. T. Matsunaga, I. Karube, and S. Suzuki, "Rapid Determination of Nicotinic Acid by Immobilized Lactobacillus Arabinosus," *Anal. Chim. Acta.* **99**, 233 (1978).
185. Y. A. Aleksandrovskii, P. K. Agasyan, A. M. Egorov, A. P. Osipov, Y. V. Rodinov, and I. V. Berezin, "Enzymic Sensor for Determining Glucose," *Zh. Anal. Khim.* **33**, 1833 (1978).
186. C. Tran-Minh and R. Guyonnet, "Enzyme Sensor for the Detection and Continuous Determination of Toxic Substances," *Compt. Rend.* **286**, 357 (1978).
187. A. S. Barker and P. J. Somers, "Enzyme Electrodes and Enzyme Based Sensors," *Enzyme Ferment. Biotechnol.* **2**, 120 (1978).
188. A. Takasaka, "Enzyme Electrodes," *Rinsho Kensa* **22**, 556 (1978).
189. M. Nanjo, "Enzyme Electrode: Application of Immobilized Enzyme Membranes as New Functional Materials," *Nippon Kinzoku Gakkai Kaiho* **17**, 1039 (1978).
190. I. Karube and S. Suzuki, "Use of Enzyme- or Microorganism-Collagen Membrane for Electrochemical Measurements," *Kagaku Kogaku* **42**, 496 (1978).
191. S. Ikeda, K. Ito, and T. Kondo, "An Enzyme-Electrode Type Glucose Sensor for an Artificial Pancreas," *Denki Kagaku Oyobi Kogyo Butsuri Kagaku* **46**, 667 (1978).
192. F. Scheller and D. Pfeiffer, "Enzyme Electrodes," *Z. Chem.* **18**, 50 (1978).

193. S. O. Enfors and N. Molin, "Enzyme Electrodes for Fermentation Control," *Process Biochem.* **13**, 9, 24 (1978).

194. S. Suzuki, I. Karube, and I. Satoh, "Fundamental Studies on Bio-Electrochemical Sensor. II. Amines Sensor," *Asahi Garasu Kogyo Gijutsu Shoreikai Kenkyu Hokoku* **32,** 355 (1978).

195. S. Suzuki and M. Aizawa, "Membrane Sensor: Use of Functional Membranes Capable of Recognizing Chemical Substances," *Maku* **4**, 37 (1979).

196. S. Hirose, M. Hayashi, N. Tamura, S. Suzuki, and I. Karube, "Poly(vinyl chloride) Membrane for a Glucose Sensor," *J. Mol. Catal.* **6**, 251 (1979).

197. M. Hikuma, T. Kubo, T. Vasuda, I. Karube, and S. Suzuki, "Amperometric Determination of Acetic Acid with Immobilized *Trichosporan brassicoe*," *Anal. Chim. Acta* **109**, 33 (1979).

198. T. Matsumaga, I. Karube, and S. Suzuki, "Electrode System for the Determination of Microbial Populations," *Appl. Env. Microbiol.* **37**, 117 (1979).

199. M. Aizawa, A. Morioka, S. Suzuki, and Y. Nagamura, "Enzyme Immunosensor. Amprometric Determination of Human Chorionic Gonadotropin by Membrane-Bound Antibody," *Anal. Biochem.* **94**, 22 (1979).

200. M. Hikuma, T. Kubo, T. Yasuda, I. Karube, and S. Suzuki, "Microbial Electrode Sensor for Alcohols," *Biotechnol. Bioeng.* **21**, 1845 (1979).

201. I. Karube, S. Mitsuda, and S. Suzuki, "Glucose Sensor Using Immobilized Whole Cells of Pseudomonas Fluorescens," *Eur. J. Appl. Microbiol. Biotechnol.* **7**, 343 (1979).

202. I. Karube, T. Matsunaga, and S. Suzuki, "Microbioassay of Nystatin with a Yeast Electrode," *Anal. Chim. Acta* **109**, 39 (1979).

203. K. Matsumoto, H. Seijo, T. Watanabe, I. Karube, I. Satoh, and S. Suzuki, "Immobilized Whole Cell-Based Flow-Type Sensor for Cephalosporins," *Anal. Chim. Acta* **105**, 429 (1979).

204. M. A. Jensen and G. A. Rechnitz, "Enzyme 'Sequence' Electrode for D-Gluconate," *J. Membr. Sci.* **5**, 117 (1979).

205. G. A. Rechnitz, "Bio-Selective Membrane Electrodes," *NBS Spec. Publ.* **519**, 525 (1979).

206. G. A. Rechnitz, M. A. Arnold, and M. E. Meyerhoff, "Bioselective Membrane Electrode Using Tissue Slices," *Natue* **228**, 466 (1979).

207. P. D'Orazio and G. A. Rechnitz, "Potentiometric Electrode Measurement of Serum Antibodies Based on the Complement Fixation Test," *Anal. Chim. Acta* **109**, 25 (1979).

208. R. L. Solsky and G. A. Rechnitz, "Antibody-Selective Membrane Electrodes," *Science* **204**, 1308 (1979).

209. M. E. Meyerhoff and G. A. Rechnitz, "Electrode-Based Enzyme Immunoassays Using Urease Conjugates," *Anal. Biochem.* **95**, 483 (1979).

210. C. R. Gebauer, M. E. Meyerhoff, and G. A. Rechnitz, "Enzyme Electrode-Based Kinetic Assays of Enzyme Activities," *Anal. Biochem.* **95**, 479 (1979).

211. M. A. Arnold and G. A. Rechnitz, "Determination of Glutomine in Cerebrospinal Fluid with a Tissue-Based Membrane Electrode," *Anal. Chim. Acta* **113**, 351 (1980).

212. L. B. Wingard, Jr., J. G. Schiller, S. K. Wolfson, Jr., C. C. Liu, A. L. Drash, and S. J. Yao, "Immobilized Enzyme Electrodes for the Potentiometric Measurement of Glucose Concentration: Immobilization Techniques and Materials," *J. Biomed. Mater. Res.* **13**, 921 (1979).

213. C. C. Liu, L. B. Wingard, Jr., S. K. Wolfson, Jr., S. J. Yao, A. L. Drash, and J. G. Schiller, "Quantitation of Glucose Concentration Using a Glucose Oxidase-Catalase Electrode by Potentiometric Measurement," *Bioelectrochem. Bioenerg.* **6**, 19 (1979).

214. L. B. Wingard, Jr., and J. L. Gurecka, Jr., "Direct Electron Transfer at an Immobilized Cofactor Electrode: Approaches and Progress," *Biotechnol. Bioeng. Symp.* **8**, 483 (1979).

215. J. L. Boitieux, G. Desmet, and D. Thomas, "An 'Antibody Electrode,' Preliminary Report on a New Approach in Enzyme Immunoassay," *Clin. Chem.* **25**, 318 (1979).

216. J. L. Romette, B. Froment, and D. Thomas, "Glucose-Oxidase Electrode. Measurements of Glucose in Samples Exhibiting High Variability in Oxygen Content," *Clin. Chim. Acta* **95**, 249 (1979).

217. C. Bourdillon, J. P. Bourgeois, and D. Thomas, "Chemically Modified Electrodes Bearing Grafted Enzymes," *Biotechnol. Bioeng.* **21**, 1877 (1979).

218. A. Malinauskas and J. Kulys, "Flow-Through NAD Sensor," *Biotechnol. Bioeng.* **21**, 513 (1979).

219. J. Kulys and A. Malinauskas, "Flow-Through Enzyme Electrode Based on Dehydrogenases with Regeneration of Cofactor (Nicotinamide-Adenine Dinucleotide)," *Zh. Anal. Khim.* **34**, 876 (1979).

220. A. Malinauskas and J. Kulis, "Use of an Enzymic Fuel Element for the Determination of Dehydrogenase Substrates," *Zh. Biol. Khim.* 1979, p. 412 (Abstr. No. 12F373).

221. P. Posadka and L. Macholan, "Amperometric Assay of Vitamin C Using an Ascorbate Oxidase Enzyme Electrode," *Collect. Czech. Chem. Commun.* **44**, 3395 (1979).

222. L. Macholan, "Determination of Beta-Glycosidases, Beta-Glucuronidase, and Alkaline Phosphatase by an Enzyme Electrode Sensitive to Phenol," *Collect. Czech. Chem. Commun.* **44**, 3033 (1979).

223. K. W. Fung, S. S. Kuan, H. Y. Sung, and G. G. Guilbault, "Methionine Selective Enzyme Electrode," *Anal. Chem.* **51**, 2319 (1979).

224. G. G. Guilbault and M. H. Sadar, "Preparation and Analytical Uses of Immobilized Enzymes," *Acc. Chem. Res.* **12**, 344 (1979).

225. C. Bertrand, P. R. Coulet, and D. C. Gautheron, "Enzyme Electrode with Collagen-Immobilized Cholesterol Oxidase for the Microdetermination of Free Cholesterol," *Anal. Lett.* **12**, 1477 (1979).

226. P. R. Coulet and C. Bertrand, "Asymmetrical Coupling of Enzymic Systems on Collagen Membranes. Application to Multienzyme Electrodes," *Anal. Lett.* **12**, 581 (1979).

227. H. Durliat, M. Comtat, and J. Mahenc, "A Device for the Continuous Assay of Lactate," *Anal. Chim. Acta* **106**, 131 (1979).

228. J. Mahenc and H. Aussaresses, "Enzyme Specific Electrode for Glucose Based on the Amperometric Detection of Ferrocyanide," *Compt. Rend.* **28**, 357 (1979).

229. S. J. Updike, M. C. Shults, and M. Busby, "Continuous Glucose Monitor Based on an Immobilized Enzyme Electrode Detector," *J. Lab. Clin. Med.* **93**, 518 (1979).

230. L. Sokol, C. Garber, M. Shults, and S. Updike, "Immobilized Enzyme Rate Determination Method for Glucose Analysis," *Clin. Chem.* **26**, 89 (1980).

231. L. C. Clark, Jr., "The Hydrogen Peroxide Sensing Platinum Anode as an Analytical Enzyme Electrode," *Methods Enzymol.* **56**, 448 (1979).

232. D. Skogberg and T. Richardson, "Preparation and Use of an Enzyme Electrode for Specific Analysis of L-Lysine in Cereal Grains," *Cereal Chem.* **56**, 147 (1979).

233. M. Mascini and C. Botre, "Enzyme Electrodes. New Advances and Procedures for Implementation," *Chim. Ind.* **61**, 542 (1979).

234. T. Yao and S. Musha, "Electrochemical Enzymic Determinations of Ethanol and L-Lactic Acid with a Carbon Paste Electrode Modified Chemically with Nicotinamide Adenine Dinucleotide," *Anal. Chim. Acta* **110**, 203 (1979).

235. C. R. Lowe, "The Affinity Electrode. Application to the Assay of Human Serum Albumin," *FEBS Lett.* **106**, 405 (1979).

236. K. G. Kjellen and H. Y. Neujahr, "Enzyme Electrode for Phenol," *Biotechnol. Bioeng.* **22**, 299 (1980).

237. B. Watson, D. N. Stifel, and F. E. Semersky, "Development of a Glucose Analyzer Based on Immobilized Glucose Oxidase," *Anal. Chim. Acta* **106**, 233 (1979).

238. H. Weise, F. Scheller, K. Siegler, and D. Pfeiffer, "Measurement of Different Carbohydrates with the Help of an Enzyme Electrode, Part I. Determination of Sucrose," *Lebensmittelindustrie* **26**, 206 (1979).

239. T.-M. Yuan, "Enzyme Electrodes," *Fen Hsi Hua Hsueh* **7**, 149 (1979).

240. S. V. Enfors and H. Nilsson, "Design and Response Characteristics of an Enzyme Electrode for Measurement of Penicillin in Fermentation Broth," *Enzyme Microb. Technol.* **1**, 260 (1979).

241. P. Abel, "Extracorporeal Continuous Glucose Measurement Using an Enzyme Electrode: Condition and Results," *Int. Wiss. Kolloq.-Tech. Hochscl. Ilmenow* **24**, 93 (1979),

242. J. E. Brady and P. W. Carr, *Anal. Chem.* **52**, 977 (1980).

FUNDAMENTALS OF CHEMICAL REACTION KINETICS IN FLOW SYSTEMS

The present chapter is concerned with some of the elementary concepts of reactions which take place in systems with fluid flow. Our goal has been to *introduce* to the reader a few of the more important concepts, such as sample dispersion, and to indicate how such factors may influence the efficiency of a reactor, without becoming deeply involved in the mathematics of such systems. Basically, we feel that the practical chemist must be aware of such factors as are introduced in this chapter so that he can empirically optimize his actual system. The interested reader should consult the references to the chemical engineering literature if a more detailed exposition is of interest.

One of the major benefits conveyed by using an immobilized enzyme is the fact that the catalyst is present on a solid phase which can easily be reclaimed. As in the case of a simple, manual, wet analysis the enzyme can be added to a convenient container, stirred to allow mixing and mass transfer, and then removed. The amount of analyte can then be determined from the change in concentration of a coreactant, for example, a coenzyme such as NADH, or from the change in concentration of a reaction product. This is a *batch mode* of operation and is analogous to a simple batch extraction in a separatory funnel or with an ion-exchange resin in a flask. By analogy to separation practices, it is much more efficient to carry out reactions in a column on a continuous basis. This permits much higher analysis rates, that is, higher sample throughput, and may provide better precision because of improved control over reaction conditions, principally time and temperature. Obviously soluble enzymes can be used in flow reactors but will be very difficult to reclaim and therefore it may be economically impossible to expose each sample to more than a small amount of the enzyme, depending upon its cost.

Consider the possibilities inherent in an insoluble preparation of enzyme. Even if its activity yield upon immobilization is only 10%, it can be reused many, many times if stored properly and if it is not poisoned by impurities in the samples. This reusability allows us to "hit" each sample with a very high level of activity. At the very least this may permit a very short reaction time to obtain a measurable signal. Longer incubation times may allow the reaction to go to equilibrium thereby greatly lessening the dependence of the analysis on precise control of time, pH, ionic strength, temperature, and all the other variables which generally have more influence on the rate of a reaction than on an equilibrium constant. Equilibrium conversions also avoid the accuracy problems involved when inhibitors or activators are present in the sample.

Real flow systems may be viewed as falling somewhere between the two extreme types of flow systems depicted in Figure 6.1. The first is generally termed a continuously stirred tank reactor (1) (CSTR). This system represents the limit of extreme sample spreading, that is, dispersion. A solid catalyst can be trapped in the system by appropriate filters or by being fixed to the wall of the reaction vessel. Samples are carried into the reactor by a flowing stream, mixed rapidly by the stirrer, and carried out to a detector by the flow of fluid. Obviously if no reaction takes place in the vessel, a very narrow pulse of sample added at the inlet would appear at the reactor outlet or exit as a decaying exponential. That is, in the absence of a chemical reaction, the system behaves as a simple exponential dilutor. Consequently, the time spent by any given molecule in the reactor vessel may be anywhere from zero to infinity and there is no simple unique residence time, but rather a distribution of residence times. This has the consequence that when a reaction is taking place in the system, some of the molecules will have almost no time to react and others will have a great deal of time. Another important analytical consequence of a CSTR is that a small volume of analyte (V_a) will be greatly diluted by being mixed into the larger volume of the reactor (V_0). This has the unfortunate effect of decreasing the sensitivity or slope of a calibration curve.

The second class of reactors (see Figure 6.1) is one in which ideally there is assumed to be *no mixing* whatsoever. Thus the sample moves through the reactor with very sharp, distinct boundaries at the front and tail of the pulse, that is the plug is transported as a set of interconnected elements moving in file through the column. This class of system is termed a *plug flow reactor* (PFR) (1). *In reality PFR's and CSTR's do not exist, they represent extreme (theoretical) limits of behavior of real reactors in*

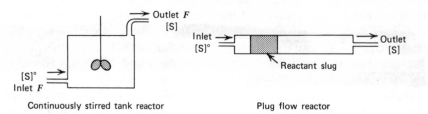

Continuously stirred tank reactor   Plug flow reactor

Fig. 6.1.   Idealized continuously stirred tank reactor and plug flow reactor.

which there is no mixing and instantaneous complete mixing respectively. As with a CSTR, the sample is carried into and out of the reaction zone by fluid flow. A PFR has a unique residence time. Consider a PFR into which a finite volume of sample is admitted. The leading and tailing edge of the sample pulse both spend exactly the same period of time in the reactor. This will greatly simplify calculation of the extent of reaction because the system will behave identically to a batch reactor. A major analytical advantage of a PFR is that the sample is *not unnecessarily diluted* and in principle there will be no undue decrease in sensitivity.

As pointed out above, plug flow and continuously stirred reactors do not exist, they are *idealized models*. Several types of real reactors are shown in Figure 6.2. These include fixed and fluidized bed reactors, cylindrical, tubular, batch, and continuous reactors. A fluidized bed is very similar to a fixed-bed system, but the bed is allowed to expand upwards as the flow rate is increased. The advantage of a fluidized bed lies in the fact that it will not clog as readily as a fixed bed; however this volume expansion permits mixing and dilution. Such systems do not appear to have any real *analytical* merit at this time and will not be considered further. Tubular reactors have been used extensively for analysis with immobilized enzymes. Hornby and his coworkers (2) were among the first to covalently attach enzymes to the walls of nylon tubes and use them in continuous segmented flow analyzers. Chapter 8 should be consulted for recent examples of the use of reactors in chemical analysis.

6.1 COMPARISON OF CONTINUOUSLY STIRRED TANK AND PLUG FLOW REACTORS

The purpose of this section is to compare the operation and use of CSTR and PFR systems and to present the assumptions and limitations in their operation. For the sake of simplicity we will assume that (1) all reactors are isothermal even though heat will be generated by reaction or by viscous forces between the fluid and the stirrer or packing material, (2) the volumetric flow rate F (ml/sec) is constant, (3) the catalyst is nonporous or diffusion in a porous catalyst is very rapid, and (4) mass transfer from the bulk of the fluid to the particle surface is very fast. In the case of a CSTR, we will assume that mixing is instantaneous; for a PFR we assume that the fluid is radially homogeneous and that no axial mixing or dilution takes place. A detailed treatment of the assumptions involved for immobilized-enzyme systems may be found in the excellent review by Vieth et al. (3). The total rate of reaction will be the product of the reaction rate $(\mathcal{R})$ and the volume in which the reaction is taking place.

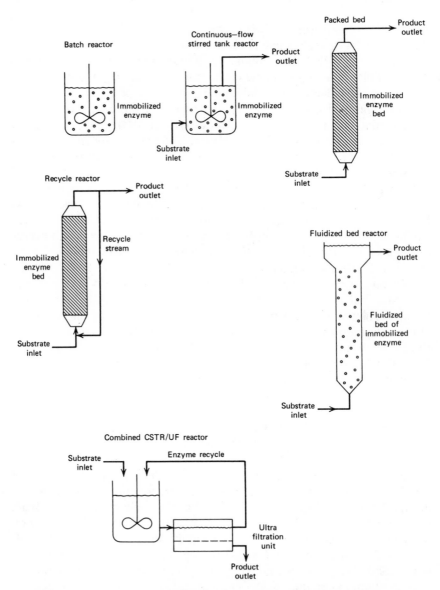

Fig. 6.2. Several types of real chemical reactors. [Reprinted from W. R. Veith, K. Venkatasubramanian, A. Constantinides, and B. Davidson, in *Applied Biochemistry and Bioengineering,* Vol. 1 (Academic Press, New York, 1976). Copyright by Academic Press.]

314

Even though the reaction is heterogeneous, that is, taking place at a surface, it can be treated, based on the mass-transfer assumptions outlined above, as if it were taking place in a homogeneous fluid. One can account for mass-transfer limitations on the reaction rate by use of so-called effectiveness factors and thereby take into account the fact that the reaction is actually heterogeneous even though we treat it as being homogeneous.

6.1.1 Mass Balance in a Continuously Stirred Tank Reactor

We will now briefly consider a continuously stirred tank reactor (Figure 6.1). Detailed treatments can be found in texts on chemical engineering (1, 4, 5). For the moment, let us assume that the reactants are being fed in continuously and that no analyte is present in the reactor when it is started up. The rate of change of analyte concentration, denoted as [S], will be:

$$V_0 \frac{d[S]}{dt} \left(\frac{mole}{sec} \right) = \text{(Flow of Analyte In)}$$

$$- \text{(Flow of Analyte Out)} - \mathcal{R} \cdot V_0 \quad (6.1)$$

The number of moles per second flowing into the reactor will be $F \cdot [S]^0$ and, since the solution is homogeneous, the concentration of analyte at the outlet will be [S]. Thus the outward flow will be $F \cdot [S]$, assuming constant density (ΔT small):

$$V_0 \frac{d[S]}{dt} = F([S]^0 - [S]) - \mathcal{R} V_0 \quad (6.2)$$

Equation (6.2) can be integrated subject to some initial condition, for example, [S] equal to zero at zero time, once the reaction-rate term has been specified. Table 6.1 should be consulted for the solutions for zero- and first-order kinetics. The use of Michaelis–Menten kinetics does not yield an *explicit* solution, so it is not presented here.

When the flow-rate terms are constant, the substrate concentration will eventually achieve a steady state, at which point its concentration ($[S]_{ss}$) in the vessel and eluent becomes:

$$[S]_{ss} = [S]^0 - \frac{\mathcal{R} V_0}{F} \quad (6.3)$$

The steady-state concentration of product will be

$$[P]_{ss} = \frac{\mathcal{R} V_0}{F} ; \quad \left(\text{only when } \frac{\mathcal{R} V_0}{F} < [S]^0 \right) \quad (6.4)$$

Table 6.1 Concentration-Time Relationships for a CSTR with Continuous Supply of Substrate

Case	Substrate concentration	Time to steady state	Fractional Conversion (X) in Steady State[d]
No reaction—flush-out only	$[S] = [S]^0\left[1 - \exp\left(-\dfrac{t}{\tau}\right)\right]$	$t > 5\tau$	—
Zero-order[a]	$[S] = ([S]^0 - k_0\tau)\left[1 - \exp\left(-\dfrac{t}{\tau}\right)\right]$	$t > 5\tau$	$X = \dfrac{k_0\tau}{[S]^0}$
First-order[b]	$[S] = \dfrac{[S]^0}{k_1\tau + 1}\left\{1 - \exp\left[-\left(\dfrac{1}{\tau} + k_1\right)t\right]\right\}$	$t > \dfrac{5\tau}{1 + k_1\tau}$	$X = \dfrac{\tau}{k_1\tau + 1}$
Michaelis–Menten[c]	Too complex to be useful	—	$[S]^0 X + \dfrac{K_M X}{1 - X} = k_2[E]\tau$

[a] k_0 = Zero-order rate constant.
[b] k_1 = First-order rate constant.
[c] K_M = Michaelis constant and $k_2[E]$ defined in Chapter 2.
[d] See reference 3 for alternate forms and more complex cases with inhibitors present.

Even though the term V_0/F has units of time in the case of a CSTR it *should not* be thought of as the residence time of the system but rather as a space–time coordinate. As will be seen this space–time parameter is very important and will be denoted as τ:

$$\tau \equiv \frac{V_0}{F} \qquad (6.5)$$

As indicated by the equation of Table 6.1, the time required to reach steady-state generation will depend upon both τ and the reaction-rate term ($\mathcal{R}$). For a simple system where no reaction takes place ($\mathcal{R} = 0$), the substrate concentration in the vessel will achieve 99% of steady state when $t > 5\tau$. In the steady state, the fractional conversion (X) of substrate to product will be:

$$X = \frac{[P]_{ss}}{[S]^0} = \frac{\mathcal{R}\tau}{[S]^0} \le 1 \qquad (6.6)$$

It is evident that for a first-order reaction the fraction conversion is independent of the inlet or feed concentration, and that for a zero-order reaction it decreases monotonically as the feed concentration increases. Equation (6.6) also points up the fact that *an increase in space–time will increase the steady-state conversion but also increases the time required to achieve steady state.* In an analytical sense, this *requires a trade-off between high sensitivity and response time.* The only way to circumvent this is to force the reaction rate term $\mathcal{R}$ to be very large, in which case the extent of conversion will approach unity and the response time can be made faster by increasing the flow rate or decreasing the reactor volume.

Rapid automated analysis may require that the sample be injected into the flow stream in a very small volume (V_a). In this instance, the first term in equation (6.1) should be dropped since there is no continuous supply of substrate. If the substrate or analyte concentration in the added volume V_a is denoted $[S]^0$, then its *initial value* immediately after mixing in the reactor will be:

$$[S]_0 = \frac{V_a \cdot [S]^0}{V_0} \qquad (6.7)$$

The integrated rate laws corresponding to the modified version of equation (6.1) are presented in Table 6.2. Once again we see that the critical parameter in determining the time response is the space–time τ. The total amount of product formed by reaction can be computed by an appropriate mass balance and integrating the product concentration multiplied by flow rate from zero to infinity. The result is simply obtained by noting that the

Table 6.2 Concentration–Time Relationship for CSTR with Impulse
Introduction of Substrate[a]

Case	Substrate concentration	Time to 1% initial concentration	Fractional conversion[b] at $t = \infty$
No reaction	$[S] = [S]_0 \exp\left(-\dfrac{t}{\tau}\right)$	$t > 5\tau$	0
Zero-order	$[S] = [S]_0 \exp\left(-\dfrac{t}{\tau}\right)$ $+ k_0\tau \left[\exp\left(-\dfrac{t}{\tau}\right) - 1\right]$	$t > 5\tau$	Not defined
First-order	$[S] = [S]_0 \exp$ $\left[-\left(\dfrac{1}{\tau} + k_1\right)t\right]$	$t > \dfrac{5\tau}{(1 + k_1\tau)}$	$X = \dfrac{k_1\tau}{1 + k_1\tau}$

[a] See Table 6.1 for definitions.

[b] Fractional conversion $= \displaystyle\int_0^\infty \dfrac{[P]F dt}{[S]_0 V_0}$

sum of the product and substrate concentration in the reactor must be given by

$$[S] + [P] = [S]_0 \exp(-t/\tau) \tag{6.8}$$

Consulting Table 6.2 for a first-order reaction and combining this with equation (6.8) indicates that

$$[P] = [S]_0 \exp\left(-\frac{t}{\tau}\right) \cdot [1 - \exp(-k_1 t)] \tag{6.9}$$

This equation indicates that to obtain complete chemical conversion of analyte to product and therefore to obtain *optimum measurement sensitivity*, the rate constant for a first-order reaction k_1) must be fast relative to the space–time parameter τ. For greater than 99% chemical conversion the value of $k_1\tau$ must be at least 5. A set of typical product-concentration–time curves for reactions with differing values of $k_1\tau$ are given in Figure 6.3.

For any finite value of the reaction-rate constant, the product concentration at the outlet builds up and then decays to zero as the fluid flow flushes through the reactor. As these plots indicate, even when the reaction generates the product very rapidly, the response time will be limited

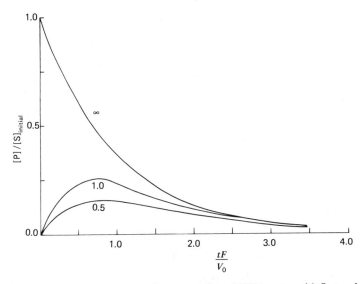

Fig. 6.3.   Product concentration transient curves for a CSTR reactor with first-order chemical kinetics. Computed from equation (6.9). The number on each curve is $k_1\tau$. Normalized time is the ratio t/τ. The product concentration is normalized by dividing it by the initial substrate concentration.

by the flush-out time of the reactor. The time required to flush out 99% of the product will at most be 5τ. The slope of a calibration curve, that is, a plot of some measurement which is proportional to the product concentration versus the analyte concentration, is related to the reaction rate. This plot will have its maximum slope when $k_1\tau$ is large.

6.1.2   Mass Balance in an Ideal Plug Flow Reactor

A complete development of an ideal PFR starts with a mass balance on a small volume element in the reactor (see Figure 6.4). (In the present treatment we ignore the presence of any packing and assume an open tube.) The substrate concentration in the plane A–B is denoted $[S]_l$ and that at the plane C–D is $[S]_{l+\delta l}$. The amount of material carried into the volume element ABCD by fluid flow will be $F \cdot [S]_l$ and that the removed by flow will be $F \cdot [S]_{l+\delta l}$. In addition to these terms, substrate is being converted to product at a rate $\Re$ (in moles/sec·ml); the net rate of chemical reaction in ABCD will be $\Re \cdot \delta V$ where δV is the volume of fluid contained in the interstices of the reactor. This volume (δV) will be equal to $a_v \cdot \delta l$ where a_v is the open area available for fluid flow. The rate of change of

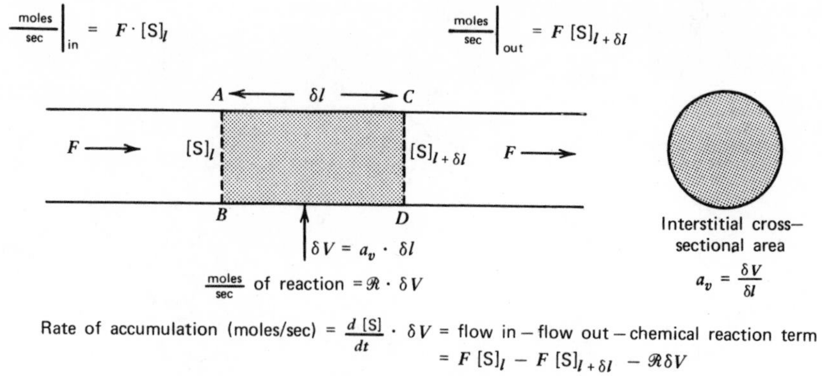

$$\left.\frac{\text{moles}}{\text{sec}}\right|_{\text{in}} = F \cdot [S]_l \qquad\qquad \left.\frac{\text{moles}}{\text{sec}}\right|_{\text{out}} = F [S]_{l+\delta l}$$

Interstitial cross–sectional area

$$\delta V = a_v \cdot \delta l$$

$$\frac{\text{moles}}{\text{sec}} \text{ of reaction} = \mathcal{R} \cdot \delta V \qquad\qquad a_v = \frac{\delta V}{\delta l}$$

Rate of accumulation (moles/sec) $= \frac{d\,[S]}{dt} \cdot \delta V =$ flow in $-$ flow out $-$ chemical reaction term

$$= F\,[S]_l - F\,[S]_{l+\delta l} - \mathcal{R}\delta V$$

Fig. 6.4.   Definition of mass balance in an ideal plug flow reactor.

concentration in the above element ABCD will be

$$a_v \delta l \cdot \frac{\partial[S]}{\partial t} = F([S]_l - [S]_{l+\delta l}) - \mathcal{R} a_v \cdot \delta l \tag{6.10}$$

In the limit as δl becomes very small, we find that

$$\frac{\partial[S]}{\partial t} = \frac{-F}{a_v}\frac{\partial[S]}{\partial l} - \mathcal{R} \tag{6.11}$$

which is the fundamental equation of a PFR. It should be noted that F/a_v is the average linear velocity (u) thus equation (6.10) may be rewritten as

$$\frac{\partial[S]}{\partial t} + u\frac{\partial[S]}{\partial l} + \mathcal{R} = 0 \tag{6.12}$$

In the *steady state,* which will be established only when the entire reactor is filled with sample, that is, is fed continuously with substrate, we find that:

$$u\frac{d[S]}{dl} = -\mathcal{R} \tag{6.13}$$

Since there is no mixing or spreading in a PFR, the sample moves down the reactor with an infinitely sharp front and tail. In the limit of a very narrow pulse of sample, it is easy to show that all of the sample and product are located at a distance along the reactor axis which is equal to the product of the average linear velocity (u) and time (t) from sample injection. Since every element of the sample spends exactly the same amount of time in the reactor, which is the residence time (τ^*) given by equation (6.14), the integrated rate law will be exactly the same as that obtained for a batch (closed) reactor:

$$\tau^* = \frac{L}{u} = \frac{V_0}{F} \tag{6.14}$$

For Michaelis–Menten kinetics, it is easy to show that the fractional conversion of substrate to products will be given by equation (6.15):

$$[S]_0 X = K_M \ln(1 - X) + k_2[E] \frac{L}{u} \tag{6.15}$$

where $[S]_0$ is the concentration of substrate injected into the reactor. Relationships for zero- and first-order reactions are summarized in Table 6.3. *A plug flow reactor may be viewed as a system in which a narrow pulse of substrate decays as it proceeds down the column and this pulse of substrate is located exactly at that axial distance equal to the time-flow rate product.* Similarly the reaction product exists only in a narrow pulse but one whose concentration *continuously increases* as the column is traversed. This situation is illustrated in Figure 6.5.

It is very instructive to compare the mathematical forms of the extent of conversion for a first-order reaction for the CSTR and PFR models. There is an exact analogy to multiple extraction separations. The fraction reacted in a CSTR bears the same relationship to that in a PFR as the fractional extracted in a single batch extraction has to an infinite number of extractions with the same volume of extractant (6). A PFR behaves as a very large number (∞) of very small CSTR's in series. One of the models for a real reactor such as a fixed-bed reactor is to treat it as several

Table 6.3   Kinetic Characteristics of a Plug Flow Reactor: Sample Injected in Narrow Pulse

Case	Concentration–distance relationship	Fraction converted[a,b]
Zero-order	$[S] = [S]_0 - \dfrac{k_0 \cdot l}{u}$	$X = \dfrac{k_0 \tau^*}{[S]_0}$
First-order	$[S] = [S]_0 \exp\left(-\dfrac{k_1 l}{u}\right)$	$X = 1 - \exp(-k_1 \tau)$
Michaelis–Menten	$[S]_0 - [S] = K_M \ln\left(1 - \dfrac{[S]_1}{[S]_0}\right) + \dfrac{k_2[E]l}{u}$	$X = \dfrac{K_M}{[S]_0} \ln(1 - X)$ $+ \dfrac{k_2[E]L}{u[S]_0}$

[a] $X \equiv ([S]_0 - [S]_L)/[S]_0$.
[b] τ^* = residence time = L/u where L is the overall length of the column.

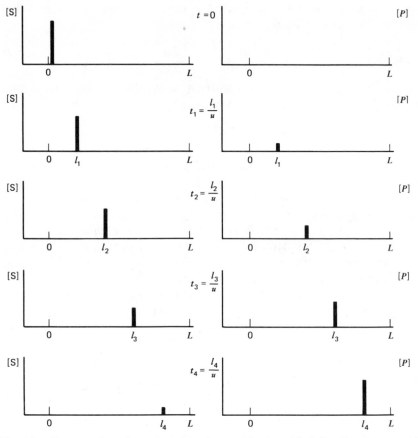

Fig. 6.5. Concentration of sample and product species in an ideal plug flow reactor with impulse input. Note that the sample concentration decreases and the product concentration increases as they move down the column.

finite volume CSTR's all acting in series (7) (see Section 6.1.5). Clearly for a first- or higher-order chemical reaction, a PFR is kinetically more efficient than a CSTR since it produces a higher conversion for equivalent operating parameter, that is, space time (τ) versus residence time (τ^*). A plot of relative efficiencies of a PFR versus a CSTR for a Michaelis–Menten kinetic system is shown in Figure 6.6. The efficiency is measured in terms of the amount of enzyme needed to provide a given fractional conversion of substrate. The number on each curve is the ratio of K_M to the feed-substrate concentration. This plot clearly indicates that a CSTR becomes less and less efficient relative to a PFR as the extent of reaction approaches 100%. The relative efficiency of these reactors for nth-order reactions has also been studied (7).

Since our intention here has been to present some of the basic ideas (mass balance, etc.) involved in flow reactors and point out some of the analytical relationships, we have avoided any lengthy development of the theory of immobilized-enzyme flow reactors. For a thorough treatment, the interested reader is advised to see the extremely complete review by Vieth and his associates, which includes a treatment of important systems such as substrate and product inhibition (3).

6.1.3   Effects of Dispersion in Flow Reactors

Since a PFR is more efficient than a CSTR, and since mixing implies dilution, it is evident that an analytical reactor should be designed to minimize those factors which induce mixing, that is, sample spreading and dispersion. An equivalent situation exists in column chromatography—those factors which cause dispersion reduce the number of plates and decrease the efficiency of the column (9).

Dispersion in a fixed-bed reactor can occur by any number of mechanisms (10). On a molecular level, the most elementary dispersion mechanism is simple longitudinal diffusion. The more time a species remains in a column the longer will be the length of column over which the species has a finite concentration. This is equivalent to the first term in the van Deemter equation of chromatography. It is not a serious problem in liquids since diffusion coefficients are much smaller (10^{-5}–10^{-6} cm^2/sec)

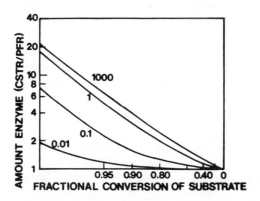

Fig. 6.6.   Comparison of efficiency of operation of CSTR and PRF reactors. Relative amount of enzyme needed indicated degree of conversion for ratio of CSTR to PFR as a function of fractional conversion and reaction order. [Reprinted from W. R. Veith, K. Venkatasubramanian, A. Constantinides, and B. Davidson, in *Applied Biochemistry and Bioengineering*, Vol. 1 (Academic Press, New York, 1976). Copyright by Academic Press.]

than in gases (10^{-1}–10^{-2} cm^2/sec). Flow around packing particles can be a significant dispersion mechanism. Mixing can be caused by obstructions in a tortuous channel in a column. Localized low-pressure regions can cause the fluid to recirculate and mix. Some flow paths will dead-end and this will necessitate slow diffusion back out of the blocked channel. If a molecule is adsorped, as in chromatography, slow adsorption–desorption kinetics can cause considerable broadening. All of these sources of dispersion exist in columns packed with *nonporous solids*. When the packing is porous, slow diffusion within the packing will cause considerable broadening. The above dispersion mechanisms and other are discussed in detail in reference 10. Giddings has developed a detailed theory of the dynamics of mixing processes (dispersion) in columns (11). The interested reader should consult this excellent source of information on the fundamentals of dispersion.

As intimated above, various flow-velocity inequalities can exist in a packed column; these processes can be described as channel flow, recycling, eddying, and stagnation in pockets. Thus the net axial dispersion coefficient will be related to both flow-length and flow-velocity inequalities. It should be pointed out that even in an unpacked tubular reactor, the viscous laminar flow profile, that is, the variation in axial fluid velocity with distance from the center of the tube, is a very serious source of dispersion (12) and exists to some extent in packed beds (11, 13). It is most pronounced in packed beds in the region nearest the column wall.

In addition to the above axial dispersion processes, radial dispersion can be quite significant if the column diameter is wide compared to the diameter of the entrance tubing, which will invariably be the case for small analytical reactors. Radial dispersion will not be very important when the ratio of the column length to diameter is high (13).

Levenspiel and Bischoff (14) identify three major classes of dispersion models for flow reactors. The first, simplest, and most important type of model is termed the axially dispersed plug flow. In this model it is assumed that *radial dispersion is negligible and the axial velocity is independent of radial position*. We will adopt this situation for general consideration. The second model of Levenspiel and Bischoff permits radial dispersion but assumes that the axial velocity is every where equal to its average value; this is termed the dispersed plug flow model. The third model (the uniform dispersion model) makes provision for both axial and radial dispersion as well as a non-uniform fluid velocity. Clearly, more complex models can exist in which the dispersion coefficients are permitted to vary along the column axis and radius. Such models are utterly intractable but are obviously more realistic since a packed bed is not truly a continuum.

Various criteria exist with which one can assess limits where radial dispersion can be neglected. Bischoff and Levenspiel (13) have shown that beyond a critical column length (L) to column diameter (d_c) ratio, radial dispersion contributes negligibly to the width of a sample peak:

$$\frac{L}{d_c} > 0.04 \frac{ud_c}{\mathcal{D}_r} \tag{6.16}$$

where u is the average linear velocity and $\mathcal{D}_r$ is the radial dispersion coefficient. The radial dispersion coefficient can be estimated from literature correlations to be a minimum of about 10^{-5} cm^2/sec (approximately equal to a molecular diffusion coefficient) at very low flow rates and it increases continuously to high values as flow rates increase. For a typical small reactor ($d_c = 5$ mm) packed with 100 μm diameter particles operated at a flow rate of 1 ml/min with an average void fraction of 0.4, radial dispersion coefficients of about 10^{-4} are estimated (15). This will require *the length to be at least 20 times the diameter*, that is, for a 5 mm diameter system, the length must be at least 10 cm. Although this is not very long for many analytical reactors, we will assume that radial dispersion can be neglected completely in order to treat the problem of dispersion in a tractable fashion, that is as a simple axially dispersed plug flow system.

6.1.4 Axially Dispersed Plug Flow and Mass Balance

The basic mass balance with dispersion is depicted in Figure 6.7. If we consider a small element of volume ABCD the net rate of accumulation of substrate will be

$$\delta V \cdot \frac{d[S]}{dt} = \text{(Accumulation by Flow In)} - \text{(Loss by Flow Out)}$$

$$+ \text{(Accumulation by Dispersion)} - \text{(Loss by Dispersion)}$$

$$- \text{(Loss by Chemical Reaction)}$$

$$\tag{6.17}$$

The dispersive accumulation and loss terms are essentially stochastic (random) processes very similar to molecular diffusion. Thus one should expect to see a term proportional to the second derivative of concentration appear in the final form. We can assume that the net flux by dispersion into the element ABCD is proportional to the concentration gradient ($\partial[S]/\partial l$) at l and the flux out by dispersion will be proportional to the concentration gradient at $l + \delta l$. Substituting such terms in equation

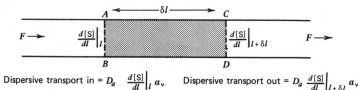

Dispersive transport in $= D_a \left. \frac{d[S]}{dl} \right|_l a_v.$ Dispersive transport out $= D_a \left. \frac{d[S]}{dl} \right|_{l+\delta l} a_v$

Rate of accumulation (mole/sec) $= \frac{d[S]}{dt}$ · δv = flow in − flow out + dispersion in − dispersion out−

chemical reaction term $= F\left([S]_l - [S]_{l+\delta l}\right) + D_a\, a_v \left(\left.\frac{d[S]}{dl}\right|_l - \left.\frac{d[S]}{dl}\right|_{l+\delta l} \right) - \mathscr{R}\delta V$

Fig. 6.7. Definition of mass balance in the axially dispersed plug flow model. The axial velocity is independent of radial position.

(6.17), using $\mathscr{D}_a$ to represent the constant of proportionality of axial dispersion, and allowing δl to approach zero, we find that

$$\frac{\partial[S]}{\partial t} = -u\frac{\partial[S]}{\partial l} + \mathscr{D}_a\frac{\partial^2 S}{\partial l^2} - \mathscr{R} \qquad (6.18)$$

The *solution of this partial differential equation is of course dependent upon the boundary conditions* imposed and has been the subject of many extensive investigations in the chemical engineering literature (16). The reaction-rate term does not effect that nature of these boundary conditions. Before considering the influence of the reaction term on the reactant or product concentration, it is important to understand the influence of the dispersion coefficient on the concentration–time–distance relationship. To do this and present the influence of the boundary conditions *we will drop the reaction-rate term* and normalize the variables by defining a new distance and time variables

$$x \equiv l/L \qquad (6.19)$$

$$\Theta \equiv \frac{u \cdot t}{L} \qquad (6.20)$$

After introducing x for l and Θ for t in equation (6.18), we obtain

$$\frac{\partial[S]}{\partial\Theta} = -\frac{\partial[S]}{\partial x} + \frac{\mathscr{D}_a}{uL}\frac{\partial^2[S]}{\partial x^2} \qquad (6.21)$$

Inspection indicates that the parameter $\mathscr{D}_a/uL$ is dimensionless; it is referred to in the engineering literature as the *Peclet number* and is defined below:

$$N_{Pe} \equiv \frac{uL}{\mathscr{D}} \qquad (6.22)$$

In order to relate the Peclet number to a concept more familiar to chemists, let us consider the behavior of a completely *unretained* (and non-reacting) species in a chromatographic column. Under ideal circumstances a perfectly Gaussian peak will be eluted when a volume of fluid (V) exactly equal to the dead volume (V_0) of the system has been pumped into the column. These ideal circumstances are really a very complex set of assumptions concerning the behavior of the column, the number of plates on it, the location and nature of the boundary conditions, the volume of sample admitted to the column, and the characteristics of the detector. A text on separations should be consulted to ascertain all of the requisite conditions (17). We assume here that a volume V_a of sample at a concentration $[S]^0$ is injected into the column. Thus the concentration of sample at the detector will be given by the well-known equation:

$$[S] = \frac{[S]^0 V_a}{V_0} \sqrt{\frac{N}{2\pi}} \exp\left(-\frac{N(V - V_0)^2}{2V_0^2}\right) \tag{6.23}$$

N in this equation is the *number of theoretical plates* and can be calculated as:

$$N = \frac{V_0^2}{\sigma_V^2} = \frac{t_0^2}{\sigma_t^2} \tag{6.24}$$

where σ_V and σ_t are the standard deviations of the peak in volume and time units respectively and t_0 is the time at which the peak maxima occurs. For a pure Gaussian peak, the standard deviation is equal to one-half of the width of the peak at its inflection points.

The *above equation* is only an ideal representation of a real peak, that is, it is only an approximate solution to the problem. In fact it is only really valid when the number of theoretical plates is rather large [according to Giddings (11), greater than 100]. *It can be shown that the Peclet number is exactly equal to 2 times the number of theoretical plates for an ideal Gaussian peak.*

Another important concept in chromatography which has direct relevance to column reactors is the height equivalent to a theoretical plate (HETP). As stated above, an ideal PFR may be considered as a collection of an infinite number of ideal, infinitesimal CSTR's. Each of these reactors has a finite length which, by analogy to a chromatographic column, could be referred to as the *height equivalent to a theoretical reactor* (HETR). It should be understood that reaction terms in the governing differential equation could well make the HETR not equal to the HETP.

The HETP is calculated as the ratio of column length (L) to number of theoretical plates thus for a Gaussian peak:

$$\text{HETP} = \frac{L}{N} = \frac{2L}{N_{\text{Pe}}} \tag{6.25}$$

As a general rule of thumb, a well-packed column under optimum conditions will have an HETP of about 3 particle diameters (18). The dependence of HETP on flow rate and particle size in liquid chromatography has been studied extensively (19–24). Snyder (25) has pointed out that for particles larger than about 10 μm, it is acceptable to use the relationship:

$$\text{HETP} = \alpha u^n d_p^m; \quad 0.3 < n < 0.6; \quad m \cong 1.8 \tag{6.26}$$

where d_p is the particle diameter. Smaller particles involve different relationships because they are operated at different linear velocities (26). When the axial dispersion coefficient is very small, or the average linear velocity is very high, the dispersive team in equation (6.21) is negligible and equation (6.12) for a pure plug flow reactor (including the reaction term) will result. Thus, as the Peclet number becomes very large, the system will behave as a pure plug flow reactor. As N_{Pe} gets smaller, more and more mixing occurs, and the system behaves as a CSTR. Clearly, the Peclet number is a very important dimensionless group. It will be a complicated function of flow rate because the dispersion coefficient (related to the HETP) will generally be a function of flow rate. At low flow the dispersion coefficient will approach the molecular diffusivity modified by a tortuosity factor; at high flow rate, other physical factors will dominate.

An analytical immobilized-enzyme reactor will generally be quite short in order to minimize dilution (spreading) and obtain high sample throughput. To avoid high pressure drops, particle sizes seldom are smaller than 30 μm, thus the number of plates will not be large. This means that the peaks obtained will be non-Gaussian. Mathematically this implies that the solution of equation (6.18) or (6.21) will depend upon the boundary conditions employed. The chemical engineering literature should be consulted for the effect of the boundary condition, the number of plates, and the sensitivity of the mathematical solution to these issues (27). These questions and the exact solution to equations (6.18) and (6.21) are well beyond the scope of this work. To summarize a complex discussion, we will make the following observations about the solution of equation (6.21):

1.   The peak mean will occur at the expected volume of eluent (V equal to V_0), but the peak maxima will occur before the mean. Thus the peak is asymmetric.

2.   The greater the number of plates on a column, the more closely the peaks resembles a Gaussian and the less sensitive they are to the precise boundary conditions.

3.   The difference between a more nearly exact solution of equation (6.21), for example, that devised by Taylor (28) in his solution to the classical laminar flow dispersion problem (29), and the true Gaussian peak is less than 10% at the peak maximum when there are more than 20 plates, that is, N_{Pe} is greater than 10 (30).

Having accepted the solution of equation (6.21) (which does not contain a chemical reaction term) as being Gaussian, we can rapidly come to several analytically vital conclusions which are illustrated in Figure 6.8.

1.   The greater the number of plates on a column, that is, the larger the Peclet number (*all else being equal*), the greater will be the concentration of material at any point in the column and at the detector, that is, the less dilution there is.

2.   Similarly, the greater N is, the sharper the signal peaks will be; thus, the greater the sample throughput will be.

3.   As pointed out above, an N_{Pe} (or N) equal to infinity corresponds to a PFR which utilizes the enzyme more effectively than a CSTR—particularly near 100% conversion.

Thus far we have only discussed the solution of equation (6.18) *without the reaction-rate term*. In general, although engineers are concerned with the transient situation equation in terms of reactor stability, only a few detailed solutions of equations as complex as the above have been devised and there development is far too tedious to present here. Most frequently simulations are employed and these use a continuous supply of substrate rather than a sharp pulse which is the *analytically* important case. The interested reader should consult the references in the review of enzyme reactors by Veith et al. (3) for an introduction to the transient behavior of reactors. Similarly, the text by Wen and Fan (1) is an excellent source. Due to the inherent nonlinearity of Michaelis–Menten kinetics, a closed-form solution will probably never be obtained. Several papers on transient analysis of enzyme reactors with Michaelis–Menten kinetics have appeared (31, 32).

To get some idea of the system behavior, consider a species undergoing a first-order reaction as it travels through a column. Suppose that it reacts very rapidly so that only product dispersion is important. Clearly, the system will behave rather as if the product were injected. Similarly, if the reaction rate is so slow that very little substrate is converted to product,

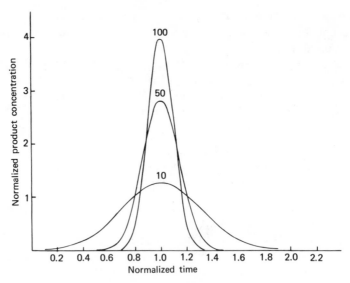

Fig. 6.8. Normalized product concentration–time curves for the axially dispersed plug flow reactor. These are normalized curves computed according to a Gaussian equation to illustrate the effect of the number of plates on the analytical sensitivity and throughput. It is assumed that the chemical reaction is instantaneous. The number of plates is given on each curve.

the sample concentration when it arrives at the detector will simply be slightly less than if no reaction at all occurred.

The only way in which the shape (not height) of the signal trace will be distorted from that of a pure Gaussian peak by the presence of the reaction term in equation (6.18) is if those molecules which elute first have substantially less opportunity to react than do those at the rear of a peak. This is contingent upon two distinct factors:

1. The net extent of *incompleteness* of the reaction must be significant (~10–90%). If the reaction is very fast or very slow, the concentration profile (shape) will be unaltered by the reaction.

2. Given the above condition, the dispersion must be sufficient to create a finite spread in residence time relative to the reaction-rate constant.

Consider an ideal PFR; clearly, reaction kinetics have no effect on the peak shape although they certainly control the peak height (see Figure 6.5). Thus we expect that there will exist a Peclet number (or number of theoretical plates, reactors, etc.) above which there will be essentially no effect of reaction kinetics on peak shape.

Simply as a matter of illustration, the data of Figure 6.9 was prepared

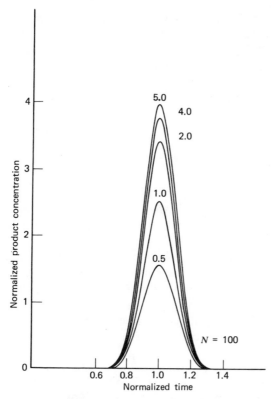

Fig. 6.9. Normalized product concentration–time curves for the axially dispersed plug flow reactor. All curves were computed assuming $N = 100$. The number of each curve is $k_1\tau$, the dimensionless first-order rate parameter.

assuming essentially no effect of reaction kinetics on peak shape. Analytically we see that the optimum sensitivity is obtained when the rate of reaction is fast so that complete conversion is obtained. In summary, analytical immobilized enzyme reactors are optimally designed to:

1.   Contain as high an enzyme activity in as little a volume as possible to drive reactions to completion and minimize dilution;

2.   Maximize the number of plates (or the Peclet number) in the column to minimize dispersion.

6.1.5   The Tanks-in-Series Model of Dispersion in Chemical Reactors

It is important to note that the types of dispersion models used above are only one possible way to simulate mixing in chemical reactors. As pointed out, a pure PFR behaves as an infinite collection of very small, well-mixed, reactors in series connection. As soon as we allow dispersion to occur in a PFR, the governing equations that describe the system become rather intractable partial differential equations. An alternative approach, which makes provision for mixing, is to consider a small number of finite-volume reactors connected in series (5). A major advantage of the tanks-in-series approach is that the governing equations are rather simple, ordinary differential, equations. Levenspiel and Bischoff presented a detailed comparison of the dispersion model and tank-in-series model and concluded that

". . . for small deviations from plug flow, both the dispersion and tanks-in-series models will give satisfactory results. Up to the present, which one is used may be largely a matter of personal preference" (33).

The number of tanks to be used (j) will vary with the extent of mixing. Evidently the less mixing, the smaller each tank must be and therefore the larger must be the number of tanks to yield a given volume (V_0). The models can be matched up by equating the variances of the two types of distributions. The substrate concentration in the jth tank as a function of time (assuming no reaction) will be given by

$$C_j(\Theta) = \frac{j^j}{(j-1)!} \Theta^{j-1} e^{-j\Theta} \qquad (6.27)$$

where Θ is a dimensionless residence time defined as:

$$\Theta = \frac{tF}{V_0} j \qquad (6.28)$$

The mean and variance of the distribution are exactly 1 and $1/j$, respectively. It can be shown that when the distribution variance is used to relate j and the dispersion coefficient, j approaches unity when the Peclet number is zero and approaches infinity as the Peclet number becomes very large. A few calculations indicate that by appropriate choice of the number of tanks one can very closely simulate any desired degree of axial dispersion. For example, values of j of 10 and 11 fall close to a curve with N_{Pe} of 10.5.

6.2 SOURCES OF DISPERSION AND ESTIMATION OF DISPERSION COEFFICIENTS AND HETP

It is clear from the preceding discussion that dispersion coefficients are as important in flow reactors as they are in chromatography. For this reason we are compelled to review some of the concepts developed by chromatographers and chemical engineers to estimate the expected magnitude of the dispersion coefficient. Most of these relationships have been developed in terms of contribution to the HETP. The concepts of number of plates, HETP, and effective dispersion coefficients are related (20) as follows:

$$\text{HETP} = \frac{L}{N} = 2\frac{\mathcal{D}_a}{u} \tag{6.29}$$

There are a great many phenomena which can contribute to dispersion in flow systems. Because of our analytical interest, we can disregard some types of reactor (e.g., fluidized beds) and can concentrate on a few distinct types of flow system in which the sample may or may not be retained (adsorbed) by a stationary phase. In terms of fluid dynamics, three classes of systems exist:

1.   Open tubes (which may be straight or coiled) but operated with one-phase flow.
2.   Open tubes operated with two-phase (air-segmented) flow.
3.   Packed (fixed) beds or columns. These may be packed with porous, nonporous, or pellicular supports.

6.2.1   Dispersion in a Straight Tube

The simplest type of system to consider is an open tube, operated with one-phase flow, which contains no adsorbent to retain a sample. We will assume that immobilized-enzyme reactors do not adsorb, in a chromatographic sense, the substrate or product. In many cases this is true (34), in others it is distinctly not the case (35). Taylor (28) and later Aris (29) studied the problem of dispersion in fully developed laminar flow in open straight tubes. Basically there are only two processes which can cause dispersion in this system. First, simple *molecular diffusion* along the axis of the column can bring about longitudinal dispersion. In liquids this will be essentially negligible. A second factor, which is much more important, is illustrated in Figure 6.10. Let us assume that a sample is injected into a flow stream in which *laminar flow is completely developed.* Under this condition one can show that the fluid velocity is a parabolic function of radial position. In fact, the center line velocity is exactly *twice* the average linear velocity, whereas the fluid at the tube wall is stagnant. Evidently, molecules in *different stream lines* will move forward at very different rates, thereby rapidly spreading the initially narrow sample slug over a progressively larger length of tube. Fortunately, this process is opposed by the fact that it rapidly introduces a *radial concentration gradient* which

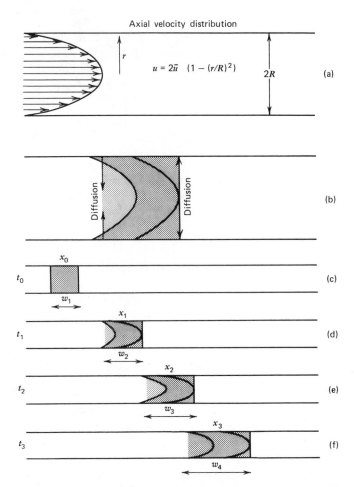

Axial velocity distribution

$$u = 2\bar{u} \; (1 - (r/R)^2)$$

Fig. 6.10. The effect of laminar flow on sample dispersion. (A) Illustration of the parabolic flow profile in a cylindrical tube. R is the tube radius, $\bar{u}$ is the average linear velocity over the tube cross section. (B) A representation of radial mixing caused by radial diffusion which opposes the axial dispersion of matter by laminar flow. (C)–(F) The gradual spreading of an initially sharp sample plug by the effective axial dispersion coefficient which results from the processes detailed in (A) and (B).

in turn causes radial diffusion. Thus a molecule at the center of the tube in the front of the slug will move to the wall, that is, into a region of low velocity. Similarly, at the rear of the sample slug the concentration is greatest near the wall. Molecules in this region will diffuse into the center, that is, a region of high fluid velocity. As shown in reference 36, it takes a finite time, which may be considerable, for this steady state to be established. **We estimate that for a $\frac{1}{16}$ in. i.d. tube, operated at a flow rate of 1 ml/min, it will take in excess of 1 min (or more than 100 cm of tubing) to establish steady state!** The dispersion coefficient (and therefore the HETP) predicted by Aris (29) is

$$\text{HETP} = \frac{2D_m}{u} + \frac{ud_t^2}{96D_m} \tag{6.30}$$

where D_m is the molecular diffusion coefficient, d_t the tube diameter, and u the average linear velocity. The first term in this equation represents the contribution of simple molecular diffusion to longitudinal dispersion. The second is due to the parabolic velocity profile. Subsequent to the development of the above result, Golay presented equation (6.31) which applies to a species which is retained on an infinitely thin-wall coated film (37):

$$\text{HETP} = \frac{2D_m}{u} + \frac{1 + 6k + 11k^2}{96(1 + k)^2} \frac{ud_t^2}{D_m} \tag{6.31}$$

where k is the chromatographic capacity ratio. Both of the above equations indicate that any difference in velocity across a tube cross-section effectively disperses (mixes) the sample longitudinally. The same holds true in a packed bed. This effect is generally referred to as *resistance to mass transfer in the mobile phase*. It would disappear completely if mass transfer in the mobile phase were very rapid. Obviously, the effects described above will take place in any open tube, including the connection tubing prior to the reactor and that which leads the sample to a detector.

6.2.2 Dispersion in Curved Tubes

It is well known that very sharp expansions and contractions in tubes can increase dispersion due to the eddy flow established. Similarly very sharp bends can have the same effect. The situation in a gently but regularly coiled tube is very complex (38–40). At least two new factors are introduced. First a molecule moving closer to the inner side of the tubing will have a shorter path to follow than one on the outside. This is referred to as the *race track effect* which obviously acts to increase dispersion. Second a distinct set of *secondary radial flows* is established by the centrifugal field due to motion in a coiled tube. This process will tend to *augment radial mixing and thereby diminish dispersion*.

Recently, Nunge, Lin, and Gill (39) carried out a detailed analysis of dispersion in coiled tubes. Their mathematical treatment is too complex to present here. The results indicate that the dispersion will always approach that of a straight tube at a sufficiently low flow velocity. As the flow velocity increases, in general the dispersion coefficient first gets larger and at a sufficiently high flow rate it drops below the value for a straight tube operated with the same flow rate. The final results, assuming reasonable (>0.1 cm/sec) flow rates for small molecules ($D_m \cong 10^{-5}$ cm^2/sec) in water (viscosity 0.01 P, density 1 g/ml), is as follows:

$$\text{HETP} = \frac{2D_m}{u} + \frac{ud_t^2}{96D_m} + \frac{ud_t^2}{D_m\lambda^2}[0.0752$$

$$+ 62.2(ud_t)^2 - 1.52 \times 10^5(ud_t)^4] \quad (6.32)$$

The term λ is the ratio of the coil radius to the radius of the tubing. The first two terms, which are identical to those in equation (6.30), are the straight-tube dispersion effects. Clearly there is very little improvement in dispersion due to coiling until reasonably high flow rate, where the fourth power of u begins to be important. For small molecules in aqueous solution, coiling will not increase dispersion significantly; thus it is always reasonable to coil tubing to compact it. The only price paid for coiling will be a slight increase in pressure drop. Recently, Moulin, Spiker, and Kolk (40) found qualitative agreement between the equation of Nunge, Lin, and Gill and measurements of axial dispersion of gases in coiled tubes. The reader is also directed to references 74–76, which are very relevant to this section.

6.3   DISPERSION IN TUBES OPERATED WITH SEGMENTED FLOW

Continuous-flow analysis, which is now a major component of all automated analysis, is generally implemented by interspersing liquid slugs, which contain the sample and reagents, with air bubbles. This technique, which was introduced by L. J. Skeggs (41), is the basis of the Technicon's Auto Analyzer system. It appears to be essential to minimizing the dispersion (42, 43) which will take place in fully developed laminar flow despite a number of contrary arguments presented by proponents of "flow injection" analysis (44–46). Most theoretical and experimental studies of dispersion in two-phase flow (47–49) indicate that sample spreading is decreased relative to unsegmented flow, but not entirely eliminated. In order for the flow pattern to be smooth and regular, the liquid must wet the walls of the tubing through which it flows; thus a given slug of sample will lay down a small but finite *film of liquid* along the entire length of tubing through which it flows. As a consequence of this, the sample will not be confined to a single liquid segment but will gradually spread over several segments, that is, it will be dispersed.

Each component (tubing, dialyzer, detector) in a continous-flow analyzer (CFA) can act to disperse a sample slug. The sample is unsegmented until it arrives downstream of the peristaltic pump at a point where a bubble is injected, and is also usually unsegmented in the photometric flow cell. Additional dispersion will be induced at tubing joints and connectors, and by a dialyzer if one is employed. Thiers, Cole, and Kirsh were among the first to experimentally assess the shape of a continuous-flow sample curve and consider the factors which control interaction between several samples (50).

6.3.1 Experimental Description of a Continuous-Flow Analyzer

The experimental curve shown in Figure 6.11 defines some important characteristics of CFA. A sampling probe is immersed in the sample for

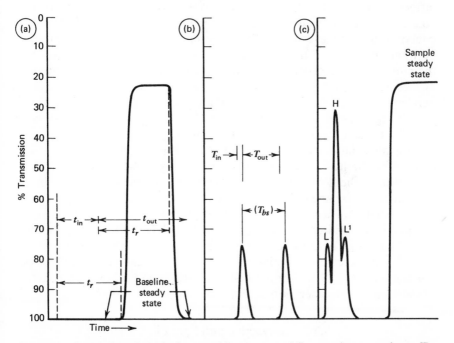

Fig. 6.11. Description of terminology used in a segmented-flow continuous analyzer. [Reproduced by courtesy of *Clinical Chemistry* (American Association for Clinical Chemistry, publisher) from R. E. Thiers, R. R. Cole, and W. J. Kirsch, *Clin. Chem.* **13**, 451 (1967).]

a period of time designated t_{in}. After a delay time, t_r, during which the specimen transverses the entire length of the flow system, an approximately exponential "rise curve" develops. The half-life of this curve, termed the "half wash time," is a function of the flow system. When the sampling interval is sufficiently long, a steady-state signal will occur and a flat trace will be observed (Figure 6.11). Under nonsteady-state conditions, the peak height is a strong function of dispersion within the system and the timing involved in sampling. Although the peak heights under this condition are still proportional to the sample concentration, the nonsteady-state signal is generally avoided when maximum precision is desired.

Eventually the signal must decrease and this exponential decay, termed the "fall curve," has the same time constant as the rise curve. If the time the sampling probe spends in the wash solution, t_{out}, is small with respect to the half wash time, the fall curve will not reach the base line before the next rise curve develops. As a result, there will be a contribution to the peak height of the second peak from the first sample. This effect is termed "interaction." It can be measured by running a standard of low concentration, followed by a high one, and finally a repeat of the first low concentration (Figure 6.11). The apparent difference in concentration between the first and third samples divided by the true concentration of the more concentrated sample is an approximate measure of the amount of interaction. This interaction on a percentage basis is generally independent of concentration. The last important characteristics of CFA is the throughput rate which is determined by the time between successive samples termed t_{bs} and is simply the sum of t_{in} and t_{out}.

With most autoanalyzer manifolds, the response curve is fairly well defined by the half wash time, $w_{1/2}$, and a "lag phase," which are characteristics of the flow system. The lag phase is a time period in the initial portion of the rise and fall curve during which the signal is not exponential. The existence of a lag phase is not surprising, considering the overall complexity of the hydraulics. More recent work has indicated that a third phase may exist near the end of the rise and fall curves which deviates from the expected exponential behavior (51). Typical values of these parameters are 9 to 21 sec for the half wash time and 6 to 23 sec for the lag phase, using first-generation equipment.

6.3.2   Interaction and Speed in CFA

The effects of these parameters on the response of the system determine the precision, interaction, and speed with which analyses can be performed. The half wash time is the key factor. For normal CFA with flat peak tops, sample plus wash time should equal at least 7 $w_{1/2}$. It is therefore important to minimize $w_{1/2}$. Portions of the flow system which have been shown to contribute to the exponential term include the unsegmented sample line from the sampler to the pump, and the flow cell and debubbler assembly. The rate at which the fluid is pumped through the flow cell should be as close as possible to the total liquid flow rate into the debubbler. The precise positioning of air segments necessary for this debubbling is attained through controlled bubble-injection techniques. Another advance in CFA has been the development of electronically debubbled flow cells which avoid the dispersion in nonsegmented portions altogether. One final point concerning the sample/wash ratio: It was shown by Thiers and his coworkers (50) that while the ratio has no influence on the interaction between specimens, it has a significant role in determining the precision of the analysis. This is because the larger the ratio of t_{in} to $w_{1/2}$, the closer to one gets to steady state. As a result, any small changes in timing of the sampler have no effect on the magnitude of the signal. Thus, if interaction is a problem, the sampling rate should be decreased rather than increasing the wash time—an all-too-common practice. The sample/wash ratio should seldom, if ever, be less than 2. It should be borne in mind that an increase in t_{in} causes an increase in sample volume.

It should be pointed out that the above considerations are true assuming the lag phase is small. If the lag phase is large, corrections must be applied in calculating the $t_{in}/w_{1/2}$ ratio. The main factors affecting the lag phase are the volume of the system, the nature of the walls containing the flowing stream, and any disparity between the donor and recipient streams during dialysis. In a well-designed flow system, the contribution of the lag phase to the sample dispersion is minor. In practice, the contribution of the unsegmented portions of the system to the lag phase is dominant.

6.3.3   Theory of CFA

Although a number of detailed experimental and theoretical studies of CFA systems have been published, the only part of the system which is simple enough to be really amenable to detailed analysis is the segmented-flow stream. Snyder and Adler have recently published studies which are experimentally quite consistent with theory (52, 53).

Most workers have assumed three conditions to simplify the mathematical model for calculating the shape of the concentration profile: (1) the film and liquid slugs mix instantaneously and completely in the contact region, (2) all segments have rigorously constant dimensions, and (3) longitudinal (axial) diffusion in the film is very small and therefore is neglected.

Based on these assumptions, one can show that if an entire sample is placed in an initial slug ($j = 0$), then its concentration is subsequent liquid slugs will be described by a Poisson distribution

$$\frac{C_j}{C^0} = \frac{q^j}{j!} e^{-q} \tag{6.33}$$

where q corresponds to a partition ratio which is the ratio of the film volume (V_f) deposited by any slug along the entire length (L) of the tube to the volume of one slug (V_s)

$$q = \frac{V_f}{V_s} \tag{6.34}$$

Geometric considerations indicate that for a tube

$$q = \frac{4\, d_f L}{L_s d_t} \tag{6.35}$$

where d_f is the film thickness and L_s is the length of a slug. In order to completely define the dispersive phenomena in terms of fundamental fluid properties, the film thickness must be obtained. Snyder and Adler (52) have used a model based on perfect wetting, that is, zero contact angle and find that

$$d_f = 0.67 d_t \left(\frac{u\eta}{\gamma}\right)^{2/3} \tag{6.36}$$

where u is the average linear velocity, γ the surface tension between the liquid and air, η is the liquid's viscosity. Since the model used is so very different from that of axially dispersed plug flow, one cannot arrive at an equation for a dispersion coefficient except in the limit of a very long tube with many slugs where the Poisson distribution becomes Gaussian. It is well known that the mean in a Poisson distribution will occur when $k \cong q$ and that the standard deviation (σ_k) will be equal to $q^{1/2}$. Thus the position and width of the peak can be specified in terms of the partition ratio which upon combining equations (6.35) and (6.36) becomes

$$q = \frac{0.67\, \pi\, L d_t^2}{V_s} \left(\frac{u\eta}{\gamma}\right)^{2/3} \tag{6.37}$$

Snyder and Adler carried out a wide variety of experiments in which all of the variables in equation (6.37) were changed over a wide range. They found very good agreement with theory particularly at low flow rate in narrow tubes if the constant term (0.67) were decreased to about 0.50. It should be noted that in order to avoid extraneous dispersive phenomena, they used a specially designed flow cuvette which did not require debubbling of the stream.

In a subsequent study Snyder and Adler studied the effect of slow mixing between the film and bulk of the slug. As is expected, this has the effect of increasing the dispersion. This increased dispersion can be calculated a priori after introducing a modified diffusion coefficient (D'_m) The total variance is described as:

$$\sigma^2 = q + \sigma_r^2 \qquad (6.38)$$

where q is given above and σ_r^2 is

$$\frac{\sigma_r^2}{q} = \frac{\pi^2}{72} \frac{d_t^4 u^{5/3} \eta^{2/3}}{\gamma^{2/3} V_s D'_m} \qquad (6.39)$$

As is expected, the increased dispersion due to slow mass transfer is more significant in large tubes at high flow rate. Considering the complexity of this situation, very good agreement was found between the model and the experimental parameter if one makes provision for the dependence of the effective diffusion coefficient on viscosity. Although the effect of coiling was not included in the theoretical model, the ratio of coil diameter to tube diameter was varied between 25:1 and 220:1 and no significant effect was observed. A straight tube was also considered and was found to produce a variance of about 3 times the variance of any of the coiled tubes. Obviously coiling is providing additional radial mixing as indicated in the previous section.

The slow mixing term is actually very significant. Substituting reasonable values in equation (6.39) and converting linear to volumetric flow rate (F) we find that:

$$\frac{\sigma_r^2}{q} = \left(\frac{5.4 \times 10^{-5}}{D'_m} \right) \frac{4^{5/3} d_t^{1/3} F^{5/3}}{V_s} \qquad (6.40)$$

Adler and Snyder found that D'_m was about 8.3×10^{-5} cm²/sec and 2.1×10^{-5} cm²/sec for potassium permanganate (M.W. = 158) and Blue Dextran (M.W. = 2×10^6), respectively. Thus the numerical coefficient in parentheses in equation (6.40) is likely to be greater than 1. A typical sample slug will be less than 0.1 ml, and since the dependence of the excess spreading on the tube diameter is rather slight, one can take the diameter to be about 1.6 mm (the central value used by Synder and Adler).

The variances presented above are all dimensionless; they can be converted to time units by multiplying by the square of the residence time $(L/u = V_0/F)$. Since the above model equation for the concentration profile is based on the assumption that the entire sample is initially contained in one slug, it will generally underestimate the actual width of the true sample which may have a volume as large as 10 to 50 slugs ($\sim$1.0 ml). When the number of slugs which contain the sample is less than half of $q^{1/2}$, then this factor will have little influence on the width of the peak and its net variance. Other factors which exist in a real system but have been disregarded in order to arrive at some simple results are: adsorption of the sample on the tubing, dispersion in tubing connectors, dispersion prior to segmentation and dispersion in the detector (flush-out). Of course, we have also neglected the effect of a chemical reaction on the peak shape and width.

It is undoubtedly true that segmented flow is absolutely essential to minimize dispersion when complex sample treatments, for example, addition of several reagents at different times and long incubation periods, are required. For this reason continuous, two-phase flow analysis has been extremely useful if not absolutely essential for modern high-speed enzymatic analysis with soluble enzymes (54).

6.3.4   Flow-Injection Analysis

Recently Ruzicka, Hansen, and coworkers introduced the technique of "flow-injection analysis" (55–63). The technique has been used with several types of detectors (ion-selective electrodes, photometry), but in most applications to date the chemical reactions used were quite rapid; lengthy delays were not needed. Despite relatively detailed discussion of the merits of flow injection versus segmented flow (42–44), the technique of flow injection appears to work better than it theoretically should. We believe that this is due to three factors:

1.   Spreading is small relative to the length of tubing occupied by the sample. Infinitely narrow sample pulses are assumed in the model equations.

2.   The effect of secondary flows has been ignored in the theoretical arguments even though coiled tubes were employed.

3.   Residence times in flow-injection systems are so short that steady state is not attained and therefore the Aris–Taylor equation does not really apply.

There is no question that under the right conditions flow-injection analysis can be analytically useful and that very high sample throughput (>300/hr) with little overlap can be achieved. In fact, a complex flow system using on-line extraction has been implemented (63).

6.4 DISPERSION IN PACKED BEDS

The systems described above are geometrically and fluid-mechanically very simple. When a packed bed of particles is employed, a number of additional complications are added, most importantly, eddy diffusion and resistance to mass transfer in the mobile phase.

6.4.1 Eddy Diffusion

Due to irregularities in packing structures, which have been summarized by Giddings (11), such as bridged regions, clusters, and non-uniformity near the wall, as well as the distribution in packing particle sizes, all paths by which a molecule can move along the packed bed will not have the same length. In classical chromatographic theory this contribution to dispersion is referred to as *eddy diffusion* (64). This term, in the classical theory, not the modern coupling theory, produces an additive contribution to the HETP which is written as:

$$\text{HETP} = 2\lambda d_p \tag{6.41}$$

where d_p is the diameter of the packing material, and λ is a constant which represents the extent of irregularity in path lengths. For a straight unpacked column, λ is obviously zero; in practice, for a well-packed column it will be approximately unity, but may be as large as 10 in a very poor column (65). It should be noted that the larger particles tend to dominate this source of dispersion, thus the upper range of the particle size distribution should be used in this equation rather than the mean particle diameter (66).

6.4.2 Mobile-Phase Mass-Transfer Dispersion

The column packing particles obviously disrupt the parabolic velocity distribution present in an open tube. In its stead, the packing generates an extremely complex set of flow streams. Giddings (11) has discussed five types of flow inequalities. In general, we expect the velocity at the tube wall to be zero, but because the wall region cannot be packed very tightly, that is, the wall interrupts the arrangement of particles, the axial velocity near the wall will be higher than average. The same is true of velocities in the center of channels in the packing structure. This process is exactly analogous to that of parabolic velocity distribution discussed above for open tubes; thus it is termed *mobile-phase resistance to mass transfer*. Clearly, if radial mixing were instantaneous, this effect would disappear. By using small particles, molecules can rapidly diffuse radially around the particles into new flow streams. Thus, we expect that this process will generate a contribution to the HETP given by an equation of the form:

$$\text{HETP} = \frac{f(k')ud_p^2}{D_M} = C_M u \tag{6.42}$$

Coupling of Eddy Dispersion and Mobile-Phase Resistance

The classical theory of chromatography treats eddy diffusion and mobile-phase resistance to mass transfer as completely independent processes. Thus, one should be able to add equations (6.41) and (6.42) to get the sum of their contributions. Giddings (11) pointed out that this is not valid since by means of molecular diffusion (and lateral dispersion) any molecule can move radially, *sampling many eddy paths,* and *moving into many flow streams.* These processes are therefore coupled. The terms are added as if they acted in parallel, thus:

$$\text{HETP} = \frac{1}{1/2\lambda d_p + 1/C_M u} \tag{6.43}$$

6.4.3 Effect of Column Diameter

All of the above radial velocity distribution considerations are complicated by the manner in which the sample is introduced onto the column. If it enters as a *point source* on the column axis, it may elute before it ever diffuses to the tube wall (see below). In contrast, if the sample enters the column as a radially homogeneous plug, it will sample all velocity distributions. These types of wall effects very obviously will depend upon the ratio of particle to column diameter (termed aspect ratio d_p/d_t). These questions have been studied extensively for unsorbed solutes ($k = 0$) (67, 68). Such studies indicate that the tube wall exerts itself to the order of 4 particle diameters. One should observe grossly different behavior as the aspect ratio is varied until the effect due to the central core of randomly packed particles completely dominates column performance. Kelley and Billmeyer (68) find that the column diameter is not at all important, provided that it is *at least 50–100 particles wide.* Their model assumes that the initial sample is radially homogeneous. Knox and Parcher (67) point out that a column will behave as if it is infinitely wide, provided that a sample is injected on the column axis *and* is eluted before it has a chance to be transported to the wall. Their analysis indicates that if

$$d_t^2 \geq 2\text{-}3d_p\cdot L \tag{6.44}$$

the column diameter is effectively "infinite."

It is generally incorrect to state that turbulent flow exists in an immobilized-enzyme reactor. Although turbulent effects have been observed to decrease dispersion, due to the onset of additional radial mass transfer (69), this phenomenon does not start in a packed bed until Reynolds numbers of at least 1-10 (70). Consider a typical enzyme reactor with a diameter of 5 mm packed with 100 μm particles. A Reynolds number of 1 requires a linear velocity of 1 cm/sec which corresponds to a flow rate of at least 4 ml/min. Complete turbulence will not occur until very high flow rates for these small reactors.

6.4.4 Diffusional Resistance to Mass Transfer

The introduction of particles into a column can create a number of additional complications. A certain fraction of the mobile phase will be trapped at contact points between particles and, if the particles are porous or pellicular, some mobile phase will stagnate in the pores. This introduces an additional contribution to the HETP known as *resistance to mass transfer in stagnant mobile phase*. It has only recently been studied in any detail (71-73). Intuitively, it will depend upon flow velocity, particle size, diffusion in the mobile phase. In addition it will be a function of the amount of stagnated mobile phase and the chromatographic capacity factor. Of course, there will be, in a chromatography column, some adsorbent present on the stationary phase which will retain the sample. Due to finite particle size, some minimum time is needed to achieve equilibrium of the solute in the stationary phase. The process of resistance to mass transfer in the stationary phase has been studied very extensively, and there is wide agreement that it contributes a term of the following form to the overall HETP:

$$\text{HETP} = q\,\frac{k'}{(1 + k')^2}\frac{ud_p^2}{D_s} \tag{6.45}$$

where k' is the capacity factor and D_s the molecular diffusivity in the stationary phase. The factor q is a term which depends upon the geometry of the stationary phase. For uniform spheres it is $\frac{1}{30}$ (11).

Why discuss this term for an immobilized-enzyme reactor since there is literally no second stationary phase? In fact, there is a thin film of protein which may retain ($k \neq k' \neq 0$) either the substrate or the product of an enzyme reaction. Thus, one may find stationary-phase resistance to mass transfer in an enzyme reactor. This phenomenon has not been systematically investigated as yet. Additional information on dispersion can be found in the references in two excellent review articles on this topic (19, 20).

2 Definition of Terms

a_v	free cross sectional area of a packed column (cm^2)
C_j	concentration of sample in the jth slug (moles/ml)
D_m	molecular diffusion coefficient (cm^2/sec)
$\mathscr{D}_a$	axial dispersion coefficient (cm^2/sec)
$\mathscr{D}_r$	radial dispersion coefficient (cm^2/sec)
d_c	column diameter (cm)
d_f	film thickness (cm)
d_p	particle diameter (cm)
d_t	tube diameter (cm)
[E]	enzyme concentration (mole/ml); k_2[E] corresponds to enzyme activity per unit volume (mole/sec·ml)
F	volume flow rate (ml/sec)
HETP	height equivalent to a theoretical plate (cm)
K_M	Michaelis constant (mole/ml)
l	axial distance from entrance to a column or tube (cm)
L	overall length of a column or tube (cm)
L_s	length of a sample slug (cm)
N_{Pe}	Peclet Number ($=\mu L/\mathscr{D}$)
N_{Re}	Reynolds Number ($=ud/\mu$)
N	number of plates on a column
[P]	product concentration as a function of time and/or distance (moles/ml)
[P]$_{ss}$	steady-state product concentration (mole/ml)
q	a partition ratio or an arbitrary constant
$\mathscr{R}$	a reaction rate (mole/sec·ml)
$\mathscr{R}_0$	reaction rate for a zero-order reaction (moles/sec·ml)
$\mathscr{R}_1$	reaction rate for a first-order reaction (moles/sec·ml)
$\mathscr{R}_M$	reaction rate in a Michaelis–Menten scheme (moles/sec·ml)

[S]	substrate or analyte concentration as a function of time and/or distance (moles/ml)
$[S]^0$	substrate concentration at a reactor inlet (moles/ml)
$[S]_{ss}$	steady-state substrate concentration (moles/ml)
t	time from the start of a reaction or addition to a system (sec)
t_c	critical time for the development of laminar flow (sec)
t_0	transit time for a plug flow reactor (sec)
u	average linear flow velocity (cm/sec) $= F/a_v$
V_a	volume of sample added (ml)
V_j	volume of film deposited by one sample slug (ml)
V_0	total available volume of a reactor, the dead volume or interstitial volume (ml)
V_s	total volume of a slug (ml)
x	dimensionless distance $= l/L$
X	fractional conversion of substrate to product
γ	surface tension (dyne/cm)
ϵ	void fraction (0.37–0.41) of a column
η	viscosity (dyne·sec/cm²)
μ	kinematic viscosity (cm²/sec)
Θ	a dimensionless time (ut/L)
λ	the eddy diffusion proportionality constant (~ 1)
σ^2	variance of a peak, σ_V^2 in volume units, σ_t^2 in time units
τ	a space–time coordinate (V_0/F) for a CSTR (sec)
τ^*	the residence time of a PFR (V_0/F, sec)
ϕ	column tortuosity (~ 0.6)

References

1. C. Y. Wen and L. T. Fan, *Models for Flow Systems and Chemical Reactors,* (Marcel Dekker, New York, 1975), p. 15.
2. W. E. Hornby, J. Campbell, D. J. Inman, and D. L. Morris, "Preparation of Immobilized Enzymes for Application in Automated Analysis," in *Enzyme Engineering,* E. K. Pye and L. B. Wingard, Jr., Eds., Vol. 3 (Plenum Press, New York, 1974), p. 401.
3. W. R. Vieth, K. Venkatasubramanian, A. Constantinides, and B. Davidson, "Design and Analysis of Immobilized Enzyme Flow Reactors," in *Applied Biochemistry and Bioengineering,* L. Wingard (ed.) Vol. 1 (Academic Press, New York, 1976).

4. O. Levenspiel, *Chemical Reaction Engineering* (Wiley, New York, 1962).
5. R. B. Aris, *Elementary Chemical Reactor Analysis* (Prentice-Hall, Englewood Cliffs, NJ, 1969).
6. H. A. Laitinen, *Chemical Analysis,* (McGraw-Hill, New York, 1960), p. 464.
7. O. Levenspiel and K. Bischoff, "Patterns of Flow in Chemical Process Vessels," in *Advances in Chemical Engineering,* T. B. Drew, J. W. Hooper, Jr., and T. Vermeulen, Eds., Vol. 4 (Academic, New York, 1963), p. 150.
8. A. R. Cooper and G. V. Jeffreys, *Chemical Kinetics and Reactor Design* (Oliver and Boyd, Edinburgh, 1971), p. 253.
9. A. B. Littlewood, *Gas Chromatography,* 2nd ed. (Academic, New York, 1970), p. 164.
10. R. A. Greenkorn and D. P. Kessler, "Dispersion in Heterogeneous Nonuniform Anisotropic Porous Media," in *Flow Through Porous Media* (American Chemical Society, Washington, DC, 1969), p. 163.
11. J. C. Giddings, *Dynamics of Chromatography* (Marcel Dekker, New York, 1965).
12. O. Levenspiel, in *Advances in Chemical Engineering* T. B. Drew, J. W. Hooper, Jr., and T. Vermeulen, Eds., Vol. 4 (Academic, New York, 1963), p. 125.
13. K. Bischoff and O. Levenspiel, *Chem. Eng. Sci.* **17,** 257 (1962).
14. O. Levenspiel, in *Advances in Chemical Engineering,* T. B. Drew, J. W. Hooper, Jr., and T. Vermeulen, Eds., Vol. 4 (Academic, New York, 1963), p. 106.
15. C. Y. Wen and L. T. Fan, *Models for Flow Systems and Chemical Reactors,* (Marcel Dekker, New York, 1975), p. 189.
16. C. Y. Wen and L. T. Fan, *Models for Flow Systems and Chemical Reactors,* (Marcel Dekker, New York, 1975), p. 127.
17. B. L. Karger, L. R. Snyder, and C. Horvath, *Introduction to Separation Science,* (Wiley, New York, 1973).
18. P. A. Bristow and J. H. Knox, *Chromatographia* **10,** 279 (1977).
19. E. Grushka, *Anal. Chem.* **46,** 510A (1974).
20. E. Grushka, L. R. Snyder, and J. H. Knox, *J. Chromatogr. Sci.* **13,** 25 (1975).
21. R. Endle, I. Halasz, and K. Unger, J. Chromatogr. **99,** 377 (1974).
22. J. N. Done and J. H. Knox, *J. Chromatogr. Sci.* **10,** 606 (1972).
23. G. J. Kennedy and J. H. Knox, *J. Chromatogr. Sci.* **10,** 549 (1972).
24. R. E. Majors, *J. Chromatogr. Sci.* **11,** 88 (1973).
25. L. R. Snyder, *J. Chromatogr. Sci.* **10,** 369 (1972).
26. L. R. Snyder, *J. Chromatogr. Sci.* **15,** 441 (1977).
27. W. N. Gill and V. Anathakrishnam, *A.I.Ch.E.J.* **12,** 908 (1966).
28. G. I. Taylor, *Proc. Roy. Soc. London* **A219,** 186 (1953).
29. R. Aris, *Proc. Roy. Soc. London* **A235,** 67 (1956).
30. O. Levenspiel and W. K. Smith, *Chem. Eng. Sci.* **6,** 227 (1957).
31. P. A. Ramachandran, *J. Appl. Chem. Biotechnol.* **24,** 265 (1974).
32. A. O. Mogensen and W. R. Vieth, *Biotechnol. Bioeng.* **15,** 467 (1973).

33. O. Levenspiel, in *Advances in Chemical Engineering*, T. B. Drew, J. W. Hooper, Jr., and T. Vermuelen, Eds., Vol. 4 (Academic, New York, 1963), p. 158.
34. R. S. Schifreen, L. D. Bowers, D. A. Hanna, and P. W. Carr, *Anal. Chem.* **49**, 1929 (1977).
35. R. E. Adams and P. W. Carr, *Anal. Chem.* **50**, 944 (1978).
36. R. J. Nunge and W. N. Gill, "Mechanisms Affecting Dispersion and Miscible Displacement", in *Flow Through Porous Media*, (American Chemical Society Publication, Washington, DC, 1970), p. 185.
37. M. J. E. Golay, in *Gas Chromatography*, V. J. Coates, H. J. Noebels, and I. S. Fagerson, Eds. (Academic, New York, 1958).
38. Z. Aunicky, *Can. J. Chem. Eng.* **46**, 27 (1968).
39. R. J. Nunge, T. S. Lin, and W. N. Gill, *J. Fluid Mechan.* **57**, 363 (1972).
40. J. A. Moulin, R. Spijker, and J. F. M. Kolk, *J. Chromatogr.* **142**, 155 (1977).
41. L. J. Skeggs, *Am. J. Clin. Path.* **28**, 311 (1957).
42. M. Margoshes, *Anal. Chem.* **49**, 17 (1977).
43. M. Margoshes, *Anal. Chem.* **49**, 1861 (1977).
44. J. Ruzicka, E. H. Hansen, H. Mosbach, and F. J. Krug, *Anal. Chem.* **49**, 1858 (1977).
45. E. H. Hansen, J. Ruzicka, and B. Rietz, *Anal. Chim. Acta* **89**, 241 (1977).
46. D. Betteridge, *Anal. Chem.* **50**, 832A (1978).
47. R. Begg, *Anal. Chem.* **43**, 854 (1974).
48. R. Begg, *Anal. Chem.* **44**, 631 (1972).
49. W. H. Walker and K. R. Andrew, *Clin. Chim. Acta* **57**, 181 (1974).
50. R. E. Thiers, R. R. Cole, and W. J. Kirsch, *Clin. Chem.* **13**, 451 (1967).
51. K. Spencer, *Anal. Clin. Biochem.* **13**, 438 (1976).
52. L. R. Snyder and J. H. Adler, *Anal. Chem.* **48**, 1017 (1976).
53. L. R. Snyder and J. H. Adler, *Anal. Chem.* **48**, 1022 (1976).
54. D. B. Roodyn, *Automated Enzyme Assays* (American Elsevier Publishing Co., Inc., New York, 1970).
55. J. Ruzicka and E. H. Hansen, *Anal. Chim. Acta* **78**, 145 (1975).
56. J. Ruzicka and J. W. B. Stewart, *Anal. Chim. Acta* **79**, 79 (1975).
57. J. W. B. Stewart, J. Ruzicka, H. B. Filho, and E. A. Zaggatto, *Anal. Chim. Acta* **81**, 371 (1976).
58. J. Ruzicka, J. W. B. Stewart, and E. A. Zaggatto, *Anal. Chim. Acta* **82**, 137 (1977).
59. E. H. Hansen and J. Ruzicka, *Anal. Chim. Acta* **87**, 353 (1976).
60. J. Ruzicka, E. H. Hansen, and E. A. Zaggatto, *Anal. Chim. Acta* **88**, 1 (1977).
61. J. Ruzicka, E. H. Hansen, and H. Mosback, *Anal. Chim. Acta* **92**, 235 (1977).
62. J. Ruzicka, E. H. Hansen, A. K. Ghose, and H. H. Mohala, *Anal. Chem.* **51**, 199 (1979).
63. B. Karlberg and S. Thelander, *Anal. Chim. Acta* **98**, 1 (1978).
64. J. J. van Deemter, F. J. Zuiderweg, and A. Klinkenberg, *Chem. Eng. Sci.* **5**, 271 (1958).

65. C. E. Ligny and W. E. Hammers, *J. Chromatogr.* **141**, 91 (1977).
66. J. J. Kirkland, *J. Chromatogr. Sci.* **10**, 129 (1972).
67. J. H. Knox and J. F. Parcher, *Anal. Chem.* **41**, 1599 (1969).
68. R. N. Kelley and F. W. Billmeyer, Jr., *Anal. Chem.* **41**, 874 (1969).
69. J. H. Knox, *Anal. Chem.* **38**, 253 (1966).
70. P. C. Carmen, *Flow of Gases Through Porous Media* (Academic, New York, 1956).
71. S. J. Hawkes, *J. Chromatogr.* **68**, 1 (1972).
72. C. Horvath and H. Lin, *J. Chromatogr.* **126**, 401 (1976).
73. C. Horvath and H.-J. Lin, *J. Chromatogr.* **149**, 43 (1978).
74. J. F. K. Huber, K. M. Jonker, and H. Poppe, "Optimal Design of Tubular and Packed-Bed Homogeneous Flow Chemical Reactors for Column Chromatography," *Anal. Chem.* **52**, 2 (1980).
75. K. Hofman and I. Halasz, "Mass Transfer in Ideal and Geometrically Deformed Open Tubes I. Ideal and Coiled Tubes with Circular Cross-Section," *J. Chromatogr.* **173**, 211 (1979).
76. I. Halasz, "Mass Transfer in Ideal and Geometrically Deformed Open Tubes II. Potential Application of Ideal and Coiled Open Tubes in Liquid Chromatography," *J. Chromatogr.* **173**, 229 (1979).

Due to the very rapid growth of the area of flow analysis we were unable to include a discussion of the following very important theoretical papers which were presented at the Amsterdam Conference on Flow Analysis (September 1979) and were published as a special issue of Analytica Chimica Acta (Volume 114, 1980):

77. J. Ruzicka and E. H. Hansen, *Flow injection analysis. Principles, applications and trends*, Anal. Chim. Acta **114**, 19 (1980).
78. R. Tijssen, *Axial dispersion and flow phenomena in helically coiled tubular reactors for flow analysis and chromatography*, Anal. Chim. Acta **114**, 71 (1980).
79. J. H. M. van den Berg, R. S. Deelder and H. G. M. Egberink, *Dispersion phenomena in reactors for flow analysis*, Anal. Chim. Acta **114**, 91 (1980).
80. J. M. Reijn, W. E. van der Linden and H. Poppe, *Some theoretical aspects of flow injection analysis*, Anal. Chim. Acta **114**, 105 (1980).
81. H. F. R. Reijnders, J. J. van Staden, G. H. B. Eelderink and B. Griepink, *A modeling approach to establish experimental parameters of a flow-through titration*, Anal. Chim. Acta **114**, 235 (1980).

MASS-TRANSFER AND DIFFUSION PROCESSES IN IMMOBILIZED-ENZYME REACTORS

In Chapter 6, all reactors were treated as if the chemical processes involved took place in homogeneous solution. Obviously, this is not at all the case when a solid-phase-supported catalyst, such as an immobilized enzyme, is employed. This chapter is essentially devoted to a treatment of how chemical engineers account for the additional mass-transfer process which may well limit the rate of product generation when a heterogeneous catalyst is used. The factor which relates the productivity of a heterogeneous process to that of an equivalent homogeneous process is the so-called *effectiveness factor,* which will be precisely defined in Section 7.1.3.

In building a small immobilized-enzyme reactor, it is essential to consider many of the factors which are vital to the design of large-scale industrial reactors. Since the enzyme is immobilized on a solid or semisolid (gel) surface and within the pores of the solid whereas the sample is contained in a fluid, no chemical reaction will take place until the substrate arrives at the surface of the particle containing the enzyme (see Figure 7.1). When small porous particles of enzyme carrier are used, the reaction may take place preponderantly in the interior of the solid. The net rate of conversion of the substrate will be dictated by the combined effect of:

1. Mass transfer from the bulk fluid to the solid surface. This is referred to as *external* mass-transfer resistance. As will be shown, this effect is most important at low flow rates, with large solid particles and with large molecules, due to their low diffusion coefficients. Exactly analogous considerations pertain in chromatography where this situation has encouraged the development of ultra-small packing materials (10 μm beads) (1).

2. The rate at which the substrate can diffuse into the catalyst pellet. Since convective flow does not take place in a small pore molecular diffusion is the sole important factor and fluid flow rate has no influence (2). This process is termed *internal* resistance to mass transfer. It should be noted that diffusion coefficients in the pores of solids can be significantly smaller than in solution. Clearly, large particle diameters can effectively decrease the availability of an enzyme which is present in the core of the

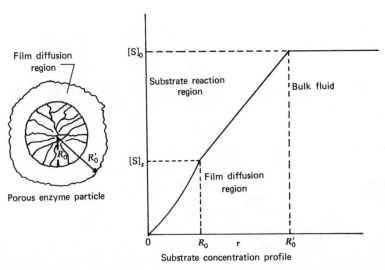

Fig. 7.1. Schematic representation of mass-transfer processes involved in immobilized-enzyme systems. Note the complex dependence of substrate concentration on distance. It is assumed in this scheme that the substrates' partition coefficient onto the enzyme particle is unity.

solid. In liquid chromatography, this problem can be avoided by using solids coated with a thin outer shell of adsorbant pellicular particles or, once again, by using very small particles. Enzyme immobilization on pellicular materials (3) have not been used extensively because this approach yields a very small amount of immobilized enzyme, which is analogous to low capacity in chromatography. Internal resistance to mass transfer can severely limit the total available enzyme activity, particularly if the reactant species is a macromolecule (protein, nucleic acid, polysaccharide). In the limit of very large molecules whose size approach the pore diameter, the net reaction rate can become very low (4).

3. The final rate-controlling process is the rate of the enzymatically catalyzed reaction. This factor can control the rate completely when the intrinsic specific activity is very low or when very little enzyme is immobilized. We should bear in mind the fact that the chemical reaction rate law will shift from first order to zero order as the substrate concentration is increased. When the substrate concentration is very high relative to its K_M, we need not be concerned with internal mass transfer as long as the reactant concentration at the center of the particle is finite, since the rate of the reaction will be independent of concentration and therefore of concentration gradients within the particle (5).

Another factor which is important in designing a reactor is the phenomenon of dispersion. As discussed previously, a plug flow reactor is more efficient than a well-mixed reactor when dealing with first- or higher-order reactions, but the two are equally efficient when a zero-order reaction is used. The difference between a plug flow and well-mixed system is the presence of dispersion, which is clearly closely related to mass-transfer considerations. Aside from the factor of reaction efficiency, excessive dispersion will cause a decrease in analytical sensitivity if sample pulses are used. Dispersion causes an increase in peak width which can result in decreased sample throughput.

The major difference between reactors for engineering purposes and those for analytical uses is a question of scale both in size and time. An industrial fixed-bed reactor is more often operated under steady-state conditions, that is, the substrate will be supplied continuously, whereas an analytical system should be able to accomodate small, short pulses of sample. In general, substrate concentrations of less than 1 M will be encountered. Although industrial reactions would include batch, fixed-bed, and fluidized-bed systems, the dispersion encountered in a fluidized bed would generally be so severe as to make such beds useless for analytical purposes. They will not be considered here. Materials of interest to clinical and biochemists will be present in the millimolar and submillimolar concentration ranges. Because the temperature changes associated with these low concentrations are so small, questions related to the effect of temperature on the reaction rate, thermal instabilities, and run away are completely irrelevant to the design of analytical reactors, but are quite crucial for large-scale industrial processors (6–8). The maximum possible temperature changes will be obtained with a plug flow reactor. Assuming a typical reaction enthalpy (10 kcal/mole) and specific heat (1 cal/°C·gm), a temperature change of 10°C is estimated for a 1 M sample if it is completely converted to products and no heat is lost to the surroundings. In a small analytical reactor immersed in a water bath, virtually all of this heat will escape. This is not the case in a large-diameter reactor whose surface-to-volume ratio is much smaller than is the case for a laboratory-scale device.

Exact and closed-form mathematical treatments of internal and external mass transfer and dispersion for the transient response of Michaelis–Menten-type reactions are not available. The steady-state behavior of such reactors when all factors are considered simultaneously is also quite complicated, particularly if the reaction-rate law is not of integral order as is the case for Michaelis–Menten kinetics. In order to establish some idea of the importance of the phenomena described above, each of the factors will be considered in turn. This will allow the estimation of the limits under which a given phenomenon can be disregarded or considered negligible.

Initially, the problem of external resistance to mass transfer to a *nonporous*, that is, solid-core catalyst pellet will be considered. A plug flow reactor will be assumed and the steady-state conversion estimated as a function of flow rate and particle size for a typical small-scale reactor. The "effectiveness factor" of a single porous catalyst particle will then be evaluated as a function of particle size. Dispersion or backmixing will be considered and its effect on the steady-state behavior will be estimated. Finally the pressure-drop requirements for a tubular flow system and a packed-bed reactor will be established.

7.1   EXTERNAL RESISTANCE TO MASS TRANSFER

Consider a nonporous catalyst pellet in which an enzyme is immobilized on the surface. In order for the substrate to react, it must be transported to the surface. In an unstirred solution, the transport will be controlled solely by diffusion. If fluid is pumped over a packed, fixed bed of catalyst, part of the transport will be due to convection and part will be due to diffusion. Therefore, the dependence of the net flux of reactant to the surface (N, in moles/sec·cm²) on diffusion coefficient will be to some power between 0 and 1.

Mass transfer in the presence of convective transport may be represented in terms of a rate process of the form

$$N = k_L([S]_0 - [S]_s)$$ (7.1)

where k_L is a mass transfer rate constant (cm/sec) which will certainly depend upon the properties of the fluid and substrate including viscosity, density, and diffusion coefficient, as well as the actual flow rate. The terms $[S]_0$ and $[S]_s$ represent the bulk and surface concentrations of the reactant (in moles/ml) as shown in Fig. 7.1.

7.1.1   Engineering Mass-Transfer Correlations

Mass transfer data (9) are usually presented in terms of a dimensionless group termed the Sherwood number (N_{Sh}), that is,

$$N_{Sh} \equiv \frac{k_L \cdot d_p}{D_m}$$ (7.2)

where d_p represents the particle diameter if it is spherical, the diameter of an equivalent sphere of the same area if the particle has some other shape, or the tube diameter when a tubular reaction is employed. Since the mass-transfer rate coefficient is a function of viscosity, density, etc., the Sherwood number will also be related to these parameters. Engineers have found it useful to summarize such dependences in terms of two dimensionless groups, the Reynolds number (N_{Re}) and the Schmidt number (N_{Sc}).

$$N_{Re} \equiv \frac{d_p G}{\mu} \tag{7.3}$$

$$N_{Sc} \equiv \frac{\mu}{\rho D_m} \tag{7.4}$$

We see that the viscosity μ (in poise) is quite important as are the fluid density ρ (g/cm^3), the particle size, and the substrate's diffusion coefficient. The term G is quite significant; it is the *superficial* mass flow rate, that is, the flow rate divided by the unpacked column geometric area (A). Phenomenologically, the Reynolds number of a fluid represents the ratio of inertial force to viscous force. It is the chief parameter which dictates the pressure required to push a fluid through a tube or packed bed. The Schmidt number interrelates momentum transfer and mass transfer. It arises from a detailed analysis of the relevant equations of motion. When N_{Sc} is equal to unity, the equations governing momentum transfer, that is, pressure drop and mass transfer, have exactly the same form and therefore have identical mathematical solutions (10), provided that the boundary conditions are similar.

The Sherwood number will be some function of both the Reynolds and Schmidt numbers

$$N_{Sh} = f(N_{Re}, N_{Sc}) \tag{7.5}$$

Experimentally it is known that k_L is proportional to $D_m^{2/3}$ over a very wide range of Schmidt numbers; thus the Sherwood number should be proportional to the one-third power of the Schmidt number:

$$N_{Sh} = N_{Sc}^{1/3} \cdot f(N_{Re}) \tag{7.6}$$

The functional relationship between the mass-transfer rate and the Reynolds number for a fixed-bed reactor defies a priori calculation. A number of workers have arrived at correlations of experimental data which work over restricted conditions. For example, the Schmidt number is said to be proportional to $G^{1/3}$ (11) and to G^1 (12) under other conditions. The results of Wilson and Geankoplis have achieved very widespread use (13). For liquids, as opposed to gases, they found that in the region of high Reynolds numbers ($55 < N_{Re} < 1500$) and over a wide range of column void fraction ($0.35 < \epsilon < 0.75$), the Sherwood number was given by the relationship:

$$N_{Sh} = \frac{0.25 G d_p}{\epsilon \rho D_m} \frac{1}{(N_{Re})^{0.31}(N_{Sc})^{2/3}} \qquad (7.7)$$

For lower Reynolds numbers ($0.0016 < N_{Re} < 55$), the following relationship was found to be valid for a wide range of Schmidt numbers ($105 < N_{Sc} < 71,000$):

$$N_{Sh} = \frac{1.09 G d_p}{\epsilon \rho D_m} \frac{1}{(N_{Re})^{2/3}(N_{Sc})^{2/3}} \qquad (7.8)$$

For a small molecule (M.W. ≈ 200) in water near room temperature one readily estimates a Schmidt number of about 1000. To determine whether equation (7.7) or (7.8) will be useful for a laboratory scale reactor such as would be used for analytical work let us estimate N_{Re} for a volumetric flow rate of 1 ml/min. The column will be packed with 100 μm particles which would be very easy to pump at low pressure and will also exaggerate the Reynolds number [equation (7.3)]. In order to estimate G, we will assume a 5 mm diameter column (d_c) whose unpacked cross sectional area (A) is about 0.2 cm². We estimate a N_{Re} of about 0.085:

$$N_{Re} = \frac{d_p G}{\mu} = \frac{4 d_p F \rho}{\pi d_c^2 \mu}$$

$$= \frac{4 \times (0.01 \text{ cm}) \times (.0167 \text{ ml/sec}) \times 1 \text{ (g/ml)}}{3.14 \times (0.25 \text{ cm}^2) \times 0.01(\text{P})} = 0.085 \qquad (7.9)$$

Both the Schmidt number and Reynolds number are well within the range of equation (7.8) so we will use that equation for a laboratory-scale analytical reactor with water as the fluid. Using the above conditions, a Sherwood number of about 12 is obtained with $\epsilon = 0.4$.

7.1.2 Criteria for Neglecting External Mass-Transfer Rate Limitations

Using the results obtained previously for the extent of reaction as a function of distance along the axis of a fixed-bed, plug flow reactor, one can estimate the length of column needed to achieve any desired degree of reaction completion.

We assume that the reaction rate is limited by the rate of mass transfer to the immobilized enzyme and not by a chemically slow step. Then a steady-state mass balance on a differential element of volume can be carried out as in Chapter 6. Realizing that the reaction is taking place at the surface, we use equation (7.1) to define the reaction rate per unit surface area. If there is a high enzyme activity on the column and mass transfer is totally rate-limiting, then the concentration of substrate on the surface will be very small relative to the bulk concentration; therefore we can neglect $[S]_s$ in equation (7.1). Integrating along the axis of the column (l), we find the following relationship between concentration and distance:

$$\ln \frac{[S]^0}{[S]_l} = \frac{N_{Sh}D_m \rho \mathbf{a} l}{d_p G} \tag{7.10}$$

where $[S]^0$ is the reactant concentration in the feed stream at the column entrance, $[S]_L$ is the concentration at some point downstream, and $\mathbf{a}$ is the surface area of catalyst per unit bed volume. It should be noted that this is the *maximum* reaction rate which might be observed for a first-order reaction.

Substituting in the previous relationship for the Sherwood number,

$$\ln \frac{[S]^0}{[S]_l} = \frac{1.09 \mathbf{a} l}{\epsilon (N_{Re} \cdot N_{Sc})^{2/3}} \tag{7.11}$$

This relationship allows us to estimate the length of column L required to achieve any degree of desired reaction. The above relationship has been employed repeatedly to assess the extent of external mass-transfer control (14, 15).

In order to do this we need to estimate the surface area of catalyst per unit bed volume (see Table 7.1). The ratio of area to volume of a single spherical particle is $6/d_p$. Since the ratio of bed volume to actual volume of catalyst is given by

$$V_{actual} = (1 - \epsilon)V_0 \tag{7.12}$$

it is easy to show that

$$\mathbf{a} = \frac{6(1 - \epsilon)}{d_p} \tag{7.13}$$

Substituting equation (7.13) for $\mathbf{a}$ and equations (7.3) and (7.4) for the Reynolds and Schmidt numbers we find that

$$\ln \frac{[S]^0}{[S]_l} = 6 \left(\frac{1 - \epsilon}{\epsilon} \right) \frac{l}{d_p^{5/3}} \left(\frac{\rho D_m}{G} \right)^{2/3} \tag{7.14}$$

Table 7.1 External Surface Area of Catalyst Per Unit Volume of Packed Bed (cm²/cm³)

d_p(cm)	100 × ϵ		
	30	40	50
0.1	42	36	30
0.05	84	72	60
0.02	210	180	150
0.01	420	360	300
0.005	840	720	600
0.001	4200	3600	3000

This equation indicates that the single most important parameter in achieving high mass-transfer rates is the particle size. Large particles will show low mass-transfer rates and will require long columns and concommitantly more immobilized enzyme for complete reaction.

Once again using water as the solvent and a molecule of average size ($D_m = 10^{-5}$ cm²/sec), we can estimate the length of column needed to achieve any satisfactory degree of completion. For analytical purposes, maximum sensitivity and precision will be obtained if we approach 99% or more conversion. Table 7.2 summarizes the calculations for a typical 5 mm diameter column (see above). Kunii and Smith found that the column voidage ranged between about 0.34 and 0.39 for glass particles between 110 to 1020 μm. Slightly larger values (0.40) were obtained with sand (16). These are quite reasonable values for a fixed-bed reactor as opposed to a fluidized-bed reactor.

Table 7.2 Column Length (cm) Required for 99% Completion[a,b]

d_p(cm)	Flow rate	
	1 ml/min	5 ml/min
0.10	4.6	13.6
0.05	1.43	4.2
0.02	0.31	0.92
0.01	0.098	0.28
0.005	3.08×10^{-2}	9.1×10^{-2}
0.001	2.1×10^{-3}	6.2×10^{-3}

[a] Computed from equation (7.14).
[b] $\epsilon = 0.40$, column diameter 5 mm, fluid is water at 25°C, diffusion coefficient of reactant is taken as 10^{-5} cm²/sec.

It is important to note the limitations of equation (7.14). Since it was derived with the assumption that mass transfer is the sole rate-limiting factor, it predicts that the reaction rate (not extent of conversion) will increase indefinitely with flow rate. Ford and coworkers (15) have designed a small-scale reactor system which uses a very thin bed of catalyst to avoid extensive reaction. They used this device to determine the effect of flow velocity on reaction rate and thereby test for the presence of external mass-transfer rate limitations. Since diffusion *within* a catalyst pellet is not dependent upon the external flow velocity, any change in reaction rate (moles/sec) with flow rate clearly indicates that external mass transfer is important. Figure 7.2 shows that the measured reaction

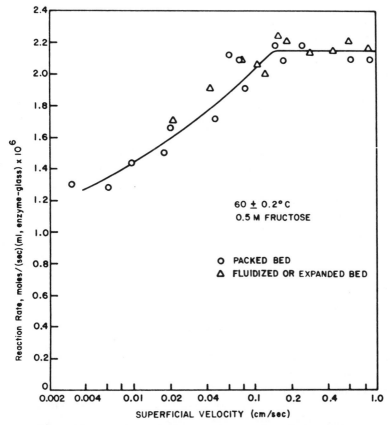

Fig. 7.2. Effect of flow velocity on apparent reaction rate. A comparison of film resistances in packed beds and fluidized beds. [Reprinted by permission of John Wiley & Sons, Inc., from Y. Y. Lee, A. R. Fratzke, K. Wun, and G. T. Tsao, *Biotechnol. Bioeng.* **18**, 389 (1976).]

rate will become independent of flow rate at high linear velocity and under these conditions, external mass transfer has very little effect.

The design of an immobilized-enzyme reactor for analytical measurements involves such questions as sensitivity (percent conversion), residence time (flow rate and column size), long-term stability, and economy of enzyme usage. Clearly it makes no sense to immobilize a large amount of enzyme on very large particles since, in a small reactor (i.e., a short one), mass transfer may limit the extent of reaction and much less of the expensive enzyme could just as well have been used.

Another serious question related to flow rate is sample throughput. Consider the now-familiar reactor (diameter 5 mm) whose voidage is 0.40. Assuming nonporous catalyst support, a 5 cm long column will contain about 0.4 ml of fluid. With a flow rate of 1 ml/min, a sample will spend about 25 sec in the column. If the catalyst is porous, for example, controlled-pore glass, the fluid will reside in the column for a considerably longer time. Unless there is very little spreading (dispersion), it is evident that the reactor design could well limit the sampling rate because it would be inadvisable to have more than one sample in a column at a given time.

There are few if any advantages to using large catalyst particles for equilibrium assays. Nonetheless, the size of the catalyst pellet is inevitably limited by the maximum tolerable pressure drop. When smaller particles are used, the pressure drop will increase dramatically. A reasonable compromise appears to be the use of particles in the 100–200 mesh range. These afford fairly short reaction lengths, good flow properties, and are not so small that they become clogged quickly. Porous-glass particles in this size range are very easily packed without elaborate equipment.

7.1.3   Kinetic Effectiveness Factor for External Resistance to Mass Transfer

The preceeding simple calculation attempts to estimate the minimum length of column required to obtain some desired degree of complete reaction. If this length is less than some small fraction of the actual column length, then it would be of interest to estimate the actual "effectiveness" of the immobilized preparation. In general, the effectiveness factor in the presence of any type of resistance to mass transfer (either external or internal) is defined as the ratio of the observed reaction rate to that which would be observed if mass transfer were infinitely fast, that is,

$$\eta \equiv \frac{\text{Rate with Finite Mass Transfer}}{\text{Rate with Infinite Mass Transfer}} \qquad (7.15)$$

Several groups have considered this problem for first-order chemical reactions (17, 18) and for Michaelis-Menten kinetics (19, 20). O'Neil has considered the effect of mass transfer in both continuously stirred tank reactors and plug flow reactors, and presented results for situations intermediate between these two cases. Kobayashi and Laidler have treated this situation where electrostatic factors and a variety of forms of substrate and product inhibition were taken into account (24).

In the complete absence of mass-transfer problems, the rate of the chemical reaction $\mathcal{R}_c$ (moles/sec·ml) will be given by Michaelis–Menten kinetics:

$$\mathcal{R}_c = \frac{k_2[E][S]}{K_M + [S]} \tag{7.16}$$

where the appropriate substrate concentration can be taken as either that at the surface ($[S]_s$) or that in the bulk ($[S]_0$) since the two are equal when mass transfer is fast. In the steady state, the rate at which matter arrives at the surface will be termed r_m (mole/sec·ml) and the rate of enzymatic reaction and mass transfer can be set equal to one another:

$$\mathcal{R}_c = r_m = k_L a([S]_0 - [S]_s) \tag{7.17}$$

The normalized substrate concentration at the surface is then calculated as:

$$[S]^* = \frac{[S]_s}{[S]_0} = \frac{1}{2}\left[\left(1 - \frac{K_M}{[S]_0} - \frac{k_2[E]}{k_L a[S]_0}\right)\right.$$
$$\left. + \sqrt{\left(1 - \frac{K_M}{[S]_0} - \frac{k_2[E]}{k_L a[S]_0}\right)^2 + \frac{4K_M}{[S]_0}}\right] \tag{7.18}$$

According to its definition the effectiveness will be:

$$\eta = \frac{[S]_s}{K_M + [S]_s} \bigg/ \frac{[S]_0}{K_M + [S]_0} = [S]^* \frac{(\beta + 1)}{(\beta + [S]^*)} \tag{7.19}$$

where β is a dimensionless Michaelis constant ($= K_M/[S]_0$) and $[S]^*$ is the ratio of surface to bulk substrate concentrations. It is evident that the effectiveness of the immobilized-enzyme preparation depends upon the relative rate of the enzymatic and mass-transfer processes ($k_2[E]/k_L a$) and on the dimensionless Michaelis constant, that is, whether the reaction is in the first- or zero-order substrate regimes. Lee et al. (19) have found it useful to rewrite equation (7.18) in the form

$$[S]^* = \tfrac{1}{2}[\{1 - \beta(\phi_f + 1)\} + \sqrt{\{1 - \beta(\phi_f + 1)\}^2 + 4\beta}] \tag{7.20}$$

where ϕ_f represents a film or external mass transfer rate modulus similar to the Thiele modulus for internal mass transfer (see below) and is defined as:

$$\phi_f \equiv \frac{[S]_0}{K_M} \cdot \frac{k_2[E]}{k_L a} \qquad (7.21)$$

Their calculated plot (19) of the effectiveness versus ϕ_f and β are shown in Figure 7.3. The reaction effectiveness or efficiency of utilization of the enzyme decreases as the rate of enzyme reaction becomes fast relative to the mass-transfer rate. For ϕ_f above 1 the effectiveness begins to drop from its maximum value. The data also indicate that the effectiveness is not as dependent upon ϕ_f in the zero-order kinetic region as in the first-order region. This results because in the zero-order region, the rate of the chemical reaction is indifferent to concentration and therefore to mass-transfer-induced concentration gradients.

Several important limiting cases are easily derived. In the limit of very low K_M or high bulk substrate concentration, the effectiveness approaches unity regardless of the film diffusion modulus:

$$\eta = 1 \text{ for } \beta = 0; \quad \text{for all } \phi_f \qquad (7.22)$$

Similarly when β is very large, the chemical reaction is first order and

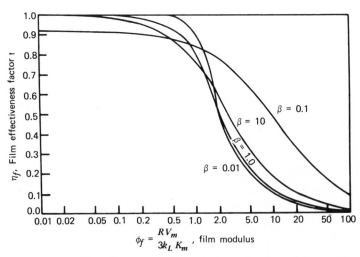

Fig. 7.3. Theoretical film effectiveness factor (η_f) as a function of the modified Thiele modulus (ϕ_f) and normalized Michaelis constant (β). [Reprinted by permission of John Wiley & Sons, Inc., from Y. Y. Lee, A. R. Fratzke, K. Wun, and G. T. Tsao, *Biotechnol. Bioeng.* **18**, 389 (1976). See equations (7.20) and (7.21).]

the factor approaches the limit

$$\eta = \frac{k_L a[S]_0}{k_2[E]}(1 + \beta) \Bigg/ \left[1 + \frac{k_L a[S]_0}{k_2[E]}(1 + \beta) \right] \qquad (7.23)$$

This is the situation studied by Rovito and Kitrell (18) in detail for a plug flow reactor. In the limit of very fast mass transfer, the film modulus approaches zero. Mathematically, the effectiveness approaches unity. For any finite Michaelis constant and bulk concentration, the effectiveness approaches zero as the film modulus increases, that is, as the rate of mass transfer decreases.

The above treatment is valid for a batch reactor and a plug flow reactor since in both of these systems, the entire sample is exposed to the catalyst for exactly the same time. The yield or conversion in a CSTR or a reactor with dispersion ($N_{Pe} < \infty$) would be different. It should be understood that the mass-transfer situations in a batch reactor and a PFR are vastly different and equation (7.8) above which relates the Sherwood number that is, the dimensionless mass transfer rate to the Reynolds number (fluid velocity) is valid only for a column reactor with plug flow. O'Neil has discussed the concept of mass transfer in a stirred-batch reactor in which the spherical catalyst particles are *suspended* in essentially stagnant fluid of roughly the same density (21). He finds that under these conditions:

$$k_L \cong \frac{2D_m}{d_p} \qquad (7.24)$$

This is equivalent to a Sherwood number of 2 [equation (7.2)]. Equation (7.8) indicates that higher Sherwood numbers can be obtained in packed reactors even at low flow rate, for example, the previous calculation with a flow rate of 1 ml/min indicated a Sherwood number of 12. Kinetically, it is therefore *more efficient to use a PFR than a batch reactor*. Use of baffles and other techniques could alter this situation. (Analytically a flow reactor is certainly easier to use for automated analysis.) The reader is advised that in some complex chemical systems, particularly where the product inhibits the enzyme, a batch reactor may be more efficient than a PFR.

7.1.4 Effect of Dispersion and Slow External Mass Transfer on the Outlet Substrate Concentration in a Flow Reactor

As discussed in Chapter 6, the inlet and outlet concentration of a PFR can be written in exactly the same form as the equation describing the soluble-enzyme reaction (Henri equation) if mass transfer and dispersion are neglected. Lilly, Hornby, and Crook (25) presented the data shown in Figures 7.4 and 7.5. The data of Figure 7.4 show excellent conformity with the form of the Henri equation, that is:

$$X[S]^0 - K_M \ln (1 - X) = \frac{k_2[E]}{[S]^0} \frac{V_D}{F} \tag{7.25}$$

where V_D/F is the residence time in the column. However the data clearly indicate that the slope of the line (K_M) is a function of flow rate (see Figure 7.5).

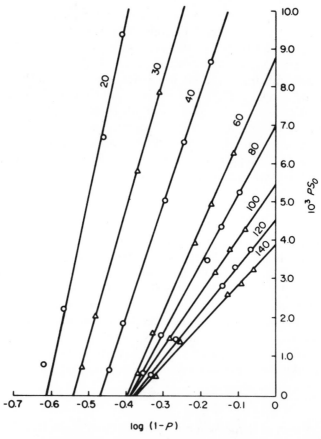

Fig. 7.4. Experimental relationship between inlet and outlet substrate concentration plotted in accord with the integrated form of the Michaelis–Menten equation. The numbers on each curve are the flow rates in ml/hr. All data are for hydrolysis of benzoyl-L-arginine ester by ficin bound to carboxymethyl cellulose. [Reproduced from M. B. Lilly, W. E. Hornby, and E. M. Crook, *Biochem. J.* **100**, 718 (1977), courtesy of the Biochemical Society.]

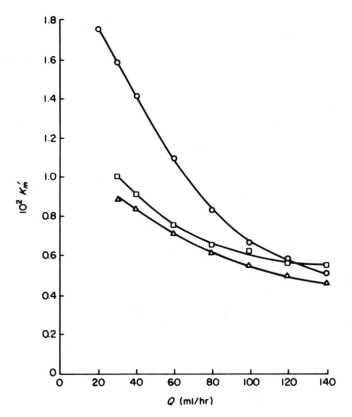

Fig. 7.5. Dependence of the apparent Michaelis constant (K'_M) on flow rate. All other conditions are the same as in Figure 7.4. [Reproduced from M. D. Lilly, W. E. Hornby, and E. M. Crook, *Biochem. J.* **100**, 718 (1966), courtesy of the Biochemical Society.]

Kobayaski and Moo-Young appear to be the first workers to derive a relationship between K'_M, the apparent Michaelis constant, and the flow rate (22, 23). Their work brings together the combined effect of slow external mass transfer and dispersion. Essentially they assume the absence of internal resistance to mass transfer and radial dispersion, an isothermal reaction, Michaelis–Menten kinetics, and a continuously fed reactor, that is, the substrate is pumped into the column continuously. They then compute the steady-state concentration of substrate at the reactor outlet. Since a steady state is assumed, their results are not exactly applicable when a narrow pulse of substrate is used. If a pulse of substrate is used, that is sufficiently wide to swamp dispersive broadening, then the peak maxima will be described by their equations.

The derivation uses a differential mass balance to obtain an equation for the *steady state*:

$$\mathcal{D}_a \frac{d^2[S]_0}{dl^2} - \frac{ud[S]_0}{dl} - \frac{k_L a}{\epsilon}([S]_0 - [S]_s) = 0 \qquad (7.26)$$

$[S]_0$, $[S]_s$, and $[S]^0$ represent the bulk substrate concentration at any point l along the column axis, the surface substrate concentrations at the catalyst pellet, and the feed supply substrate concentration, respectively. $\mathcal{D}_a$ is an axial dispersion coefficient. In the steady state, the mass-transfer reaction rate can be equated to the chemical reaction rate and the surface substrate concentration can be estimated as per equation (7.18). The second-order differential equation requires two boundary conditions.

Obviously the final equation is very difficult to solve in closed form, therefore Kobayashi and Moo-Young (see Table 7.3) sought the limiting behavior at high and low Peclet numbers, that is, plug flow reactor and continuously stirred tank reactor behavior as well as several intermediate cases. Two of the more significant results are obtained when the mass-transfer rate is fast but not infinitely fast. Under this condition for a continuously stirred reactor, the reaction rate ($\mathcal{R}_c$) becomes:

$$\mathcal{R}_c = \frac{k_2[E](1 - \epsilon)}{\epsilon} \frac{[S]_0}{K_M' + [S]_0} \qquad (7.27)$$

with

$$K_M' = K_M + \frac{3}{4} \frac{k_2[E](1 - \epsilon)}{k_L a} \qquad (7.28)$$

As is evident, this is the same form as the Michaelis–Menten rate law and predicts that K_M' will be a function of the flow rate. Under similar conditions for a plug flow reactor (fast but finite mass transfer)

$$[S]^0 - [S]_e = K_M' \ln \frac{[S]_e}{[S]^0} + \frac{k_2[E](1 - \epsilon)V_0}{F} \qquad (7.29)$$

where K_M' is given by equation (7.28). Equation (7.29) is of precisely the same form as the integrated Michaelis–Menten rate law. Clearly equation (7.28) predicts that the Michaelis constant will be related to flow since k_L the mass transfer coefficient will depend on the Reynolds number [see equation (7.5)]. This is in accord with the results of Lilly et al. (25) (see Figure 7.5). As the flow rate increases, k_L increases, thereby reducing K_M'. In the limit of very high flow rate, the apparent Michaelis constant should approach K_M, the intrinsic constant for the soluble enzyme, provided that the substrate partition coefficient is unity. Although the data of Lilly et al. asymptotically approach a limit at high flow rate, this is

Table 7.3 Equations of Kobayashi and Moo-Young for Outlet Substrate Concentration with Dispersion and Slow Mass Transfer[a]

CSTR	$N_{Pe} = 0$	$\dfrac{[S]_e}{[S]^0} = \dfrac{1}{2}\left[\dfrac{\lambda + 2\alpha}{\lambda + \alpha}\lambda - \lambda - \dfrac{\beta\lambda}{\lambda + \alpha} + \sqrt{\left(\dfrac{\lambda + 2\alpha}{\lambda + \alpha} - \lambda - \dfrac{\beta\lambda}{\lambda + \alpha}\right)^2 + \dfrac{4[(\alpha + \beta)\lambda - \alpha]}{\lambda + \alpha}}\,\right]$
CSTR	$\begin{array}{l}N_{Pe} = 0 \\ \alpha = 0\end{array}$	$\dfrac{[S]_e}{[S]^0} = \dfrac{1}{2}[1 - \lambda - \beta + \sqrt{(1 - \lambda - \beta)^2 + 4\beta}]$
PFR	$\begin{array}{l}N_{Pe} = \infty \\ \alpha = 0\end{array}$	$[S]^0 - [S]_e = K_M \ln\dfrac{[S]_e}{[S]^0} + k_2[E]\dfrac{(1 - \epsilon)}{\epsilon}\tau$
PFR	$\begin{array}{l}N_{Pe} = \infty \\ \alpha\ \text{small}\end{array}$	$[S]^0 - [S]_e = (K_M + \tfrac{3}{4}\alpha)\ln\dfrac{[S]_e}{[S]^0} + \dfrac{k_2[E](1 - \epsilon)}{\epsilon}\tau$

[a] Definition of terms $\alpha \equiv k_2[E](1 - \epsilon)/k_L a[S]^0$; $\beta = K_M/[S]^0$; $\lambda = k_2[E](1 - \epsilon)\tau/[S]^0$; $\tau = V_D/F$.

seldom found to coincide with the Michaelis constant of the soluble enzyme due to chemical effects related to immobilization, as well as microenvironmental and electrostatic factors, which produce substrate partitioning. Nonetheless, it is observed that *slow mass transfer invariably acts so as to increase the apparent Michaelis constant.*

Qualitatively the effect of slow mass transfer to increase K'_M can be rationalized by the fact that the rate of mass transfer is always proportional to concentration. Thus as mass transfer becomes more important, that is slower, the net reaction rate should appear to be first order at higher concentrations.

7.1.5 Effect of Slow External Mass Transfer in an Open Tubular Reactor

Several papers have appeared on the use of reactors in which an enzyme is immobilized on the wall of a tube (26–36). Three modes of immobilizing the enzyme on the inner wall of a tube can be envisioned: (1) direct covalent immobilization, (2) covalent immobilization on a thin porous crust of solid coated on the wall, (3) immobilization in a gel to form an annulus of enzyme. There are a number of significant advantages to such a system which are similar to those obtained in open tubular gas chromatography; most notable is the low pressure drop of such systems. In addition, this type of reactor is ideal for use in continuous-flow analysis. Because pressure drops will be much lower than in a packed-bed reactor, and this type of device can be operated with segmentation gas bubbles to inhibit sample spreading (dispersion), quite long reactors can be used. For example, Hornby and coworkers (32) have employed 3–10 m nylon tubes to bring about a high fractional conversion with only a small amount of enzyme.

Although Kobayashi and Laider (37) have carried out a detailed theoretical investigation of mass transfer in a one-phase (i.e., unsegmented) tubular reactor of immobilized enzyme, their results will not be detailed here because for analytical purposes the system will generally be operated in two-phase flow to prevent spreading and preserve sample integrity. Their results are nonetheless important because they include the entire range of concentrations relative to K_M as well as the effect of substrate and product inhibition. Such a tubular reactor would undoubtedly be formed into a coil for compactness. The secondary flows (see Figure 7.6) created by coiling and gas segmentation (38, 39) can have a profound influence on radial mass transfer and this factor was not included, for the sake of simplicity, in the derivation of Kobayashi and Laider. Horvath and coworkers (38) have carried out an extensive experimental evaluation

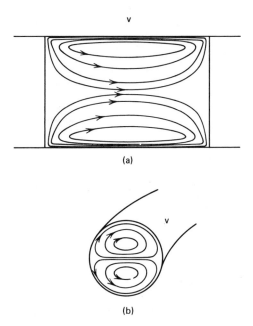

Fig. 7.6. Illustration of the secondary radial flows created by use of segmented-flow systems and coiling and tubes. (a) Flow pattern due to segmentation with a gas in a straight tube. (b) Flow pattern due to use of a coil without segmentation. [Reprinted with permission from C. Horvath, B. A. Solomon, and J. M. Engasser, *Ind. Eng. Fund.* **12,** 431 (1973). Copyright by the American Chemical Society.]

of tubular enzyme coils operated under conditions of two-phase flow, that is, segmented flow.

The extreme effects of coiling and segmentation on the fractional conversion to product are shown in Figure 7.7, which is from the data presented by Horvath and his coworkers (36) for the hydrolysis of *N*-benzoyl-L-arginine ethyl ester by trypsin immobilized on the walls of a circular tube. It should be noted that the enzyme activity per unit area is so high that the substrate concentration at the wall is essentially negligible, that is, the net reaction rate is *totally mass-transfer controlled.* For a straight tube operated with one-phase flow, that is, no gas segments, it is observed that the fractional conversion decreases monotonically with flow rate. However, for an uncoiled tube operated with segmented flow, the extent of conversion rises very rapidly with flow rate. *With a coiled tube operated with segmented flow, the conversions are even greater.* Since the residence time decreases as flow rate increases, it is evident that the rate

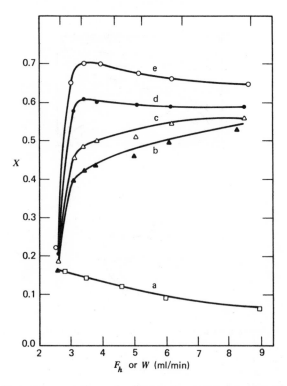

Fig. 7.7. The effect of segmentation and coiling on the extent of reaction (X) as a function of flow rate (F_h or W). The flow rate is F_h for homogeneous flow and W for slug flow. Curve a: straight tube—no segmentation; curve b: straight tube—with segmentation; curves c–e. Coiled tube—with segmentation. δ = 0.005, 0.009, and 0.017 for curves a, b, and c–e, respectively. [Reprinted with permission from C. Horvath, B. A. Solomon, and J. M. Engasser, *Ind. Eng. Fund.* **12**, 431 (1973). Copyright by the American Chemical Society.]

of mass transfer must increase dramatically with slug flow and is strongly coupled to the radius of curvature of the coil helix.

The data collected by Horvath et al. were correlated in terms of a *logarithmic mean* Sherwood number (40), $\overline{N_{Sh}}$, for which equation (7.30) is taken essentially as a definition:

$$X = 1 - \exp(-4\overline{N_{Sh}}\cdot Z) \qquad (7.30)$$

where X is the steady-state conversion defined as the fractional change in substrate concentration:

$$X = \frac{[S]^0 - [S]_e}{[S]^0} \qquad (7.31)$$

The term Z is a dimensionless reaction time

$$Z = \frac{\pi D_m L}{4W} \qquad (7.32)$$

The variables, N_{Sh}, and Z are subscripted with an s or h where appropriate to indicate either segmented or homogeneous flow, respectively. We expect the mass transfer rate to depend upon the Reynolds number, the Schmidt number, the reactor length (L) since it is not packed, the reactor diameter (d_t) and the slug length (l_s). If the tube is coiled, then it will depend upon the diameter of the helix denoted (H):

$$\overline{N_{Sh}} = f(N_{Re}, N_{Sc}, \psi, \gamma, \delta) \qquad (7.33)$$

where for a tubular reactor Horvath takes:

$$N_{Re} = \frac{4W}{\pi d\mu\rho} \qquad (7.34)$$

$$\psi = \frac{l_s}{d_t} \qquad (7.35)$$

$$\gamma = \frac{L}{d_t} \qquad (7.36)$$

$$\delta = \frac{d_t}{H} \qquad (7.37)$$

W is the combined flow rate of the liquid and gas. The dependence on all of the parameters except the tube length (γ) is quite strong, as shown in Figure 7.8. The mass-transfer process acts as if it were intermediate between homogeneous laminar and turbulent flow, that is, the Sherwood numbers for short slugs approach those characteristic of turbulent flow even though the Reynolds numbers are well below that required for turbulence in cylindrical tubes ($N_{Re} > 2000$) with homogeneous flow.

The maximum mass-transfer rates occur at small slug lengths ($\psi < 3$) and with tightly coiled tubes. As shown in Figure 7.8(d), tube coiling is not as significant an effect as is segmentation. There is certainly a price to be paid for the improved mass-transfer coefficient, that is, an increased pressure drop. At $\psi < 3$, the pressure drop required to achieve a Reynolds number of 220 with a column voidage of 0.5 is somewhat higher than that required for homogeneous flow. As ψ decreases further, the pressure drop increases quite rapidly. Engasser and Horvath (41) have compared the relative mass-transfer coefficients of packed beds, and open tubular columns operated with and without slug flow. They conclude that open tu-

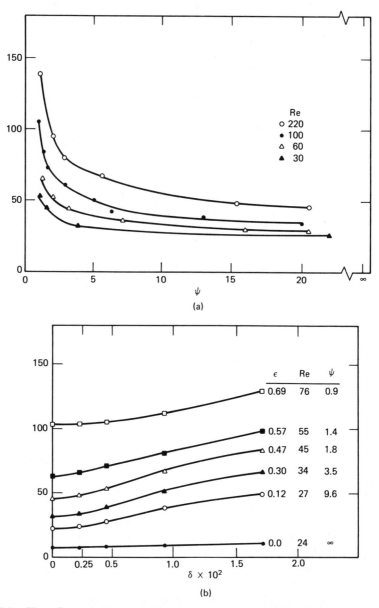

Fig. 7.8. These figures indicate the effect of the experimental parameters on the measured logarithmic mean Sherwood number ($\bar{N}_{Sh}$). See equations (5.33)–(5.40) for the definition of all terms used. The subscript h refers to homogeneous flow and s to segmented flow. (a) Plots illustrating the dependence of the slug flow Sherwood number, N_{Sh_u}, on the dimensionless slug length, ψ, at fixed void fraction, $\epsilon = 0.5$ and at different Reynolds numbers, $\bar{N}_{Re}$. The values of the corresponding Sherwood number for homogeneous flow, $\bar{N}_{Sh_h}$, are also indicated. (b) Plots of the slug flow Sherwood number, $\bar{N}_{Sh_u}$, versus the ratio of tube diameter to coil diameter, δ, for different values of the void fraction, ϵ. At the fixed liquid

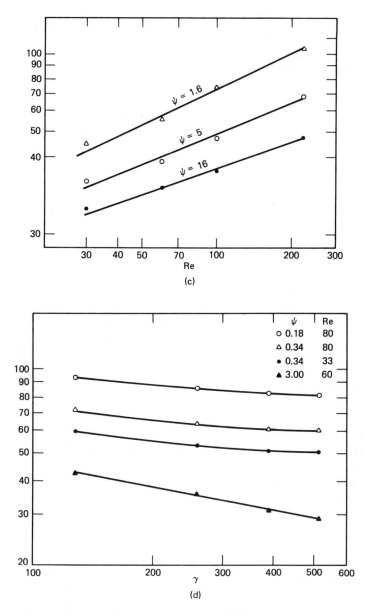

(c)

(d)

flow rate, F_s = 2.6 ml/min, different values of ϵ represent also different ψ's and $\bar{N}_{Re}$ as indicated. The dependence of $\bar{N}_{Shh}$ on δ in homogeneous flow (ϵ = 0) at the same flow rate is also illustrated. (c) Plots of the slug flow Sherwood number, $\bar{N}_{Shu}$, versus the Reynolds number, $\bar{N}_{Re}$, at different values of the dimentionless slug length, ψ. (d) Plots of the slug flow Sherwood number, $\bar{N}_{Shu}$, against the dimensionless tube length, γ, at different values of the dimensionless slug length, ψ, and Reynolds number, $\bar{N}_{Re}$. [Reprinted with permission from C. Horvath, B. A. Solomon, and J. M. Engasser, *Ind. Eng. Fund.* **12,** 431 (1973). Copyright by the American Chemical Society.]

373

bular reactions permit higher mass-transfer rates with a given pressure drop than do packed beds.

Equation (7.30) in conjunction with the graphical correlations of Figure 7.8 can be used to estimate the length of tube required to bring about any desired degree of reaction. The nylon tube reactors used by Hornby and his coworkers have a typical diameter of 1 mm and are operated at flow rates of 7 ml/min (32). This corresponds to a Reynolds number of about 150. Choosing ψ as 1.6 (the lowest relative slug length studied) yields a Sherwood number of about 80. The minimum reactor length for the conversion X is obtained by combining equations (7.30) and (7.32)

$$L_{min} = \frac{W}{\pi D_m N_{Sh}} \ln \left(\frac{1}{1-X} \right) \qquad (7.38)$$

Choosing a generous value for D_m of 10^{-5} cm^2/sec, we see that a minimum length of 210 cm is required to achieve 99% conversion. This means that mass-transfer effects can be quite severe with open tubular reactors. It should be noted that the total volume of the reactor is only 1.65 ml and at a flow rate of 7 ml/min, the sample is in residence for only about 15 sec.

The above calculation suggests not surprisingly that the optimum tubular reactor should have a very narrow bore, be very long, and operated with small slug lengths. This will obviously increase the pressure required to operate the system. Complete conversion can be obtained by going to very low flow rates. These will be far too low to obtain an acceptable analysis rate. Horvath's data indicates that for a slug flow system, an optimum flow rate will be at Reynolds number of 30–40 (see Figure 7.7).

Both gas segmentation and column coiling set up secondary flow fields in addition to the flow in the direction along the column axis. Horvath concludes that although each of the secondary flows are due to separate factors, they both act in unison to enhance radial mass transfer, that is, they provide convective transport toward the inner wall of the tube in addition to that provided by diffusion and the normally small radial component of flow velocity which exists in the laminar-flow regime. Obviously, turbulent flow can also aid radial mass transfer, but it is generally not efficient to go to flow rates where turbulence sets in since turbulence is accompanied by a dramatic increase in pressure drop, and will not generate much of an improvement in mass transfer over that provided for by slug flow in coiled tubes.

Another very real problem in using immobilized-enzyme tubes is the low surface area available for the catalyst. A 1 mm diameter tube will have an area of 0.314 cm^2 per cm of length. Thus a typical column of 2–10 m will have a total surface of a few hundred cm^2. This is contrasted with the several hundred m^2 per gram of small porous catalyst beads. This problem can be counteracted by etching nylon or glass tubes or by coating the tubes with colloidal silica (32, 38).

Despite the above considerations, the analytical advantages of a segmented flow tubular reactor, such as the simplicity in it placing in an automated analyzer and the inhibition of spreading, are so significant that these devices must be viewed as very useful for routine analysis. Specific applications of these are detailed in Chapter 8.

7.2   EFFECT OF SLOW INTRAPARTICLE MASS TRANSFER ON THE EFFICIENCY FACTOR

Probably the majority of all supports for immobilization which have found significant use (polyacrylaride, porous glass, ion-exchange resins, Agarose, etc.) are highly porous solids or semi solids. The interior surface area of such materials is generally very much greater than their exterior area. Thus the bulk of the catalyst will be present in a region *which is not directly accessible by convective mass transfer* from the bulk of the fluid. Diffusion is the only process which can bring a substrate into contact with the interior surface of the solid and can limit the effectiveness of the preparation. Qualitatively, we want the substrate to be able to diffuse into the entire particle very rapidly. Thus, the effectiveness will be directly related to the substrate diffusion coefficient and inversely related to the radius of the particle or the thickness of a membrane.

It is important to note that diffusion coefficients inside a porous solid may be smaller than in free solution due to the tortuousity of the path. Various geometric models of pores such as series of hyperbolas of revolution, sinusoidal tubes, series of capillaries of various diameters, etc., have been proposed. The effective diffusion coefficient may be as much as a factor of 10 smaller than the true molecular diffusion coefficient. In addition to modified bulk-diffusion, material can be transported by Knudsen diffusion and diffusion on the outer surface of the solid. A detailed discussion of all of these processes is beyond the intent of this book. The interested reader is referred to the book by C. N. Satterfield (2). The use of highly cross-linked resins and polyacrylamide gels should be avoided because of restricted internal diffusion in these media.

A great many studies of the effect of intraparticle diffusion and membrane diffusion have appeared in part because these preparations are geometrically similar to living cells. Intraparticle diffusion or resistance to mass transfer has been studied theoretically with many different complicating factors including micro-environmental and electrostatic effects (42, 43), two limiting substrates, that is, reactant and coenzyme diffusion (44), slow diffusion of buffer (45), substrate and product inhibition (46), the geometry of the particle or membrane (46–48), the nature of boundary conditions (46, 49, 50), and the simultaneous occurrence of external resistance to mass transfer (50, 51). The effectiveness factors for these systems have been calculated by many groups (51–55) and the influence of diffusion on the shape of Lineweaver–Burke plots (56, 57) and the validity of the experimental kinetic parameters has been studied. The coupling of internal diffusion and the apparent thermal stability has also been studied (58).

Three distinct situations will be considered. First, diffusion in an immobilized-enzyme particle, which is immersed in a very well-stirred solution, will be considered for a spherical pellet and for a thin slab (see Figure 7.9). The purpose here will be to obtain an estimate of the effectiveness factor of the catalyst. Second, enzyme membranes will be considered in terms of the flux of substrate and product into and out of the membrane as well as to determine the shape of the concentration profile. It is not our purpose to review all of this work. To illustrate the nature of the problems, simple Michaelis–Menten kinetics will be assumed and electrostatic factors, which are not at all negligible in dilute buffers, will be ignored. The reader is referred to the cited works for more detailed treatments of the complicating factors.

7.2.1 Spherical Particles

The effect of internal diffusion can be very severe. O'Driscoll and coworkers (52) carried out a series of experiments using the recirculating reactor system of Ford et al. (15) on the hydrolysis of α-N-benzoyl-L-arginine ethyl ester with trypsin and of o-nitro-β-D-galactopyranoside with β-galactosidase trapped in a cross-linked polymer gel. They adjusted the flow rate for large particles such that the reaction rate was independent of flow and therefore external resistance to mass transfer was negligible. Their data shown in Figure 7.10 indicate that the reaction rate is strongly coupled to particle size, and that the rate drops rapidly as the particle size increases above some critical value. This is quite unambiguous evidence for the existence of internal diffusion as a rate-limiting factor. Their data as a function of concentration is plotted in Figure 7.11. Conventional Lineweaver–Burke plots are indeed linear for small particles but are concave downward for big particles. Even for small particles where the plots are linear, the slope of the line is a function of particle size. This can evidently lead to quite erroneous estimates of Michaelis constants when

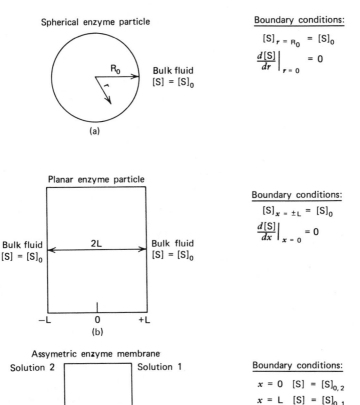

Fig. 7.9. Generalized models and boundary conditions for diffusion within catalyst particles and membranes. (a) Spherical particle; (b) planar particle; (c) assymetric membrane.

conventional graphical analyses are used and has been the subject of extensive research (55–57).

It will be assumed that external mass transfer can be ignored in all cases so that the substrate concentration on the outer surface of the solid will be essentially equal to that in the bulk fluid. Blaedel and coworkers have treated membrane systems in which external resistance to mass transfer was not negligible (49, 50). This allows us to establish one boundary con-

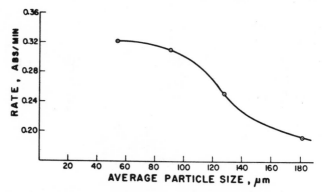

Fig. 7.10. Plot of measured trypsin activity as a function of average particle size. Conditions: 0.1 M phosphate buffer (pH 8.0), temperature 22°C, 0.5 mM α-N-benzoyl-L-arginine ethyl ester. [Reprinted with permission of John Wiley & Sons, Inc., from K. F. O'Driscoll, S. Hinberg, R. Korus, and A. Kaporilas, *J. Polymer Sci.* **46**, 227 (1974).]

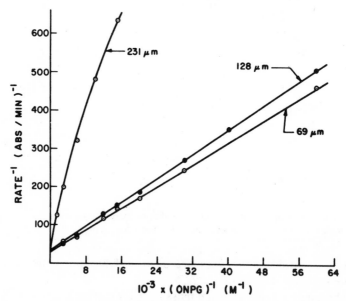

Fig. 7.11. Measurement of the effect of particle size on a Lineweaver–Burke plot. Note that although small particles give linear plots, the slopes are different. [Reprinted with permission of John Wiley & Sons, Inc., from K. F. O'Driscoll, S. Hinberg, R. Korus, and A. Kaporilas, *J. Polymer Sci.* **46**, 227 (1974).]

dition for enzyme particles in a stirred fluid: diffusion fields are symmetric (see Figure 7.9 for a definition of the coordinates) consequently we can show that the substrate gradient at the center of the sphere is identically zero; thus,

$$\frac{\partial [S]}{\partial r}\bigg|_{r=0} = 0 \tag{7.39}$$

The differential equation describing the *steady-state* diffusion system will be (52, 56):

$$D_m \left(\frac{d^2[S]}{dr^2} + \frac{2}{r}\frac{d[S]}{dr} \right) = \frac{k_2[E][S]}{K_M + [S]} \tag{7.40}$$

where [E] is the enzyme concentration per unit volume. Sundaram, Tweedale, and Laidler (59) have indicated that several seconds or longer may be required to establish steady state. The net reaction rate in the catalyst particle in the steady state can be obtained by taking the integral of the flux of material into the particle over the external surface area ($4\pi R_0^2$), that is, in the steady state the rate of the chemical reaction is consuming substrate as fast as it enters the sphere. Thus the actual reaction rate ($\mathcal{R}_c$) per unit area (in moles/sec·cm^2) will be:

$$\mathcal{R}_c = D_m \frac{d[S]}{dr}\bigg|_{r=R_0} \tag{7.41}$$

For the sake of simplicity, we will assume that both the substrate and its product have identical diffusion coefficients (D_m). To obtain $\mathcal{R}_c$, equation (7.40) must be solved and then the derivative at R_0 evaluated.

Unfortunately, a general solution of equation (7.41) has not been found. Analytical solutions exist only for the case in which [S] is much greater or much less than K_M (17):

$$\mathcal{R}_c = \frac{k_2[E]R_0}{3} \qquad \text{when} \quad [S] \gg K_M \quad \text{(zero order)} \tag{7.42}$$

and

$$\mathcal{R}_c = \frac{D_m[S]_0}{R_0}[\sqrt{\alpha R_0^2}\coth(\sqrt{\alpha R_0^2}) - 1] \quad \text{(first order)} \tag{7.43}$$

where

$$\alpha \equiv \frac{k_2[E]}{D_m K_M} \qquad \text{when} \quad [S] \ll K_M \tag{7.44}$$

As will be seen α is a very important parameter and appears repeatedly. We will refer to it here as the *enzyme loading factor*. Intermediate cases can be handled only by means of numerical solution of equation (7.40) (51, 55). An enzyme effectiveness (η) can be calculated as in the case of film diffusion. Again defining η as the ratio of the observed net rate (in moles/sec) to that which would occur if there were no diffusional limitations:

$$\eta = \frac{4\pi R_0^2 \cdot D_m d[S]/dr \,|_{r=R_0}}{\pi R_0^3 (k_2[E][S]_0)/(K_m + [S]_0)} \tag{7.45}$$

At low substrate concentration ($[S]_0 \ll K_M$), the effectiveness factor becomes

$$\eta = \frac{3}{\alpha R_0^2} (\sqrt{\alpha R_0^2} \coth \sqrt{\alpha R_0^2} - 1); \quad [S]_0 \ll K_M \tag{7.46}$$

In the limit of very small particles ($R_0 \to 0$), there will be no diffusional resistance to mass transfer and we can show that the effectiveness approaches unity. Miyamoto et al. (53) have calculated the effectiveness factor for both limiting cases discussed above, that is, high and low substrate concentration as well as intermediate cases which were obtained by a modified Runge–Kutta integration of equation (7.40) by computer. Others have carried out similar calculations (51, 55).

Since diffusion will have the greatest influence on a first-order reaction ($[S]_0 \ll K_M$) and no effect on a purely zero-order reaction, provided that the reactant concentration remains finite, ($[S]_0 \gg K_M$), equation (7.46), which is plotted in Figure 7.12 can be used to obtain a *conservative* estimate of the actual catalyst effectiveness regardless of the ratio of $[S]_0$ to the Michaelis constant. The data in this figure indicate that the catalyst will be at least 50% effective provided that $\sqrt{\alpha R_0^2}$ is greater than about 8.2. Since the parameter α is related to the enzyme characteristics and R_0 is the particle size, one can calculate the maximum particle size for 50% effectiveness for any given set of enzyme parameters. A set of results is given in Table 7.4 for $D_m = 10^{-5}$ cm²/sec, and a variety of reaction parameters. It is obvious that very active enzyme preparations (units/ml) require quite small particles to be more than 50% effective.

The data of O'Driscoll (Figure 7.10) are seen to follow the general shape of the curve in Figure 7.12. Several other groups have shown that the apparent enzyme activity can be improved by grinding up the particles. Any improvement in reaction rate, which is independent of flow rate, upon use of smaller particles is clear-cut evidence for slow internal mass transfer. The best experimental method for evaluating the effectiveness is to take the ratio of the observed rate to that measured with very small particles which show no further increase in rate upon decrease in size.

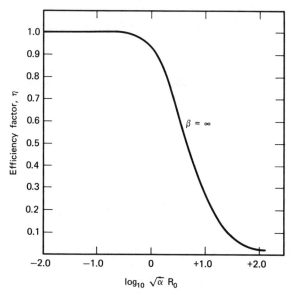

Fig. 7.12. Theoretical effectiveness factor (η) for first-order reaction as a function of the enzyme loading factor (α) with spherical particles. Calculated from equation (7.46).

7.2.2 Enzyme Slab Model and Reaction Effectiveness

Many workers (54) have used a planar slab model in place of spherical particles in order to simplify the attendant differential equations. In this situation the slab is assumed to be bathed on both sides by a common solution. The mathematics are simplified by a shift in coordinates so that the center of the slab is at $x = 0$ and the surfaces are located at $x = \pm L$. Under these circumstances the appropriate boundary conditions become:

$$[S] = [S]_0 \qquad \text{at } x = \pm L \tag{7.47}$$

and

$$\frac{d[S]}{dx} = 0 \qquad \text{at } x = 0 \tag{7.48}$$

In the steady state

$$D_m \frac{d^2[S]}{dx^2} = \frac{k_2[E][S]}{K_M + [S]} \tag{7.49}$$

The boundary-value problem can be solved in closed form only when the substrate concentration is high or low with respect to K_M. In the first-

**Table 7.4   Maximum Particle Size^a for 50%
Effectiveness for a First-Order Reaction^b**

Enzyme activity (μ mole/min·ml)	10^{-5}	K_M (mole/ml) 10^{-6}	10^{-7}
1.0	0.63	0.20	0.063
5.0	0.28	0.089	0.028
10	0.20	0.063	0.020
50	0.089	0.028	0.0089
100	0.063	0.020	0.0063
500	0.028	0.0089	0.0028
1000	0.020	0.0063	0.0020

^{a} Radius in centimeters.
^{b} For $D_m = 1 \times 10^{-5}$ cm^2/sec.

order kinetic region, various groups have shown that

$$\eta = \frac{\tanh(\sqrt{\alpha}L)}{\sqrt{\alpha}L} \qquad [S]_0 \ll K_M \qquad (7.50)$$

A plot of η versus $\sqrt{\alpha}L$ is similar in shape and assymptotes as the previous case for spherical particles. The effectiveness will be less than 50% when $\sqrt{\alpha}L$ is greater than about 1.9. This should be compared with the parameter $\sqrt{\alpha}R_0$ for spheres which must exceed 8.2 for the effectiveness to be reduced to less than 50%.

7.2.3   Immobilized-Enzyme Membranes

Enzyme membranes can be used for analytical purposes in many different ways. For example an enzyme could be immobilized on the surface or within a thin dialysis tube in a conventional Auto Analyzer. The most important case here would be that situation in which one side of the membrane is exposed to a substrate while the other side of the membrane is continuously flushed to collect product for analysis. Katchalski's group and Blaedel's group have studied the problem extensively. Ignoring external mass transfer (50) and electrostatic factors (42), the relevant differential steady-state equations are:

$$D_m \frac{d^2[S]}{dx^2} = \frac{k_2[E][S]}{K_M + [S]} \qquad (7.51)$$

$$-D_m \frac{d^2[P]}{dx^2} = \frac{k_2[E][S]}{K_M + [S]} \qquad (7.52)$$

Regardless of the nature of the boundary conditions, these equations cannot be solved in closed form unless [S] is either much greater or much smaller than K_M. When the bulk product concentration is held quite low and substrate is present on only one side of the membrane, then the substrate- and product-concentration profiles in the membrane as obtained by Katchalski's group (54) are:

$$[S] = [S]_0 \frac{\sinh\sqrt{\alpha}(L - x)}{\sinh\sqrt{\alpha}L} \qquad \text{when } [S]_0 \ll K_M \qquad (7.53)$$

$$[P] = [S]_0 \left(\frac{\sinh\sqrt{\alpha}(L - x)}{\sinh\sqrt{\alpha}L} + \frac{x}{L} - 1 \right) \qquad (7.54)$$

The flux of product (mole/cm^2·sec) at $x = L$ and $x = 0$, i.e. the rate at which the product escapes from the membrane are:

$$J_P \bigg|_{x=0} = \frac{D_m[S]_0}{L} [\sqrt{\alpha}L \coth\sqrt{\alpha}L - 1] \qquad (7.55)$$

$$J_P \bigg|_{x=L} = \frac{D_m[S]_0}{L} [\sqrt{\alpha}L \operatorname{csch}\sqrt{\alpha}L - 1] \qquad (7.56)$$

The substrate- and product-concentration profiles corresponding to equation (7.53) and (7.54) are plotted in Figure 7.13 and clearly indicate that the product builds up most rapidly on that side of the membrane where the substrate is present in the bulk, that is, at $x = 0$. It is also important to note that the product flux out of the membrane is greatest into the compartment containing the substrate. For example when $\sqrt{\alpha}L$ is large, the flux ratio is equal to $\sqrt{\alpha}L - 1$.

$$\frac{J_p(x = 0)}{J_P(x = L)} = \sqrt{\alpha}L - 1; \quad \sqrt{\alpha}L \text{ large} \qquad (7.57)$$

When $\sqrt{\alpha}L$ (the total enzyme activity in the membrane) is low

$$\frac{J_P(x = 0)}{J_P(x = L)} = 2 \qquad (7.58)$$

Therefore for analytical purposes there is an inherent disadvantage in using the flow-collector stream as opposed to the donor stream since the original stream will contain more of the product. Furthermore the receptor stream will have less and less product as the enzyme activity increases. An effectiveness factor can be computed in the usual fashion. For a first-

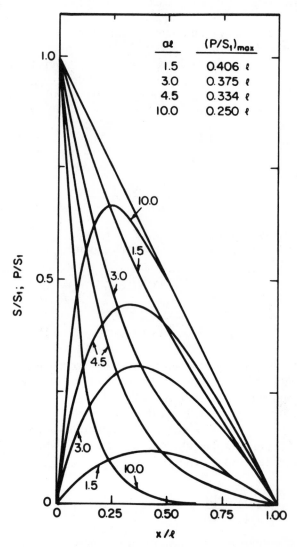

Fig. 7.13. Calculated concentration profiles for substrate in an enzyme membrane. The product was not present in the bulk and the substrate was present on only one side of the membrane ($x = 0$, $[S]_0$). The values on the curves correspond to $\sqrt{\alpha}\, L$ and all results are computed from equations (7.53) and (7.54). [Reproduced with permission from R. Goldman, L. Goldstein, and E. Katchalski, in *Biochemical Aspects of Reactions on Solid Supports*, G. R. Stark, Ed. (Academic P, New York, 1971). Copyright by Academic Press.]

order reaction,

$$\eta = \frac{2[\cosh\sqrt{\alpha}L - 1]}{\sqrt{\alpha}L \, \sinh\sqrt{\alpha}L} \qquad (7.59)$$

This result once again indicates that η will decrease as the membrane thickness is increased or when a high enzyme activity is used (α). When $\sqrt{\alpha}L$ is greater than about 5.0 the effectiveness will be less than 50%.

7.3   DETAILED CALCULATION OF EFFECTIVENESS FACTOR FOR SIMULTANEOUS INTERNAL AND EXTERNAL RESISTANCE TO MASS TRANSFER WITH COMPLETE MICHAELIS–MENTEN KINETICS

The two major shortcomings of the steady state a priori effectiveness factors presented above are that (1) resistance to mass transfer was localized exclusively to within the catalyst particle or to the external solution around the particle and (2) the calculations were restricted to purely zero-order or first-order kinetic regimes. Several groups have attempted to overcome either one or the other of these problems. For example, Rony has considered both external and internal mass in the first-order region (60); Blaedel and his coworkers (50) have considered both factors in either the zero- or first-order, but not intermediate regions; Kobayashi in two separate studies derived effectiveness factors for internal mass transfer (56) and external mass transfer (22–23) over the entire range of substrate concentrations. Fink, Na, and Schultz (51) developed a new numerical and theoretical analysis which permits the presentation of effectiveness-factor curves which include both types of resistance to mass transfer and cover the entire range ratios of substrate concentration to Michaelis constant. For the sake of ease of presentation, we will assume that the partition coefficient of substrate between phases is unity and that the enzyme is distributed homogeneously in a sphere of radius R_0. The following parameters arise in their work:

$$\beta = \frac{K_M}{[S]_0} \qquad (7.60)$$

$$\alpha = \frac{k_2[E]}{D_m K_M} \qquad (7.61)$$

$$\omega = \frac{k_L R_0}{D_m} \qquad (7.62)$$

All of the terms have their usual significance. The term ω is equivalent to the Sherwood number for the rate of external mass transfer. The results of their calculations are plotted in terms of an effectiveness factor η and a modified Thiele modulus ϕ which is defined as

$$\phi = \sqrt{\frac{k_2[E]R_0^2}{D_m(K_M + [S]_0)}} = \sqrt{\frac{\alpha \cdot \beta \cdot R_0^2}{1 + \beta}} \tag{7.63}$$

Two types of effectiveness factors were computed: those in which external mass-transfer effects are negligible ($\omega = \infty$), these are denoted η_i (see Fig. 7.14) and an overall effectiveness where both forms of resistance to mass transfer are significant (see Fig. 7.15). The effectiveness is plotted versus the Thiele modulus (ϕ) at constant αR_0^2. Thus for a given set of enzyme and size parameters, we can follow the effectiveness as a function of bulk substrate concentration ($[S]_0$) or Michaelis constant. All of the plots (Figures 7.14 and 7.15) indicate that when the bulk substrate concentration is very large, ϕ is small and the effectiveness approaches unity. These plots can be used in conjunction with mass-transfer correlations, that is, Sherwood number as a function of Reynolds number, and estimates of the amount of enzyme and particle size to determine the overall effectiveness of the catalyst.

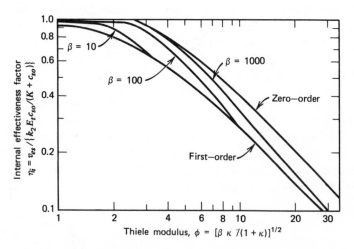

Fig. 7.14. Theoretical plot of internal effectiveness factor (η_i) as a function of the modified Thiele modulus ($\emptyset$). All calculations are for spherical particles and Michaelis–Menten kinetics. The effectiveness factor is defined by equation (7.15) and the Thiele modulus in equation (7.63). The effect of film diffusion has been neglected. [Reprinted by permission of John Wiley & Sons, Inc., from D. F. Fink, T. Y. Na, and J. S. Schultz, *Biotechnol. Bioeng.* **15**, 879 (1973).]

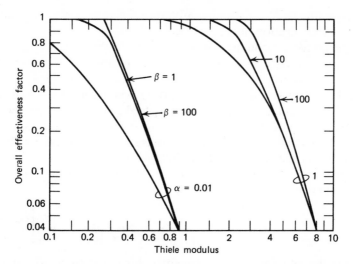

Fig. 7.15.   Plot of overall effectiveness factor (η) as a function of a modified Thiele modulus. All calculations are for spherical particles and Michaelis–Menten kinetics. The results include both slow external and internal diffusion. The Thiele modulus is defined by equation (7.63). [Reprinted by permission of John Wiley – Sons, Inc., from D. F. Fink, T. Y. Na, and J. S. Schultz, *Biotechnol. Bioeng.* **15**, 879 (1973).]

Marrazo, Merson, and McCoy obtained closed-form solutions to steady-state reactor problems when axial dispersion, slow film (external) and slow internal diffusion coexist (61). Their results were limited to the pure zero-order or pure first-order rate law regions due to the nonlinear behavior when Michaelis–Menten kinetics are used. Because of the nature of the problem, two distinct sets of boundary conditions, one which pertains to long columns and one for short columns, were employed. Their results are summarized below.

The behavior of the reactor under zero-order conditions is fairly simple:

1.   The nature of the boundary condition does not make much difference. When N_{Pe} is greater than 10, there is only a 10% difference between the exact solution for a short column and that for a long column.

2.   The extent of conversion in a long column is completely independent of axial dispersion. Obviously any practical short column, with N_{Pe} greater than 10, will also show little effect on the extent of conversion.

3.   Neither intraparticle (internal) or film (external) diffusion have any effect at all on the extent of conversion. This results because a zero-order reaction is insensitive to concentration gradients. This will be true provided that the substrate concentration never decreases to the point where

it is comparable to K_M. Consequently Marrazo et al. found that the inlet substrate concentration must be at least $10K_M$.

4. The fractional conversion of substrate to product is linearly proportional to residence time and inversely proportional to the inlet substrate concentration.

The extent of conversion in the first-order kinetic regime is more complicated. Their results indicate that increased axial dispersion and slow internal and external mass transfer act to decrease the fractional conversion. The extent of conversion decreases almost exponentially with flow rate but is independent of the inlet substrate concentration. Their results for a long plug flow reactor ($\mathscr{D}_a = 0$) agree with those of Rovito and Kittrell (18) thus:

$$[S]_1 = [S]^0 \exp\left[- (1 - \epsilon)\left(\frac{R_0}{3k_L} + \frac{K_M}{\sigma k_2[E]\eta}\right)^{-1} \cdot \frac{l}{u_0}\right] \quad (7.64)$$

where σ is the internal particle porosity. This equation clearly indicates the exponential dependence of the extent of conversion on the residence time in a PFR system even in the presence of slow mass-transfer processes. In the limit of fast external mass transfer ($k_L \to \infty$) and very small particles ($R_0 \to 0$; therefore $\eta \to 1$) equation (7.64) reaches the limiting form

$$[S]^{\ddagger} = [S]^0 \exp\left(- (1 - \epsilon)\frac{\sigma k_2[E]}{K_M} \frac{l}{u_0}\right) \quad (7.65)$$

where $[S]^{\ddagger}$ denotes the ideal substrate concentration obtained under these limiting conditions. This equation is what we should expect for an ideal PFR packed with porous particles. We can now see very clearly the importance of the results given in equations (7.2)–(7.7), which allow an a priori prediction of k_L (the mass transfer rate constant), and the data of equation (7.46) and Figure 7.12 which allow estimation of the internal effectiveness factor (η). Taken together, these permit estimation of the overall extent of conversion via equation (7.64) and of the overall effectiveness of the reaction.

The above is of major analytical significance because for a PFR ($\mathscr{D}_a = 0$) we have already shown that the extent of conversion for a steady-state reactor, that is, a continuously fed system, is identical to that obtained with a pulsed addition of sample. Thus *equations (7.64) and (7.65) allow us to estimate the upper limit of sensitive of an enzyme column with pulse sample injection as well as with continuous addition of sample.* If enough sample is added to give flat top peaks, then the column is being operated essentially as an ideal steady-state PFR and equations (7.64) and

(7.65) are exactly applicable and therefore the slope of the calibration curve can be computed a priori.

One last point which Marrazzo, Merson, and McCoy make very succinctly is that the *internal effectiveness factor (η) should be based on the diffusion coefficient in the porous particle.* As eluded to before (2) this will depend upon the internal porosity (σ) of the particle and the tortuosity (θ) of the diffusion path. Thus the internal diffusion coefficient will be:

$$D_i = \frac{\sigma D_m}{\theta} \qquad (7.66)$$

The porosity may be as high as 0.5 to 0.8 for porous glass and tortuosities are generally in the range of 2 to 10. They estimated that the internal diffusion coefficient of sucrose is in the range 2.9 to 14.7×10^{-7} cm^2/sec, which is much smaller than the value in aqueous solution.

7.4 CONCLUSION

The present chapter would have been much more extensive if we had considered detailed treatment of micro-environmental and electrostatic effects, substrate partitioning, non-Michaelis–Menten kinetics, diffusion of coreagents, buffer diffusion, and a myriad of real-world complexities. The most serious limitation, to our mind, of the theory of enzyme reactors is that the bulk of the work to date really pertains to steady-state systems. Analytically, we are much more interested in pulsed catalytic reactors whose theory is now starting to be developed (62). The ideas presented here do permit a first approximation of the required particle size, amount of enzyme, etc. needed to achieve a desired steady-state conversion. As emphasized in Chapter 6, these results will apply with good accuracy to a well-designed pulsed reactor in which dispersion is minimized. Thus, from an analytical viewpoint, if sample spreading is sufficiently low to use a reactor, it will more closely resemble a PFR than a CSTR. Under these conditions the error made in assuming the use of steady-state design equations will not be serious, particularly if we acknowledge their limitations and allow experiments to define the final operating conditions.

Definition of Terms and Symbols

a   surface area of catalyst per unit bed volume (cm^2/cm^3)

A geometric cross-sectional area of an unpacked column (cm^2)

A_c a characteristic area (cm^2)

d_p particle diameter (cm)

d_t diameter of a tube (cm)

$\mathscr{D}_a$ axial dispersion coefficient (cm^2/sec)

D_m diffusion coefficient in solution (cm^2/sec)

D_i diffusion coefficient in a porous solid (cm^2/sec)

[E]   enzyme concentration (moles/ml)

F volumetric flow rate (ml/sec)

G superficial mass flow rate (g/sec·cm^2) = mass flow rate/A

H diameter of a helical coil (cm)

J_p flux of product (moles/cm^2·sec)

K fluid kinetic energy per unit volume (ergs/ml)

K_M Michaelis–Menten constant (moles/ml)

K'_M apparent Michaelis–Menten constant (moles/ml)

k_2 product formation rate constant of the enzyme substrate complex (sec^{-1})

k_L mass-transfer rate constant (cm/sec)

k' enzyme activity per unit volume (mole/ml·sec); $k' \equiv k_2[E]$

l axial distance from inlet of reactor (cm)

l_s length of a sample slug in a segmented flow reactor (cm)

L total length of a reactor or half thickness of a flat particle (cm)

L_{min} minimum length for a given fractional conversion (cm)

N molar flux to a surface (moles/sec·cm^2) $\equiv r_m$/total surface area

N_{Re} Reynolds number ($d_p/G/u$, etc.)

N_{Sc} Schmidt number ($\mu/\rho D_m$)

N_{Sh} Sherwood number ($k_L d_p/D_m$)

N_{De} Dean number; $N_{De} = (H/dt)^{1/2}N_{Re}$

$\overline{N_{Sh}}$ logarithmic mean Sherwood number

$\mathscr{R}_c$ rate of a *chemical* reaction (moles/ml·sec)

r_m mass transfer rate (moles/ml·sec)

R_0 radius of a tube or particle (cm)

r distance along a radius from the center or axis (cm)

[S]   substrate concentration (moles/ml)

[S]^0 inlet substrate concentration (moles/ml)

[S]$_0$ bulk fluid substrate concentration (mole/ml)

[S]$_s$ surface substrate concentration (moles/ml)

[S]$_e$ exit substrate concentration (moles/ml)

$[S]_l$ substrate concentration along axis of a column (moles/ml)

$[S]^*$ normalized surface substrate concentration; $[S]^* \equiv [S]_s/[S]_0$

$[S]^\ddagger$ substrate concentration in the absence of all complications including dispersion and slow mass transfer

u average linear flow velocity (cm/sec)

u_0 superficial linear flow velocity (cm/sec)

V_{actual} real volume occupied by particles in a column (ml)

V_0 total geometric volume of an unpacked reactor (ml)

V_D interstitial or void volume (ml)

W net volume flow rate of a liquid and a gas in a segmented flow system (ml/sec)

X fractional conversion; $X = ([S]^0 - [S]_e)/[S]^0$

Z $\dfrac{\pi D_m L}{4F}$ (dimensionless length)

α enzyme loading factor (cm^{-2}), Thiele modulus; $\alpha \equiv k_2[E]//K_M D_M$

β dimensionless Michaelis constant; $\beta \equiv K_M/[S]_0$

γ L/d_t;

δ d_t/H

ϕ_f film or external mass transfer modulus

ϕ modified Thiele modulus; $\phi \equiv \{R_0^2 k_2[E]/D_m(K_M + [S]_0)\}^{1/2}$

ϵ void fraction $= V_D/v_\phi$

ρ density of fluid (g/ml)

μ viscosity of fluid (poise)

σ internal porosity of solid particle

ψ l_s/d_t

ω $k_L R_0/D_m$; (a Sherwood number)

η an effectiveness factor either internal, external or both, see equation (7.15) for definition

θ tortuosity

References

1. R. Endele and I. Halasz, *J. Chromatogr.* **99**, 377 (1974).
2. C. N. Satterfield, *Mass Transfer in Heterogeneous Catalysis* (MIT Press, Cambridge, MA, 1970), p. 33.
3. J. M. Engasser and C. Horvath, *J. Theoret. Biol.* **42**, 137 (1973).
4. D. L. Eaton, "Optimization of Porous Materials for Immobilized Enzyme Systems," in *Immobilized Enzymes and Affinity Chromatography,* R. B. Dunlap, Ed. (Plenum Press, New York, 1974), p. 246.

5. C. N. Satterfield, *Mass Transfer in Heterogeneous Catalysis* (MIT Press, Cambridge, MA, 1970), p. 134.
6. S. P. O'Neil, *Biotechnol. Bioeng.* **14**, 473 (1972).
7. S. H. Lin, *Biophysik* **8**, 302 (1972).
8. S. H. Lin, *Biophysik* **10**, 235 (1973).
9. C. N. Satterfield, *Mass Transfer in Heterogeneous Catalysis* (MIT Press, Cambridge, MA, 1970), p. 80.
10. C. O. Bennett and J. E. Meyers, *Momentum, Heat and Mass Transfer* (McGraw-Hill, New York, 1962), p. 473.
11. D. Kunii and M. Suzuki, *Int. J. Heat Mass Transfer* **10**, 845 (1967).
12. R. Pfeffer, *Ind. Eng. Fund.* **3**, 380 (1964).
13. E. J. Wilson and C. J. Geankoplis, *Ind. Eng. Fund* **5**, 9 (1966).
14. W. H. Pitcher, Jr., "Design and Operation of Immobilized Enzyme Reactors," in *Immobilized Enzyme for Industrial Reactors,* H. H. Weetall, Ed. (Academic, New York, 1975), p. 162.
15. J. R. Ford, A. H. Lambert, W. Cohen, and R. P. Chambers, "Recirculation Reactors Systems for Kinetic Studies of Immobilized Enzymes," in *Symposium 3 on Biotechnology and Bioengineering,* E. L. Gaden, Ed. (Wiley, New York, 1972), p. 6.
16. D. Kunii and J. S. Smith, *A. I. Ch. E. J.* **7**, 29 (1961).
17. A. D. Traher and J. R. Kitrell, *Biotechnol. Bioeng.* **16**, 44 (1974).
18. B. J. Rovito and J. R. Kittrell, *Biotechnol. Bioeng.* **15**, 143 (1973).
19. Y. Y. Lee, A. R. Fratzke, K. Wun, and G. T. Tsao, *Biotechnol. Bioeng.* **18**, 389 (1976).
20. Y. Y. Lee and G. T. Tsao, *J. Food Sci.* **39**, 667 (1974).
21. S. P. O'Neil, *Biotechnol. Bioeng.* **14**, 675 (1972).
22. T. Kobayashi and M. Moo-Young, *Biotechnol. Bioeng.* **13**, 893 (1971).
23. T. Kobayashi and M. Moo-Young, *Biotechnol. Bioeng.* **15**, 47 (1973).
24. T. Kobayashi and K. J. Laidler, *Biotechnol. Bioeng.* **16**, 77 (1974).
25. M. D. Lilly, W. E. Hornby, and E. M. Crook, *Biochem. J.* **100**, 718 (1966).
26. H. Filippuson and W. E. Hornby, *Biochem. J.* **120**, 215 (1970).
27. P. V. Sundaram and W. E. Hornby, *F.E.B.S. Letters* **10**, 325 (1970).
28. J. P. Allison, L. Davidson, A. G. Hartman, and G. B. Kitto, *Biochem. Biophys. Res. Commun.* **47**, 66 (1972).
29. D. J. Inman and W. E. Hornby, *Biochem. J.* **129**, 255 (1972).
30. D. J. Inman and W. E. Hornby, *Biochem. J.* **137**, 55 (1974).
31. W. E. Hornby and D. L. Morris, "Modified Nylon Tubes in Enzyme Immobilization and Their Use in Analysis," in *Enzymology,* H. H. Weetal Ed., Vol. 1 (Marcel Dekker, Inc., New York 1975), p. 141.
32. W. E. Hornby, J. Campbell, D. J. Inman, and D. L. Morris, in *Enzyme Engineering,* E. K. Pye and L. B. Wingard, Jr., Ed., Vol. 2 (Plenum Press, New York 1974), p. 401.
33. C. Horvath and B. A. Soloman, "Open Tubular Heterogeneous Enzyme Reactors," in *Enzyme Engineering,* E. K. Pye and L. B. Wingard, Jr., Eds. (Plenum Press, New York, 1974), p. 259.
34. A. H. Emery, "Annular Column Enzyme Reactors," in *Enzyme Engineer-*

ing, E. K. Pye and L. B. Wingard, Jr., Eds., Vol. 3 (Plenum Press, New York, 1974) p. 271.

35. C. Horvath and B. A. Soloman, *Biotechnol. Bioeng.* **14**, 885 (1972).
36. C. Horvath, L. H. Shendalmon, and R. T. Light, *Chem. Eng. Sci.* **28**, 375 (1973).
37. T. Kobayashi and K. J. Laidler, *Biotechnol. Bioeng.* **16**, 99 (1974).
38. C. Horvath, B. A. Solomon, and J. M. Engasser, *Ind. Eng. Fund.* **12**, 431 (1973).
39. L. R. Snyder and H. J. Adler, *Anal. Chem.* **48**, 1017 (1976).
40. R. B. Bird, W. E. Stewart, and E. N. Lightfoot, *Transport Phenomena*, (Wiley, New York 1960), p. 644.
41. J. M. Engasser and C. Horvath, *Ind. Eng. Fund.* **14**, 107 (1970).
42. B. K. Hamilton, L. J. Stockmeyer, and C. K. Cotton, *J. Theoret. Biol.* **41**, 547 (1973).
43. M. L. Shuler, H. M. Tsuchiya, and R. Aris, *J. Theoret. Biol.* **41**, 347 (1973).
44. B. Atkinson and D. E. Lester, *Biotechnol. Bioeng.* **16**, 1321 (1974).
45. J. M. Engasser and C. Horvath, *Biochim. Biophys. Acta* **358**, 178 (1974).
46. M. Moo-Young and T. Kobayashi, *Can. J. Eng.* **50**, 162 (1972).
47. R. Goldman, O. Kedam, and E. Katchalski, *Biochemistry* **7**, 4518 (1968).
48. R. Goldman, O. Kedam, and E. Katchalski, *Biochemistry* **10**, 165 (1971).
49. R. C. Boguslaski, W. J. Blaedel, and T. R. Kissel, "Kinetic Behavior of Enzymes Immobilized in Artificial Membranes," in *Insolubilized Enzymes*, M. Salmona, C. Saronio, and S. Garrattinio, Eds. (Raven Press, New York, 1974), p. 87.
50. W. J. Blaedel, T. R. Kissel, and R. C. Boguslaski, *Anal. Chem.* **44**, 2030 (1972).
51. D. F. Fink, T. Y. Na, and J. S. Schultz, *Biotechnol. Bioeng.* **15**, 879 (1973).
52. K. F. O'Driscoll, I. Hinberg, R. Korus, and A. Kapoulas, *J. Polymer Sci.* **46**, 227 (1974).
53. K. Miyamoto, T. Kujii, N. Tamasraki, M. Ozazaki, and Y. Miura, *J. Ferment. Technol.* **51**, 566 (1973).
54. R. Goldman, L. Goldman, and E. Katchalski, "Water Insolubilized Enzyme Derivatives and Artificial Membrane," in *Biochemical Aspects of Reactions on Solid Supports*, G. R. Stark Ed. (Academic, New York, 1971).
55. B. K. Hamilton, C. R. Gardner, and C. K. Cotton, *A. I. Ch. E. J.* **20**, 503 (1974).
56. T. Kobayaski and K. J. Laidler, *Biochim. Biophys. Acta.* **302**, 1 (1973).
57. S. Gondo, S. Isayama and K. Kusanoki, *Biotechnol. Bioeng.* **17**, 423 (1975).
58. D. F. Ollis, *Biotechnol. Bioeng.* **14**, 871 (1972).
59. P. V. Sundaram, A. Tweedale, and K. J. Laidler, *Can. J. Chem.* **48**, 1495 (1970).
60. R. P. Rony, *Biotechnol. Bioeng.* **13**, 431 (1971).
61. W. N. Marrazzo, R. L. Merson, and B. J. McCoy, *Biotechnol. Bioeng.* **17**, 1515 (1975).
62. A. M. Sica, E. M. Valles, and C. E. Gigola, *J. Catalysis* **51**, 115 (1978).

ANALYTICAL APPLICATIONS OF IMMOBILIZED ENZYMES: REACTORS AND FREE MEMBRANES

The analytical applications of immobilized enzymes have been extensively reviewed (1–5). The recent advances in the analytical applications of these materials have made the systematic classification and discussion of their use difficult. The first use of an immobilized enzyme for analysis, other than in batch form, was in a column. Subsequently, "enzyme electrodes" were reported followed by numerous other forms of application. In order to present some of the principles involved in different types of applications, we have arbitrarily divided these analyses into three categories: the immobilized-enzyme reactor (IMER) approach, in which the effluent of the reactor is coupled to some type of detection system such as a spectrophotometer or electrode system; the immobilized-enzyme membrane systems, in which a thin layer or membrane of enzyme is used to catalytically convert the substrate to product prior to a separation or detection step; and transducer-bound enzymes such as the "enzyme electrode," in which an artificial enzyme membrane is fixed directly to the transducer. The latter systems are discussed in Chapter 5. The former two classifications will be reviewed here with particular emphasis on clinical and bioanalytical determinations.

Of the two classifications to be discussed, the IMER is perhaps the most versatile area of application because of its ability to be interfaced with virtually any detection system, its compatibility with flow systems and therefore the potential of high sample throughput rates, and its ability to be used, in certain cases, for total conversion of substrate to product in an equilibrium type of analysis. Certainly, the IMER has received more widespread use to date. As a result, the following discussion will involve some of the theoretical aspects of IMER's, as well as an overview of applications.

8.1 THEORETICAL CONSIDERATIONS OF IMER

In designing an IMER for analysis, it would be beneficial to have a model so that a logical approach to optimization could be made. Any immobilized-enzyme flow analyzer can be thought of as having three significant processes which affect the analysis: enzymatic conversion of the analyte (the enzyme's substrate) to product; dispersion of the sample and product as they flow through the system; and detection of the reaction product for quantitation. Enzyme kinetics in both fixed-bed and tubular reactors were discussed in Chapter 6 and will not be considered in detail here. Since dispersion and the detection system both have an influence on the utility of an analytical system, we shall consider them in some detail.

Snyder (6) and Snyder and Adler (7) have recently presented a model for dispersion in air-segmented tubular systems. Equations analogous to those below can be written for this type of system. These were presented in Section 6.3.3. For fixed-bed reactors, dispersion can be considered in a manner analogous to that employed in high-performance liquid chromatography (HPLC). In the following discussion, we will assume that the substrate is instantaneously converted to product (an equilibrium analysis) and deal with the dispersion of the substrate and product. It should be pointed out that differential equations can be written to consider both reaction kinetics and dispersion simultaneously (See Chapter 6). They are, however, difficult to solve and, as will be seen, are not required for a semiquantitative treatment as presented here.

The principle criteria of quality for an automated analytical IMER system are sample volume requirements, sample throughput rate, and adequate sensitivity and limit of detection. The importance of the criteria is somewhat dependent on the application. For example, sample volume is more critical in a clinical setting than in monitoring an industrial production line. Nevertheless, the purpose of this chapter is to relate these criteria to operational factors such as flow rate, support size, and reactor volume so that appropriate action can be taken to improve the analysis.

8.1.1 Effects of Column Dilution

Subject to the assumptions of the plate theory of chromatography, an infinitely narrow sample input function, and instantaneous total conversion of substrate to product, the concentration of product eluting from the column will be given by:

$$[P]_t = [P]_{max} \exp\left(-\left[\frac{N(t - t_e)^2}{2t_e^2}\right]\right) \tag{8.1}$$

where

$$[P]_{max} \equiv \frac{C_a V_a}{V_e} \sqrt{\frac{N}{2\pi}} \tag{8.2}$$

and $[P]_t$ is the concentration of product at any time t, $[P]_{max}$ is the maximum concentration of product, t_e is the elution time (analogous to the retention time in chromatography), N is the number of "theoretical plates" for the system, C_a is the concentration of analyte in the sample, V_a is the injected volume, and V_e is the volume required to elute the peak. The most important result of the Gaussian peak shape model is the relation between the maximum product concentration ($[P]_{max}$) and the sample volume (V_a), and the magnitude of the dispersion factors (N) and the size of the reactor and associated volume (V_e).

If we assume that the support material does not retain the substrate or product, which appears not to be true for many bound-enzyme systems, then the elution volume, V_e, can be represented by the relation

$$V_e = V_d + K_d V_i \tag{8.3}$$

where V_d is the dead volume, K_d is the distribution coefficient analogous to that used in gel-permeation chromatography, and V_i is the interstitial volume of the support. For small substrate molecules such as glucose and urea, K_d should be unity. Since relatively small reactors are used in IMER work (<1.0 ml), the connecting lines between various components may contribute significantly to the elution volume.

The spreading factors which contribute to N in liquid chromatography are well known (8, 9). The number of plates, N, in a system can be related to the variance of the peak, that is,

$$N = \frac{V_e^2}{\sigma_v^2} \tag{8.4}$$

The contribution to the overall variance (σ_v^2) may be due to the sample injector ($\sigma_{v,s}^2$), the connecting tubing ($\sigma_{v,t}^2$) column processes ($\sigma_{v,c}^2$), or connections between the various components. Making use of the additivity of variances,

$$\sigma_v^2 = \sigma_{v,s}^2 + \sigma_{v,t}^2 + \sigma_{v,c}^2 \tag{8.5}$$

The contribution of the connecting tubing can be estimated using the Golay equation for open tubes (10). As in HPLC, it is important to keep connecting tubing short and of small internal diameter. The contribution of the injector and the column are much more difficult to assess a priori. If we assume, as in HPLC, that the optimum height equivalent to a theoretical plate will be about three times the support diameter (11), then for column processes

$$\sigma_{v,c}^2 = \frac{L V_c}{3 d_p} \tag{8.6}$$

where L is the column length and d_p is the particle diameter. We have not found this to be the case, which is not surprising considering the crude packing technique and large-bore columns which have been generally employed. The values for these variances in an IMER are shown in Table 8.1 (12). It is apparent that the equilibration coil required for a thermal detection system contributes greatly to the dispersion. The other major contributors are the injector and the column.

The object of minimizing dispersion is to optimize the ratio of the peak height to peak width. This would allow both a higher sample throughput rate without sample overlap and an improved limit of detection. Again, if we attain a situation analogous to that in HPLC, we can employ the functional relationship between the other parameters such as flow rate and column radius and the maximum product concentration. Karger, Martin, and Guiochon (13) have substituted empirical values into equation (8.2) and obtained the following relation for an unretained peak:

$$[P]_{max} = \frac{C_a V_a}{\sqrt{2\pi\epsilon_T \pi r^2 L^{1/2} d_p^{0.9} u^{0.2}}}$$

where ϵ_T is the total column porosity, r is the inside radius of the column,

Table 8.1 Estimation of Sources of Sample Dispersion[a]

System component	Component geometric volume[b]	Total apparent volume[b,c]		Calculated component half-width[b]	
		40 μl[d]	120 μl[e]	40 μl[d]	120 μl[e]
Sample valve[f]	—	0.27	0.18	0.14	0.15
Connection tubing	0.03	0.32	0.27	~0	~0
Thermal equilibration	0.20	0.64	0.53	0.19	0.18
Reference thermistor junction	0.08	0.65	0.57	~0	~0
Column[g]	0.42[h]	1.09	0.96	0.24	0.23
Total	—	1.09	0.96	0.34[i]	0.33[i]
Thermal system[j]	—	0.95	0.84	0.31	0.32

[a] All measured with a low dead-volume refractive index detector at a flow rate of 1.22 ml/min. All results taken from reference 12.
[b] All in milliliters.
[c] Calculated from time to peak maxima and the measured flow rate.
[d] Measured with a 40 μl sample volume.
[e] Measured with a 120 μl sample volume.
[f] This includes the detector.
[g] Column (4.0 × 33 mm) packed with 200–400 mesh porous glass.
[h] Total volume of unpacked column.
[i] Total measured half-width.
[j] Same conditions as refractive-index detector.

and u is the solvent velocity. It is apparent from this relation that the column volume should be as small as possible, the velocity should be relatively low (although the functionality is low), and the particles should be as small as possible (see below). We have observed a slight decrease in the peak height with increasing flow rate for several different detection systems (12) which is probably the result of increased dispersion.

Up to this point we have considered only the case of very rapid, complete conversion of the substrate. Certainly this is not always the case. As mentioned before, differential equations can be written for this situation, but they are complex and have no simple solution. We can consider two extremes of reactor behavior to get a "feel" for the effect of incomplete reaction on the peak height and width. In an ideal plug reactor, which is characterized by the complete absence of longitudinal mixing or spreading, we would expect incomplete reaction to result in a peak of less magnitude than a corresponding peak reflecting total conversion. An important feature in this case is the absence of an influence on the peak width by incomplete reaction. Thus, for a plug flow reactor, less than complete conversion results in decreased sensitivity and may give a non-linear calibration curve if the reaction proceeds under zero-order kinetics, but does not influence the sample throughput rate.

If one considers a continuously stirred reactor under the same conditions, that is zero-order kinetics, a different result is obtained. If the reaction is not sufficiently rapid, both the rate of removal of unreacted substrate from the reactor and the rate of conversion to product will be significant processes (See Section 6.1.1). It can be shown that the product concentration during this period will be

$$[P] = k'\tau \left[1 - \exp\left(-\frac{t}{\tau} \right) \right] \tag{8.7}$$

where k' is the product of the enzyme concentration and the rate constant for the decomposition of the enzyme–substrate complex, that is, V_{max}, and τ is equal to the reactor volume divided by the volumetric flow rate. The substrate will be depleted at time t' where

$$t' = \tau \ln \left[\frac{[S]_0}{k'\tau} \right] + 1 \tag{8.8}$$

As can be seen from Figure 8.1(C) an exponential dilution of product occurs after t'. The instantaneous product concentration can be obtained

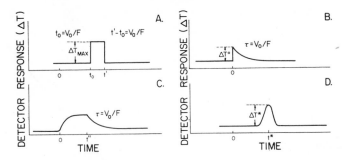

Fig. 8.1. Product concentration as a function of time for (A) a plug flow reactor; (B) a continuously stirred reactor with instantaneous conversion of substrate; (C) a continuously stirred reactor under conditions of zero-order kinetics; and (D) an axially dispersed plug flow reactor. See list of symbols at the end of the chapter for definition of the terms in the figure.

from the relation

$$[P] = [P]_{t'} \exp\left[-\left(\frac{t - t'}{\tau}\right)\right] \qquad (8.9)$$

where $[P]_{t'}$ is the concentration at time t'. Since the peak dispersion for a Gaussian peak can be measured by determining the width at half the peak height, it would be useful to express the theoretical spreading in these terms, that is,

$$\Delta t_{1/2} = \tau \ln\left(2 + \frac{[S]_0}{k'\tau}\right) \qquad (8.10)$$

where $\Delta t_{1/2}$ is the peak width at half height expressed in time units.

The variation of the dispersion with increasing substrate concentration is shown in Figure 8.2. The theoretical curve was calculated from equation (8.10) assuming a constant enzyme concentration, reactor volume, sample size, and flow rate. From the figure, it should be noted that once the substrate concentration exceeds a certain value $[S]_0 \cong 1.3 \, k'\tau$), peak spreading due to incomplete reaction becomes significant. Obviously, the substrate concentration where this change of spreading mechanism takes place will be a function of those parameters which affect k' and τ as well as the K_M of the enzyme. Since zero-order reaction conditions in enzyme reactions are encountered only with high substrate concentrations, spreading due to incomplete reaction should only be observed at saturating concentrations of substrate. It should be borne in mind that once

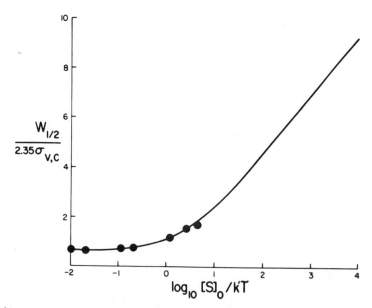

Fig. 8.2. Dispersion due to incomplete conversion of substrate in the reactor. The solid curve is calculated from equation (8.10). The circles are data from an immobilized urease reactor with a k' value of 800 μmoles/min, a residence time, τ, of 0.35 min, and a column volume of 0.42 ml ($\sigma_{v,c}$). [Reprinted with permission from R. S. Schifreen, D. A. Hanna, L. D. Bowers and P. W. Carr, *Anal. Chem.* **49**, 1929 (1977). Copyright by the American Chemical Society.]

the threshold concentration is passed, measurement of the peak variance will not yield an authentic estimate of the column or system dispersion but will also reflect "chemical" broadening. This situation is at least partially analogous to the broadening of a chromatographic peak which occurs when the nonlinear region of an adsorption isotherm is reached. The data in the figure were obtained using an analytical urease reactor developed in our laboratory (12). It is apparent from the figure that the reactor approximates a continuously stirred reactor rather well.

A second effect related to the enzyme which may influence peak shape is the interaction of the substrate or reaction products with the enzyme and/or support matrix. Injection of urea into a urease–CPG reactor gave rise to very broad and skewed peaks when an electrochemical pH stat was used as the detector (14). This was in dramatic contrast to the peak shape observed for immobilized-enzyme preparation using other detection systems.

Injection of an equivalent amount of the reaction product, ammonium carbonate, also differed markedly from the urea injection as shown in

Figure 8.3. Although some adsorption of the ammonium carbonate occurred, increased levels of the background electrolyte, sodium perchlor-
ate, did little to decrease the peak tailing for urea. This indicates that both
the ionic strength and the nature of the anion are important in limiting
adsorption. The successful use of 1,3-diamino-propane, a competitive inhibitor for urease, to sharpen the urea peaks is a strong indication that
the dispersion is caused by the enzyme–substrate interaction. As shown
in Figure 8.3, the urea peak is actually narrower than an injection of
ammonium carbonate. Since a similar effect has been observed in another

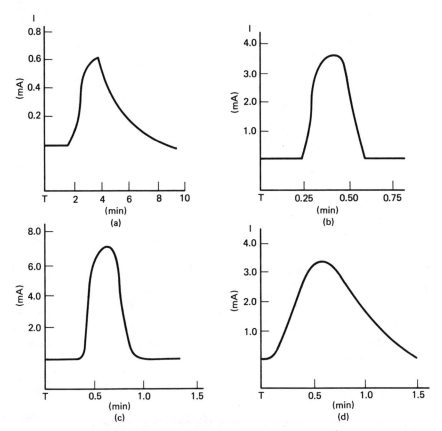

Fig. 8.3. Current–time curves from a coulometric analyzer. (A) Urea injected through
urease column 3 ml/min; (B) ammonium carbonate injected through urease column, 3 ml/
min; (C) urea injected through column in presence of 1, 3-diaminopropane, 1.6 ml/min; (D)
ammonium carbonate injected under same conditions as (C). [Reprinted with permission
from R. E. Adams and P. W. Carr, *Anal. Chem.* **50,** 944 (1978). Copyright by the American
Chemical Society.]

system requiring low buffer capacity (15), this may be a general effect which is buffer-capacity-dependent. Nevertheless, the use of an inhibitor to decrease dispersion is an interesting phenomenon worthy of additional study.

8.1.2   Effect of Sample Size

The influence of the sample volume on the sensitivity and peak width can also be computed. It is apparent from equation (8.2) that as long as the sample width does not interfere with the overall peak width, that is, the peak is Gaussian, the sensitivity increases with increasing sample volume. Sternberg (14) has shown that as the sample volume becomes large relative to the dispersion of the system, the effluent from the column can be represented by

$$[P]_t = \frac{[P]_{max}}{2} \left\{ \text{erf} \left(\frac{t - t_e}{\sqrt{2}\sigma_{t,c}} \right) - \text{erf} \left(\frac{t - t_e - \Theta}{\sqrt{2}\sigma_{t,c}} \right) \right\} \qquad (8.11)$$

where Θ is the plug width in time units, that is,

$$\Theta = \frac{V_a}{F} \qquad (8.12)$$

The maximum concentration is given by

$$[P]_{max} = \text{erf} \left(\frac{\Theta}{2\sqrt{2}\sigma_{t,c}} \right) \qquad (8.13)$$

It is easily shown that the peak maximum and symmetry center occur when

$$t = t_e + \tfrac{1}{2}\Theta \qquad (8.14)$$

The total variance of a peak resulting from a large volume injected as a plug was shown by Steinberg to be

$$\sigma^2_{v,\,total} = \sigma^2_{v,c} + \frac{V_a^2}{12} \qquad (8.15)$$

Thus, for very large sample volumes, the peak width approaches V_a. The effect of sample volume on the peak height and width is shown in Figure 8.4 which is a series of curves calculated from equation (8.11). The curves indicate that the peak height initially increases quite rapidly with increasing sample volume but ultimately becomes independent of Θ as Θ in-

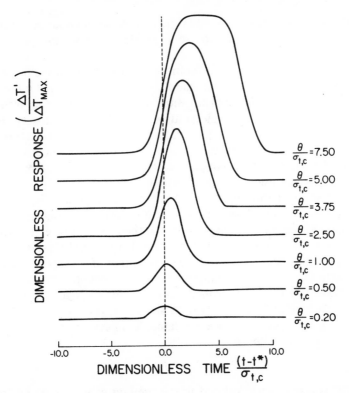

Fig. 8.4. Peak shape as a function of the initial substrate volume as calculated from equation (8.11) assuming a value of 0.20 min for $\sigma_{t,c}$ and 1.0 min for t^*.

creases. It should be evident from this function that an increase in sensitivity can be obtained via an increase in the sample volume without affecting the sample throughput only under some conditions. It can be seen from equation (8.15) that if $\sigma^2_{v,c}$ predominates, changes in the sample volume will not affect the peak width ($V_a \leq 3\sigma_{v,c}$). From Figure 8.4 and equation (8.13) it can be observed that the sensitivity is a function of the sample volume as well. Thus, there is an trade off when $V_a \cong 3\sigma_{v,c}$ where both the sample throughput and the sensitivity of the assay are acceptable. An increase in V_a above this level will still increase the peak height but at the cost of a decrease in sample throughput.

Equation (8.13) specifies the dependence of the peak height on the sample volume. Inspection of Figure 8.4 indicates that as V_a is increased, a flat-topped peak is eventually obtained when $V_a > 5\sigma_{v,c}$. Further increases in sample volume have no effect on peak height but do increase the peak half-width and therefore decrease the sample throughput. Consequently there is no analytical reason to use a sample whose volume is larger than about $5\sigma_{v,c}$.

Several important practical features can be obtained from the above discussion. First, the optimum sample size can be obtained quite simply by injecting a small volume of substrate solution into the reactor to obtain an estimate of $\sigma_{v,c}$. The sample volume for optimum sensitivity can then be calculated. Second, if sample volume is limited or if a high throughput rate is important, the parameter to optimize is the column or system dispersion. This is exactly analogous to the improvement of resolution in high-performance liquid chromatography.

In the discussion thus far, we have assumed that the sample was injected as a plug which, in all probability, is not correct. A discussion of the interaction of the injector dispersion and the column dispersion for chromatography has been presented by Sternberg (16). In essence, the resultant peak has been found to be a convolution of the input function and the column function. For a plug injection, this concept gives rise to equations (8.11) and (8.15). Sternberg has derived analogous equations for several other input functions (16). A simpler approach to sample dispersion during the injection process relates the injection volume, V_a, to the dispersion by the relation

$$V_a = K\sigma_{v,s} \qquad (8.16)$$

where K is a parameter related to the method of injection and the conditions of the injection (13, 16). For plug injection, K has been shown to be $\sqrt{12}$, while for other injection profiles, the value of K decreases (16). Another factor which will affect the relationship between peak dispersion and sample volume is the functional relationship between V_a and K. We have assumed that K is independent of V_a. If this is not the case, equations (8.11)–(8.15) are not a good estimate of the effect of V_a on the dispersion or sensitivity. Thus the performance of the injection system may have a significant effect on the total system dispersion, and more specifically on the optimum sample volume. It should be noted that values between 2 and 2.7 have been reported for K for on-column liquid chromatography injections (13).

8.1.3   Effects of the Detector System

The ability of the detection system to authentically reproduce the concentration of the incoming column effluent is also an important consideration. There are a number of factors which influence the fidelity of the peak profile including mixing in the detector volume, slow mass or heat transfer (if a thermal device is used) to the detector, and slow response characteristics of the associated electronics. All of these processes can be represented by a simple exponential function and, as shown above, the output of the detector can be represented as a convolution of the column effluent function, that is, a Gaussian curve, and the detector function (16). There are essentially three effects of interest: the peak-height maximum is diminished, the maximum is also displaced, and the peak is broadened. Several authors have presented quantitative treatments of these phenomena for chromatographic peaks (16–18) and the situation for IMER's is again analogous. If the variance of the detector in time units is given by $\sigma_{t,d}$, then the movement of the peak is given by a relation similar to equation (8.11) and the increased dispersion is given by equation (8.5). A slow detector has no effect on the peak area provided the peak is integrable and therefore if peak area is used for quantitation, there is no diminution of sensitivity.

All chromatographic detectors can be divided into two general classes (19): concentration-sensitive detectors and mass-flow-sensitive detectors. Most of the detectors currently in use in enzymatic determinations (absorbance, refractive index, conductivity, thermal, potentiometric, and amperometric) can be classified as concentration-sensitive detectors. A characteristic of this type of detector is that the instantaneous magnitude of the signal is independent of the flow rate but the peak area is inversely related to the flow rate (17). Thus, at the relatively low flow rates employed in IMER's, this type of detector is preferable to the mass-flow-sensitive system. This also agrees with the conclusion of Bakalyar and Henry that better precision is obtained by measurement of the peak height if flow fluctuations are encountered (18). Precision data for an IMER system are presented in Table 8.2. It is important to note that the peak area is independent of the sample dispersion factors.

The response of any of the above mentioned detectors can be predicted from equations (8.1) and (8.2) and the appropriate relationship between concentration and detector response, such as Beer's law. For most of these detectors, good descriptions of the response functions in liquid chromatographic systems are available (21–23). For the thermal detectors, however, a number of additional parameters are involved. First, both mass- and heat-transfer rates must be considered. We have estimated that heat, which has a much higher diffusivity than matter, equilibrates thoroughly with the interstitial fluid and fluid within the internal pores within 0.2 sec. Second, heat can be lost from the reactor and detection system. The peak height as a function of flow rate are shown for a refractive index and thermal sensor in Figure 8.5. The roll-off at low flow rates for the thermal detector is due either to heat-transfer effects or heat loss from

Table 8.2   Effect of Sample Concentration on Peak Height and Area[a]

Concentration (mmol/l)	Peak height[b] $(10^{-3}\,°C)$	CV[c] (percent)	Peak area[d] $(10^{-3}\,cal)$	CV[c] (percent)
Blank	0.00	—	0.00	—
1	3.96	1.7	1.32	2.9
2	7.90	0.7	2.62	1.7
5	19.7	0.4	6.40	0.8
10	38.9	0.6	12.5	0.6
20	78.5	1.2	25.1	1.0
50	197	0.4	63.0	0.1
100	416	0.0	134	0.8
200	816	0.4	259	0.8

[a] Conditions: Column 4.0 × 33 mm, 0.120 ml samples, flow rate = 1.22 ml/min.
[b] Least-squares analysis of a plot of peak height versus concentration indicates that: y (peak height, m °C) = 4.1 ± 0.02 × (concentration, mM) − 1.3 ± 1.6 with a correlation coefficient of 0.999.
[c] Coefficient of variation = standard deviation/mean; N = 5.
[d] Least-squares analysis of a plot of area versus concentration indicates that y (area, millicalories) = 1.30 ± 0.01 × (concentration, mM) − 0.17 ± 0.7 with a correlation coefficient of 0.999.
[e] All results taken from reference 12 with the permission of the authors.

the column. Since continuous introduction of sample solution resulted in the same roll-off, heat leak is the most likely explanation. This can be further confirmed by changing the surface area to reactor volume ratio, a maneuver which increases heat leak and decreases the observed peak height at a particular flow rate. Third, since the diffusivity of heat is almost the logrithmic mean of the mass diffusivity in gases and liquids, it might be anticipated that longitudinal thermal spreading might be a significant contributor to the dispersion process. It does not, however, appear to be a major consideration (12). Despite these disadvantages, the universality, simplicity and inexpensiveness of the thermal detector make it a useful device.

8.1.4   Optimization of Reactor Design

In conclusion, there are a number of factors which can be optimized for improving the sensitivity and sample throughput of an IMER. In the previous sections, some of the parameters which influence these operating characteristics have been described. The reactor itself may also be designed for optimum performance. Much of the work which has been re-

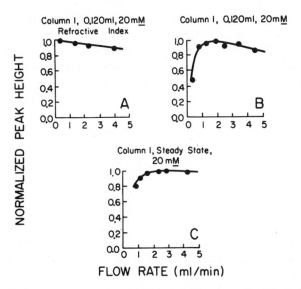

Fig. 8.5. Dependence of normalized peak height on flow rate. (A) 120 μl of 20 mM urea, refractive index detector; (B) 120 μl of 20 mM urea, thermal detector; (C) Steady-state peak height for 20 mM urea, thermal detector. Column 1 dimensions: 4.0 × 33 mm. [Reprinted with permission from R. S. Schifreen, D. A. Hanna, L. D. Bowers and P. W. Carr, *Anal. Chem.* **49**, 1929 (1977). Copyright by the American Chemical Society.]

ported has not involved an optimization since the original attempt has provided analytically acceptable results. As IMER's are applied to assays which require more sensitivity and speed, these parameters will become more important.

One of the ways in which the reactor can be improved is optimization of enzyme-support size. The smallest particles which have been used in enzyme reactors are 400 mesh (37 μm in diameter). The size distribution of CPG of this type ranges from 200 mesh (74 μm in diameter) to 400 mesh. Both of these figures are quite large relative to the supports used in modern liquid chromatography. Substantial decreases in peak width should be possible using smaller particles of a narrow size distribution and using better packing procedures. Smaller particles in a narrower range of sizes have recently become available. Hamilton has announced a line of spherical styrene–divinylbenzene beads in various degrees of cross-linking (which affects pore size) and particle sizes from 7–10 to 50 μm. Any of the various manufacturers of HPLC gel-permeation supports would also be a source of high-performance particles. Two other advantages would also result from smaller particles: the larger surface area to volume ratio would increase the rate of external mass transport, and the smaller diameter would decrease the internal diffusion. Both of these factors would lead to more efficient use of the enzyme (see Chapter 7).

The actual effect of this change is difficult to assess quantitatively. It is important to realize that as the activity of the immobilized-enzyme preparation is increased, the internal and external mass transfer become rate-limiting. Thus more effective use of the enzyme in the column could result in less support required for the column and thus, in smaller reactors. The smaller particles also require a larger driving force with its increased mechanical complexity. Since the pressure drop and column dimensions are related, the final system parameters must be determined by the specific application requirements. Another consideration is the fact that in short columns the validity of some of the above relations has been questioned (24). Studies on the effects of column dimensions on enzyme reactors have not been reported. It should be apparent from this discussion that there are no clearcut boundaries in the optimization of an IMER. The economic and practical aspects of the application should be considered with the preceding factors in mind.

8.2 APPLICATIONS OF IMMOBILIZED-ENZYME REACTORS

The immobilized-enzyme reactor (IMER) is perhaps the most important area of application because of its ability to be interfaced with virtually any detector, its compatibility with flow systems and therefore with high sample throughput, and its ability to be used, in some cases, for total conversion of substrate to product, thus avoiding the complexities of kinetic analysis. In the following section, we will discuss some of the systems which have been developed using various detection systems. As with any attempt to survey a rapidly expanding field of science, keeping a comprehensive listing of the latest advances is nearly a futile endeavor. Hopefully, between the text and Tables 8.3 and 8.4, the reader will have a firm and relatively complete survey of the field.

8.2.1 Spectrophotometric Detection

The first successful analytical application of an immobilized-enzyme preparation was the determination of urea in serum and urine using immobilized urease and a colorimeter (25). The ammonium produced by the hydrolysis was quantified using a manual Nesslerization procedure. Although a good correlation was reported between the immobilized-enzyme reactor and the Berthelot method, the primary goal of the paper was the characterization of the bound urease. As a result, analytical details were rather sparse.

In 1966, Hicks and Updike reported a system using polyacrylamide gel-entrapped lactate dehydrogenase (LD) and glucose oxidase (GO) for the determination of lactate and glucose respectively (26). Once again, the major emphasis of the paper was a study of the characteristics and preparation of the immobilized enzyme. There are, however, several interesting conclusions which arise. First, using the automated system shown in Figure 8.6, up to 20 samples/min could be run through the system which coupled the products of the reactions, NADH or H_2O_2, to a dye which absorbs in the visible region, 2,6-dichlorophenolindophenolphenazine methosulfate or o-tolidine. The more interesting feature, however, was that as the amount of glucose oxidase entrapped in a given amount of gel increased, the linear range of the assay decreased. This was found to be true even when the total column activity was identical, that is, for variable amounts of gel in the column. The limited linear range was the result of the diffusional limitations in the relatively large (20–40 mesh) support.

Colorimetry has also been used in conjunction with an immobilized enzyme system to measure part per billion quantities of nitrate in envi-

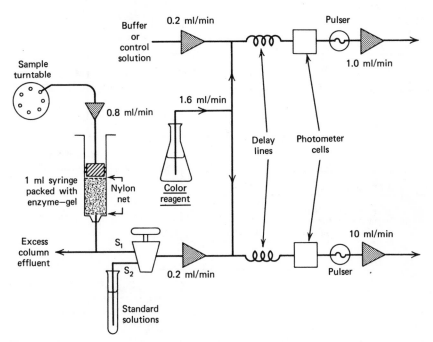

Fig. 8.6.   Original apparatus designed for the spectrophotometric determination of glucose. [Reproduced with permission from G. P. Hicks and S. J. Updike, *Anal. Chem.* **36**, 726 (1966). Copyright by the American Chemical Society.]

ronmental water samples (27). The nitrate was reduced to nitrite in the presence of methyl viologen by the reaction

$$2H^+ + NO_3^- + 2MV^{\ddot{+}} \xrightarrow{\text{nitrate reductase}} 2MV^{2+} + NO_2^- + H_2O$$

where MV^{2+} and $MV^{\ddot{+}}$ represent the oxidized form of methyl viologen and its one-electron reduction product, respectively. The nitrite produced in the reaction was then quantified by a classical azo dye reaction monitored at 543 nm. A schematic drawing of the continuous-flow-based device is shown in Figure 8.7. A limit of detection of 17 ppb nitrate was postulated for total conversion of the nitrate although this condition was not achieved. The system was rapid and simple to operate and showed improved specificity over the other commonly used nitrate techniques. As an example, the determination of nitrate in a sample effluent which

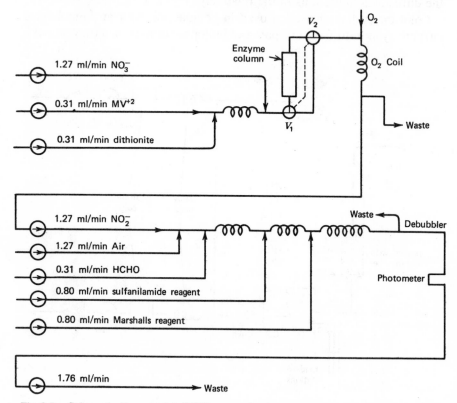

Fig. 8.7. Schematic diagram of an IMER system for the trace-level determination of nitrate. [Reproduced with permission from D. R. Senn, P. W. Carr, and L. N. Klatt, *Anal. Chem.* **48,** 954 (1976). Copyright by the American Chemical Society.]

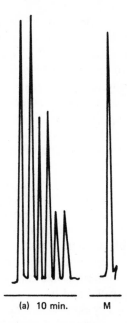

(a) 10 min.   M   (b) 10 min.

Fig. 8.8.   Detector response for (a) glucose and (b) ethanol standard runs using a high-performance liquid-chromatography variable-wavelength detector in conjunction with an IMER. Duplicate injections of various volumes of a standard solution are shown. Peak M corresponds to a 1.5 μl injection of a diluted malt syrup sample. Note the narrow peaks in (b). [Reproduced with permission of John Wiley & Sons, Inc., from P. Cremonesi and R. Bovara, *Biotechnol. Bioeng.* **18**, 1487 (1976).]

contained organic nitrogen gave much higher results for the Cu–Cd reductor method compared to the enzyme technique. The authors postulated the difference arose from the conversion of organic nitrogen to nitrate or nitrite in the Cu–Cd reductor. It is interesting to note that the limit of detection was a result of flow-stream noise and not the stabilized photometer detection system.

A recent paper outlines the advantages of coupling a high performance liquid chromatography detector with a small immobilized-enzyme reactor (28). An example of some of the data obtained for glucose standards, a malt syrup sample, and ethyl alcohol standards is shown in Figure 8.8. Using on-column injection and small (2–3 mm inner diameter × 5 cm) reactors these authors were able to achieve relatively low limits of detection for glucose, ethanol, and testosterone. The glucose system used glucose oxidase and horseradish peroxidase immobilized on CNBr-activated mercerized cellulose linters, the latter reaction to oxidize I^- to form a colored species. Using the liquid chromatography detector, a limit of detection and linear range of 2.5×10^{-9} moles and 2.5×10^{-9} to 1 $\times 10^{-7}$ moles, respectively ($\sim$2.5–100 mM glucose) was obtained as opposed to a 1×10^{-8} moles limit of detection with an 80 μl spectrophotometer cell. Ethanol and testosterone were determined by injecting the sample into a stream of NAD^+ at the optimum pH for alcohol dehydrogenase or 3β, 17β-hydroxysteroid dehydrogenase, respectively. The limit of detection for the testosterone was reported to be 50 ng. Although this would require the injection of 10 ml of serum to be clinically useful, it does demonstrate rather good sensitivity for this type of system. Numerous examples of absorbance measurements coupled to immobilized-enzyme systems can be found in Table 8.3.

A new device, called an immobilized-enzyme pipette or impette, has been recently reported. The principle of the device is conversion of substrate to product in a short length of nylon tubing containing immobilized enzyme attached to an automatic adjustable volume pipette. Immobilization technology developed for continuous segmented flow analyzers is directly applicable. In the case of a urease impette, 0.25 ml of standard or deproteinized and diluted serum was aspirated and held in the tube for 5 min. An aliquot of the sample was then expelled and a colorimetric analysis performed to measure the ammonia produced. Good correlation was obtained with both solution-phase urease methods and with the diacetyl monoxime method (163). Since the enzyme activity in the tubes used is quite low, the technique is essentially a single-point kinetic method. The dependence of the method on such factors as reaction time, tubing length and specific gravity, and analyte concentration have been discussed (164). The simplicity and inexpensiveness of the technique may make it of value in laboratories which lack automation.

8.2.2 Electrochemical Detection

Electrochemical transducers have also proven to be very useful as detectors for IMER systems. Once again, electrochemical sensors can be divided into amperometric and potentiometric classes. Although both of these types of systems have been used, the first electrochemical device was an amperometric one designed for the determination of glucose in blood by Updike and Hicks (29). A Clark oxygen electrode was used to measure the oxygen content of the effluent stream from a small column of polyacrylamide entrapped glucose oxidase. The depletion of O_2 was taken as a measure of the glucose concentration. In their study, Updike and Hicks varied both the activity per support particle and the total activity in the column, but did not observe a change in the linear range. This was the first evidence of diffusion control of the reaction which might be

Table 8.3 Immobilized-Enzyme Reactor Applications

Analyte	Enzyme (E.C. number)	Reactor type (and specifications)	Detection system	Comments	Reference
L-Amino acids	L-amino acid oxidase (1.4.3.2); catalase (1.11.16)	Column (3.2 mm × 45 mm) CPG–glutaraldehyde	Ammonia gas sensor	Reactor at 37°C; catalase used to recycle O_2 for increased conversion; linearity for L-leucine 3×10^{-5} to 10^{-3} M	66
L-Arginine	Arginase (3.5.3.1) and urease (3.5.1.5)	Tubular (0.25 cm × 8 m-nylon) (0.25 cm × 6 m-nylon)	Ammonium ion-selective electrode	Linear 5×10^{-5} to 5×10^{-2} M for either amino acid (Colorimetric system—reference 61)	67, 68
L-Asparagine	Asparaginase (3.5.1.1)	Glutaraldehyde			
L-Aspartic acid	Aspartate aminotransferase (2.6.1.1) Malate dehydrogenase (1.1.1.37)	Column (Sephadex CNBr) (0.7 ml volume)	UV absorbance (340 nm)	NADH, pyridoxyl phosphate cofactors; sample introduced, left in reactor 5 min, eluted, and absorbance measured. 10-fold increase in sensitivity experienced with co-immobilization	69
Creatine kinase	Hexokinase Glucose-6-Phosphate Dehydrogenase	Column (Agarose) (2 × 17 mm)	UV absorbance (340 nm)	Effluent from DEAE–Sephacel column mixed with substrate before passage through enzyme reactor.	70
	Hexokinase Glucose-6-Phosphate Dehydrogenase	Tubular (nylon)	UV absorbance	Enzymes co-immobilized. Good correlation with soluble enzyme Auto Analyzer method.	71

413

Table 8.3 *(Continued)*

Analyte	Enzyme (E.C. number)	Reactor type (and specifications)	Detection system	Comments	Reference
Creatinine	Creatinine amidohydrolase Creatine kinase Pyruvate kinase Lactate dehydrogenase	Tubular (nylon–polyethylene imine polymer–glutaraldehyde)	UV absorbance (340 nm)	Pyruvate kinase–Lactate dehydrogenase co-immobilized to measure ATP. Various combinations of immobilized and co-immobilized IMER's tested.	154
Cholesterol	Cholesterol oxidase (1.1.3.6) Catalase (1.11.16)	Column (1 ml)	Thermal	Linear to less than 0.03–0.1 mM. Cholesterol oxidase precolumn, catalase thermistor	58
Cyanide	Rhodanese Injectase	Column (Zirconia-clad CPG)	Thermal	Assay range 20–600 μM cyanide with injectase. Range 20–1000 μM with rhodanese.	72
Glucose	Glucose oxidase (1.1.3.4.)	Column (Polyacrylamide entrapped)	Colorimetric	o-tolidine redox dye. Linear range dependent on enzyme concentration 10 mg/100 ml glucose oxidase (GO) linear to 10 mg/100 ml	26
Glucose	Glucose oxidase	Tubular (polystyrene) (nitrate, reduction)	Colorimetric	40 samples/hr. Useful, but nonlinear in the range 4–16 mM glucose.	73
Glucose	Glucose oxidase	Tubular, Column	Colorimetric	Detect I_2 color. Compare tube, column, and dialysis membrane reactor in continuous flow.	74

414

Analyte	Enzyme	Support	Detection	Comments	Ref.
Glucose	Glucose oxidase	Tubular (nylon by dimethyl sulfate activation)	Colorimetric and amperometric	Colorimetric system linear to 1 mM; electrode to 2.5 mM. Rate for amperometric—20/hr, colorimetric—20/hr. Study effect of spacer from surface.	75
Glucose	Glucose oxidase	Tubular (30 cm)	Colorimetric	Color system of 3-methyl-2-benzothiazolin hydrazone and dimethyl analine (DMA) for clinical samples. 60 samples/hr. 25,000 assays on a single tube.	71
Glucose	Glucose oxidase	Column (polyacrylamide entrapped (0.15 × 4 cm)	Amperometric (O_2)	Linear range 0–40 mg/100 ml. Observe extent of linear range varies with enzyme concentration.	29
Glucose	Glucose oxidase	Column (CPG–diazo)	Amperometric	Small reactor (1 ml syringe). Operated in either kinetic or end-point mode—30/hr in former. Catalase causes nonlinearity above 10^{-4} M.	30
Glucose	Glucose oxidase	Column (CPG–glutaraldehyde)	Amperometric	40/50 mesh titania-coated glass. Using 9 μl sample, linear to 5 g/l. Good analysis of curve factors.	77
Glucose	Glucose oxidase	Column (0.5 cm inner diameter, 1.2 ml) poly (HEMA) gel immobilization	Amperometric	Linear for serum samples between 0.01–0.15 mg/ml. In vivo monitoring of dog blood.	78
Glucose	Glucose oxidase Catalase	Column (mixed bed)	Amperometric		31

Table 8.3 (*Continued*)

Analyte	Enzyme (E.C. number)	Reactor type (and specifications)	Detection system	Comments	Reference
Glucose	Glucose oxidase	Column	Amperometric	End-point mode. Linear in range 0–50 ppm glucose; higher ranges by dilution.	79
Glucose	Glucose oxidase	Column (Sephadex)	Chemiluminescence	H_2O_2–luminol-Fe$(CN)_6^{3-}$ system.	38
Glucose	Glucose oxidase	Column	Chemiluminescence	H_2O_2-luminol-Fe$(CN)_6^{3-}$ system.	40
Glucose	Glucose oxidase	Column (CPG-)	Chemiluminescence	H_2O_2-luminol-Fe$(CN)_6^{3-}$ system.	39
Glucose	Glucose oxidase	Column	Thermal	Modified LKB heat-flow calorimeter; low flow rates yield poor throughput; limit of detection is 0.5 mM	47
Glucose	Glucose oxidase	Column	Thermal	Linear range 0.03–0.5 mM glucose	53
Glucose	Glucose oxidase Catalase	Column	Thermal	Extend range to 5 mM with catalase. 5 min peak width.	58
Glucose	Glucose oxidase Catalase	Column (CPG-GO)	Differential conductivity	15 min assay time. Recycle substrate through the reactor.	80
Glucose	Glucose oxidase	Packed bed (CPG)	Thermal	Dual-column–dual-thermistor split-flow system. 20 µl sample volume–1 ml injection (50 × dilution). Linear range 0.01–0.45 mM glucose. Range extended to 0.90 mM by co-immobilization of catalase.	60

416

Analyte	Enzyme	Support	Detection	Comments	Ref.
Glucose	Glucose oxidase	Tubular (nylon)	Absorbance	No loss of activity in 4 months at 37°C. 25,000 specimens run on tube.	71
Glucose	Glucose oxidase	Open tubular (nylon–glutaraldehyde)	Absorbance	Evaluation of Carlo Erba cartridge. 20% conversion of glucose in 0.3–3 g/l range. 20,000 assays reported for single tube.	81
Glucose	Glucose oxidase	Tubular (nylon–glutaraldehyde)	UV absorbance (365 nm)	Stopped-flow analyzer to increase reaction time.	82
Glucose	Glucose oxidase	Column (Alumina–cross-linked enzyme)	Amperometric (H_2O_2)	Use cellulose dialysis membrane to remove oxidative ascorbate, urate interference. Linear to 4 g/l. Good correlation.	155
Glucose	Glucose dehydrogenase (1.1.1.47)	Tubular (nylon–glutaraldehyde)	UV absorbance (340 nm)	Complete conversion of 8 g/l in 3 m reactor. 20–30% activity retained after 6 weeks. Good correlation with other methods.	156
Glucose	Glucose dehydrogenase (1.1.1.47)	Tubular (nylon–polyethylene imine polymer –glutaraldehyde)	UV absorbance (340 nm)	Linear to 4 g/l. Good correlation with hexokinase/glucose-6-phosphate dehydrogenase method. Enzyme retained 20% activity after 3500 analyses regardless of time interval.	157
Glucose	Hexokinase Glucose-6-Phosphate Dehydrogenase	Tubular	UV absorbance		83

Table 8.3 (*Continued*)

Analyte	Enzyme (E.C. number)	Reactor type (and specifications)	Detection system	Comments	Reference
Glucose	Hexokinase	Column (CPG—Glycophase Schiff's base)	Thermal		62
Glucose	Hexokinase Glucose-6-phosphate dehydrogenase	Tubular (nylon)	UV absorbance	Evaluation of Technicon tubular reactor for clinical laboratory. Stable for one month—12,000 assays.	84
Galactose	Galactose oxidase	Column	Amperometric	Determine lactose in milk and blood. Linear range in blood is 0–0.6 mM	85
Glucose	β-Galactose dehydrogenase	Column (Sephadex–CNBr)	UV absorbance	A "filter tube" of enzyme gel was notated by a motor for a specified interval. Linear for at least the range 10–50 mg/100 ml.	86
Glutamate–oxaloacetate	Malate dehydrogenase Alcohol dehydrogenase	Tubular	UV absorbance		87
Lactic acid	Lactate dehydrogenase	Column (Sepharose 4B)	Fluorescence	Small bed attached to syringe—sample flushed from syringe into bed for 1 min incubation. Linear 10^{-4} to $10^{-2}M$ lactate.	88
Lactic acid	Lactate dehydrogenase/ Alanine transaminase	Tubular (nylon–polyethyl-eneimine–glu-taraldehyde)	UV absorbance	Need ALT to "pull" pyruvate. Stability limited by ALT to 4 weeks or 2,000 assays.	89
Lactic acid	Lactate Dehydrogenase	Polyacrylamide entrapment	Chemiluminescence	Detect 10–100 pmoles lactate	90

418

Analyte	Enzyme	Configuration	Detection	Comments	Ref.
Lactose	β-galactosidase, Glucose oxidase	Tubular	Colorimetric	Linear over 0.1–0.6 mM. Used for assay in milk. Separate reactors.	91
Lactose	Glucose oxidase catalase	Column	Thermal	Linear 1 to 3 mM. Immobilized—galactosidase used before reactor. Analysis in milk.	58
Lactose	β-galactosidase, Glucose oxidase	Column (2 column system)	Amperometric		92
Lecithin	Phospholipase D, Choline oxidase	Column (CNBr–octyl Sepharose 4B)	Amperometric	Linear to 3 g/l. Concentration of Ca^{2+} and Triton X-100 critical, 70% activity retention after 9 days.	158
Maltose	Amyloglucosidase, Glucose oxidase	Tubular	Colorimetric	Useful over range of 5–30 mM with separate tubes or 1–6 mM with co-immobilized tube	91
Mercury	Urease	Column (CPG)	Potentiometric	Hg(II) inhibition of urease. 0–30 nM is 25 ml sample solution used. Regenerate column with thioacetamide. Cu(II) and Zn(II) also detected.	93
Mercury	Urease	Column (CPG)	Thermal	Hg(II), Ag, Cu inhibition. 30 sec pulse of sample. Regeneration with NaI and EDTA.	63
Neutral Lipids	Lipoprotein lipase	(Polystyrene-γ-aminopropyl-triethoxycilane-glutaraldehyde) Packed column	Potentiometric (pH electrode)	Extract lipids with isopropanol, silicic acid; inject supernatant. Stable for 200 analyses over 10 days.	159

Table 8.3 (*Continued*)

Analyte	Enzyme (E.C. number)	Reactor type (and specifications)	Detection system	Comments	Reference
Nitrate	Nitrate reductase	Column (CPG-diazo)	Colorimetric	Detection limit of 17 ppb. Linear range 0.05–5 ppm. Used in EPA samples with good correlation with reference method.	27
Nitrate	Nitrate reductase	Column	Potentiometric	Effluent from column collected—analyzed by air gap NH_3 electrode. 5–7 min for 10^{-4} M response.	94
Nitrate	Nitrite reductase			Detection limit $2 \times$ for 10^{-5} M NO_3^-.	
Nitrite	Nitrite reductase (BSA copolymer)	Batch	Potentiometric	Methyl viologen electron donor. 5 min incubation, filter, add NaOH, close air-gap electrode. Linear 0.1–100 mM.	95
Organophosphorus pesticides	Cholinesterase (3.1.1.7)	Membrane reactor	Electrochemical	Useful for all pesticides in air and water. Commercially available.	37
Penicillin	Penicillinase	Tubular			87
Penicillin	Penicillinase	Column (CPG glutaraldehyde)	Thermal	Linear, 1 ml sample pulse—about 6 min peak width, Stable at least 3 weeks	53
Penicillin	Penicillinase	Column	Potentiometric		33
Penicillin	Penicillin acylase				94
Peroxide	Catalase	Membrane reactor of	Thermal	Response time—20 min for 3 mg H_2O_2.	95

420

Analyte	Enzyme	Configuration	Detection	Comments	Ref.
Peroxide	Peroxidase	Column	Fluorometric		96
Peroxide	Peroxidase	Tubular (nylon)	Chemiluminescence	Clear nylon tube placed directly on PMT.	41
Peroxide	Peroxidase	Column (polystyrene)	Fluorescence	Homovanillic acid used as fluorometric hydrogen donor. Tris best buffer. Some metals enhance or inhibit fluorescence and/or enzyme.	97
Phosphate	Alkaline phosphotase	Column	Colorimetric	pNO_2 phenylphosphate substrate. Inhibition by phosphate gives linear curve from $1-10 \times 10^{-4}$ M	98
Phosphoenolpyruvate	Pyruvate kinase / Lactate dehydrogenase	Column (CPG-diazo)	UV absorbance	Separate reactors. Linear over range from $0.8-10 \times 10^{-5} M$. Also could be used for ADP. 4 min analysis time.	99
Phosphoenol pyruvate (ATP)	Pyruvate kinase / Lactate dehydrogenase	Tubular (nylon–polylysine)	UV absorbance		162
Pyruvate / Pyruvic acid	Lactate dehydrogenase	Column (CPG-diazo)	UV absorbance	Linear range $0.5-10 \times 10^{-5}$ M, 4 min analysis time, dependent on geometry, flow rate.	99
Pyruvic acid	Lactate dehydrogenase	Tubular (nylon–polyethylene imine–glutaraldehyde)	UV absorbance	Useful for serum pyruvate. "Maintained considerable activity" at 4°C for 18 months.	89
Sucrose	Invertase (3.2.1.26)	Column	Colorimetric	Determine total reducing sugars.	100

Table 8.3 (*Continued*)

Analyte	Enzyme (E.C. number)	Reactor type (and specifications)	Detection system	Comments	Reference
Sucrose	Invertase Glucose oxidase	Tubular	Colorimetric	Linear over the ranges 20/hr 0.5–2.5 mM or 5–30 mM depending on use of co-immobilized or separate tubes and lag between them.	91
Sucrose	Invertase Mutarotase Glucose oxidase	Column	Amperometric		101
Sulfate	Arylphosphatase	Column	Colorimetric	One of a number of anions which nonspecifically inhibit the enzyme at concentrations below 1 mM.	98
Testosterone	3β, 17β-Hydroxy steroid dehydrogenase	Column	UV absorbance	HPLC detector. Linear in range from	28
Triglycerides	Glycerol kinase	Tubular (nylon) (15 cm reactor)	UV absorbance	Lipase, pyruvate kinase, and lactate dehydrogenase in solution. Reactor stable for at least 3000 assays (1 week).	71
Triglycerides	Glycerol dehydrogenase	Tubular (nylon–polyethylene imine)	UV absorbance	10 min residence time in column. Total conversion of substrate in linear range of 0–50 nmole/2 ml.	160
Tryptophan	Tryptophanase or tryptophanase/LDH co-immobilized CNBr Sepharose	Column	UV spectrophotometric (NADH)		102

422

Substrate	Enzyme	Reactor	Detection	Comments	Ref.
Urea	Urease	Column	Colorimetric	Nesslerization of effluent of column. First application of immobilized enzyme to analysis.	25
Urea	Urease	Tubular	Colorimetric	Chaney–Marbach method in AA system. Comparison of column, tubular, and membrane reactors.	74
Urea	Urease	Tubular	Colorimetric	Study of stability and enzyme properties of immobilized urease.	103
Urea	Urease	Tubular	Colorimetric	Chaney–Marbach method in AA system. Linear range 0.3–3.0 mM urea.	104
Urea	Urease	Tubular	Colorimetric	Chaney–Marbach method in AA system. Linear to at least 160 mg/100 ml urea.	105
Urea	Urease	Column	Potentiometric		34
Urea	Urease	Column	Potentiometric		35
Urea	Urease	Column	Thermal (single)		52
Urea	Urease	Column	Thermal (Differential)		61
Urea	Urease	Column	Thermal	Linear 2.5–7 mM. Single thermistor system.	53
Urea	Urease	Column	Thermal (single)	1 hr equilibration time. 30 samples/hr maximum throughput. Linear 0.02–200 mM/l. 1 ml sample pulses with or without dilution.	54

Table 8.3 (*Continued*)

Analyte	Enzyme (E.C. number)	Reactor type (and specifications)	Detection system	Comments	Reference
Urea	Urease	Column (alumina–cross-linked)	Potentiometric	Evaluation of Kimble BUN Analyzer	106
Urea	Urease	Tubular (nylon)	Absorbance	Evaluation of Miles IMER	107
Urea	Urease	Tubular (nylon–glutaraldehyde)	Absorbance	Evaluation of Carlo Erba cartridge. 10,000 assays claimed.	81
Urea	Urease	Tubular (nylon)	Absorbance	Direct coupling yields more activity than annulus. Stable for 4 months and 2000 tests. Measure citrulline by difference.	108
Urea	Urease	Column (polyacrylamide)	Potentiometric		109
Uric acid	Uricase (1.7.3.3)	Column (nylon powder)	Colorimetric	I_2 detection. Nonlinear even at 25×10^{-6} M urate. 30 samples/hr.	104
Uric acid	Uricase Catalase	Column	Thermal	Linear in 1–4 mM region.	58
Uric acid	Uricase	Tubular (nylon)	Absorbance	Good stability of reactor. MBTH–Anthranilic acid color system. 0–12 mg/dl range.	111
Uric acid	Uricase	Tubular (nylon–polyethylene imine–glutaraldehyde)	Absorbance	Linear from 0–12 mg/dl at 0.33 ml/min with 2 m reactor. 23 days (4000 samples) shows 60% activity loss.	112

424

Analyte	Enzyme	Support (immobilization)	Detection	Comments	Ref.
Uric acid	Uricase	Tubular (nylon–diamine spacer–glutaraldehyde)	Absorbance	Evaluation of Carlo Erba cartridge. 50% activity loss in 70 hr continuous running. 20,000 assays performed.	81
Uric acid	Uricase	Column (alkylamine CPG–glutaraldehyde)	Amperometric		110
Uric acid	Uricase		Chemiluminescence	Coil of transluscent Teflon on PMT to detect luminol $Fe(CN)_6$–H_2O_2 system. Linear 10^{-7}–10^{-5} M.	161

expected for 20–40 mesh particles. In spite of this observation, many investigators have continued to use large particles for an immobilized-enzyme support.

Subsequently, several groups have studied the glucose oxidase IMER–platinum-electrode system. Weibel and his coworkers (30) reported a system in which the effluent from the glucose oxidase–CPG column is directed into a microtangential cell. A 0.1 ml sample was introduced via an automatic dilutor which performed a 1:100 or 1:200 dilution. As a result of the large sample size, the sample throughput was relatively low. It should be pointed out that this instrument was generally operated in the end-point mode. A more detailed study of a very similar system was reported by Kunz and Stasny (31). Using a 9 μl sample volume and the column dilution to assist in adjusting the concentration to the appropriate range, these workers were able to reduce the peak width from 90 sec in the Weibel system to less than 30. It was observed that the slope of the response curve, the peak maximum, and the peak area were proportional to the glucose concentration. The area was found to give a slightly greater linear dynamic range than the peak height, although the precision was influenced by the fact that the peak area was a function of the flow rate.

Further study by Weibel and his group indicated that the technique was limited by the depletion of O_2 in the interior of the support. As a result, glucose oxidase and peroxidase were co-immobilized on CPG (32). The reaction scheme is given by

$$\text{glucose} + O_2 + H_2O \xrightarrow{\text{glucose oxidase}} H_2O_2 + \text{glucuronic acid}$$

$$H_2O_2 \xrightarrow{\text{peroxidase}} H_2O + \tfrac{1}{2}O_2$$

where the effect of peroxidase is to recycle O_2 in the system and to reduce the overall stoichiometric relationship of glucose and O_2 to one-half. The result of this procedure was a more efficient use of the immobilized enzyme and much greater stability of the glucose oxidase. Another interesting feature of this system is its use with both plasma and aerated whole blood. Formation of clots in the column is apparently suppressed by the use of EDTA.

As mentioned in Chapter 3, a number of systems are available for anslysis with amperometric detection. Other systems with analytes such as uric acid and galactose have been reported using the IMER–platinum-electrode approach and are listed in Table 8.3.

As mentioned earlier, potentiometric transducers are also applicable to monitoring column effluents and have seen substantial expansion recently. Watson and Keyes have described a dedicated system for the analysis of urea using a column of immobilized urease as shown in Figure 8.9 (33). A stream of Tris buffer (pH 7.5) is passed continuously through the column. A 5–20 μl aliquot of serum is introduced into the stream and totally converted to ammonium and carbonate ions. A stream of NaOH is then added to the effluent to liberate the ammonia which is measured with an ammonia ion-selective electrode. The NaOH is used to improve the system sensitivity so that such small samples can be used. Excellent agreement was obtained between the urease IMER system and commonly used urea methodology for both quality assurance sera and patient specimens. One potential limitation of potentiometric sensors is their rather slow response time. This is particularly true of gas membrane electrodes which require 23 to 150 sec to attain a stable steady state. In the case of this device, a sample throughput rate of from 45 to 60 samples/hr has been reported. For an instrument of similar design, a sample throughput rate of 8 samples/hr was necessitated by the slow response of the electrode system (34).

A continuous-flow analyzer making use of a flow-through pH electrode has been described for the determination of penicillin in fermentation broths (15). A buffer solution was passed continuously through the glass-

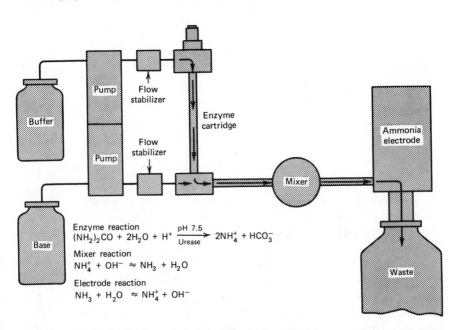

Enzyme reaction
$$(NH_2)_2CO + 2H_2O + H^+ \xrightarrow[\text{Urease}]{\text{pH 7.5}} 2NH_4^+ + HCO_3^-$$

Mixer reaction
$$NH_4^+ + OH^- \rightleftharpoons NH_3 + H_2O$$

Electrode reaction
$$NH_3 + H_2O \rightleftharpoons NH_4^+ + OH^-$$

Fig. 8.9. Schematic diagram of a potentiometric immobilized-enzyme instrument. [Reprinted from B. Watson and M. H. Keyes, *Anal. Letter* **9**, 217 (1976), by courtesy of Marcel Dekker, Inc.]

immobilized penicillinase column and electrode. Samples were introduced via a sampling valve. A dialysis attachment was required for the fermentation broth specimens both for sample clean-up and for dilution of the specimen to the appropriate concentration range (1–5×10^{-4}M). Above this range, the pH change across the column decreases the enzyme activity and thus reduces the efficiency of the reactor. Although the sample throughput was only 6 samples/hr, the method has significant advantages over the established hydroxamic acid colorimetric method and may herald the application of more enzyme reactors in pharmaceutical analyses and process control.

One of the most interesting electrochemical immobilized-enzyme systems which has been developed is for the monitoring of environmental air and water for the presence of organophosphorus pesticides. The first such device was based on the pesticides' inhibition of cholinesterase (35). The principle of the device is that cholinesterase, immobilized by starch gel on the surface of an open pore urethane foam, could catalyze the hydrolysis of butylthiocholine iodide, i.e.

$$C_3H_8COSCH_2CH_2\!-\!\overset{\overset{\displaystyle CH_3}{|}}{\underset{\underset{\displaystyle CH_3{}^{\oplus}}{|}}{N}}\!-\!CH_3\; I^{\ominus} \xrightarrow{\text{cholinesterase}} C_3H_8COOH$$

$$+\; HSCH_2CH_2\!-\!\overset{\overset{\displaystyle CH_3}{|}}{\underset{\underset{\displaystyle CH_3{}^{\oplus}}{|}}{N}}CH_3\; I^{\ominus}$$

The pad containing cholinesterase is placed between a pair of Pt electrodes and a stream of substrate and sampled air passed continuously through the electrodes and pad. In the absence of cholinesterase inhibitors, the formation of thiol by the enzyme reaction is sufficient to carry the 2 μA current between the electrodes with a low applied voltage. In the presence of an inhibitor, no thiol is present to be oxidized and the voltage required to maintain the current flow increases and the inhibitor thus is detected.

Application of the device to water monitoring required the use of a two-cycle system (36). First, the water to be tested is passed through the pad. Since most cholinesterase inhibitors are irreversibly bound, this in effect is a preconcentration step. In the second cycle, air and substrate–buffer solution are passed through the electrodes and pad system. During the second cycle, voltage measurements analogous to those described above are made to determine the presence of pesticides. This device, called the Cholinesterase Antagonist Monitor (CAM-1), is commercially available for air and water monitoring and recently received an $IR^{2} \cdot 100$ award for product excellence. This device, with an automatic enzyme pad changer and alarm system, can detect concentrations of pesticides (36) and zinc (37) in the ppm range.

8.2.3   Chemiluminescent Detection

Another new detection system using chemiluminescent reactions appears to have significant potential. Bostick and Hercules (38) have reported a system in which the hydrogen peroxide produced by the catalytic action of glucose oxidase is quantitated by measuring the light emitted from a ferricyanide-catalyzed luminol reaction. In their study of the peroxide–luminol system, these authors found a number of metals such as Cu(II), Co(II) and Ni(II) which catalyzed the reaction at pH 10.5 or greater. $K_3Fe(CN)_6$ was chosen as a catalyst because it provided the best reproducibility. The light intensity was found to be a strong function of the $Fe(CN)_6^{-3}$ and luminol concentrations, the pH, and the flow rate into the mixing cell. A small (0.4 × 16 cm) reactor packed with Sepharose-bound glucose oxidase was placed in a continuous-flow system as shown in Figure 8.10.

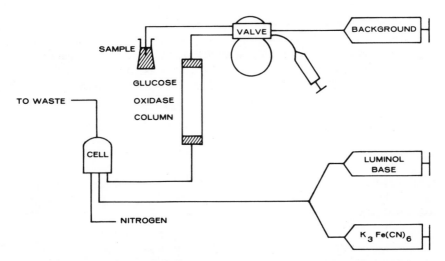

Fig. 8.10.   Schematic diagram of apparatus designed for IMER based chemiluminescent determination of serum glucose. [Reproduced with permission from D. T. Bostick and D. M. Hercules, *Anal. Chem.* **47**, 447 (1975). Copyright by the American Chemical Society.]

Serum samples (0.1 ml) were prepared for analysis by dilution 1:20 with water, deproteinated by the Somogyi technique, and the resultant solution diluted 1:50 with acetate buffer. The glucose injected into the column was completely converted to its products. The lower limit of detection was $10^{-8}M$ peroxide and a linear response over four decades of concentration was observed. The precision at $10^{-6}M$ peroxide was better than 1%. Attempts to measure urine glucose levels were unsuccessful due to a uric acid interference. Williams, Huff, and Seitz (39) modified the system and were able to obtain good correlation with the reference method for urine specimens. A similar system was reported at the same time by Maloy (40). Recently Hornby has reported on a transparent open tubular reactor coiled on a photomultiplier tube for extremely sensitive detection of hydrogen peroxide generating systems (41).

There are several drawbacks to the luminol chemiluminescence system for biochemical measurements, including the pH of the reaction, the necessity for metal-free solutions, and the peculiarities of the luminol itself. Two new chemiluminescent reactions have recently been reported for use with enzyme systems which produce hydrogen peroxide (42) and NADH (43). These systems in large measure eliminate the previously mentioned deficiencies of luminol and make the further application of chemiluminescence imminent.

8.2.4   Thermal Detection

Of the many measurement techniques which have been used in conjunction with immobilized enzymes, one which has enjoyed particular expansion has been solution enthalpimetry. The fundamental measurement parameter is, of course, temperature. For an adiabatic system, the temperature change is related to the number of moles of analyte, n, by the equation

$$\Delta T = \frac{n\Delta H}{C_p} \tag{8.17}$$

where ΔH is the enthalpy of the reaction or reactions being measured and C_p is the heat capacity of the solution and the surrounding vessel (container). Excellent reviews of solution-phase calorimetry have been published (44–46).

The first application of thermal methodology to immobilized-enzyme coupled analysis, however, involved the use of a heat-flow calorimeter to measure glucose via glucose oxidase reaction (47). Since the total amount of heat generated in the reaction was measured by the system and since flowing-stream heat-flow calorimetry requires low flow rates, this technique had a small limit of detection (10 μg glucose), but a very low rate of analysis. A modified flow cell for the LKB microcalorimeter was later reported (48).

The first reported application of a thermistor in conjunction with immobilized enzymes consisted of a thermistor covered by a cross-linked trypsin–BSA membrane (49). The steady-state temperature change was predicted by derivation of an equation assuming substrate diffusion limitations. Differential temperature measurements with a reference and immobilized-enzyme membrane thermistor gave quantitative results but required about 12 min for establishment of steady-state conditions. Subsequently, probes coated with hexokinase or glucose oxidase have been reported (50). Although these authors were able to demonstrate the feasibility of a thermal enzyme probe, limitations caused by diffusional resistance and reduced sensitivity made the device of limited analytical value. A detailed discussion of these devices was presented in Chapter 5.

An enzyme thermistor was reported by Mosbach and Danielsson (51) again making use of immobilized trypsin. In this application, the thermistor is immersed in paraffin oil inside of a cup formed by a coil of polyvinylchloride tubing. The paraffin oil, due to its low heat capacity, is used to increase the observed temperature change. The substrate conversion occurs in the polyvinylchloride tubing which contains glass-bead-immobilized trypsin. Thus, an integrated heat was measured over the entire conversion in the tubing. This fact, however, greatly decreased the sample throughput of the system since a 1 ml sample of 6 mM TEAA caused a peak approximately 56 m°C high and over 15 min wide. Part of the slow response problem may arise from the relatively poor thermal conductivity of polyvinylchloride. The low flow rate (10 ml/hr) gave a relatively high steady-state conversion (80%), but the low sample throughput makes the system of limited analytical use.

Canning and Carr reported a system in which CPG-bound urease was used to determine urea in serum (52). The important features of the system were the use of a well-controlled temperature bath and a second "inner bath" in which the sample was equilibrated before entry into the IMER. A single thermistor mounted in the end of the Dewar-type column was used to measure the temperature change corresponding to the total conversion of urea to ammonia and carbon dioxide. The peaks obtained were significantly tailed, but peak-height determinations were useful for quantifying urea in the clinically significant range at a rate of about 40 samples/hr. The major problem was the relatively expensive and cumbersome temperature control required to use the single-thermistor approach.

A very similar system has subsequently been reported by Mosbach and his coworkers. A thin-walled glass column of 0.8–1.0 ml volume was enclosed in an inner bath which was further immersed in a controlled temperature bath (53). A single detection thermistor was placed at the column exit. A stainless-steel heat exchanger was used to equilibrate the sample with an "inner bath." Urease, glucose oxidase, and penicillinase were bound to glass and used in the system for analysis of their substrates. For penicillin a 6 min wide, 80 m°C high peak was observed for a 1.5 ml sample. A slightly modified version of this system has been reported for the analysis of urea using glutaraldehyde-activated glass-bound urease (54).

Recently, three groups have reported similar approaches to flow enthalpimetry: differential temperature measurement across an adiabatic IMER. Schmidt, Krisam, and Grenner (55, 56) reported a system in which gold capillaries were used to transfer the heat generated in the reaction to the thermistors which were not in direct contact with the flow stream. Using commercially available immobilized enzymes, a glucose oxidase–catalase coupled enzyme system was used for determination of glucose. A determination of total temperature change was made every 20 min; although the authors imply that a higher sample throughout is possible, no data is presented. A later report from this group indicated an analysis time of 5 min with tubular reactors for glucose and urea (57). Mosbach and coworkers also devised a differential temperature system using the same column mentioned above (58), but one equipped with a thermistor at the head of the column. Lactase, glucose oxidase, urate oxidase, and cholesterol oxidase were coupled to CPG and used for analysis. Glucose analyses were linear to 0.5 mM due to the limited solubility of oxygen and thus were not useful for clinical analysis. The lactose system, coupled to glucose oxidase and catalase, and urate system coupled to catalase, were linear to 3 and 4 mM, respectively. A split-flow thermistor system for glucose has also been reported which may be useful in the clinical range (59). This group has also developed a dual-column dual-thermistor split-flow system for the analysis of serum glucose (60). Serum must be diluted 50-fold and 1 ml must be injected into the system to achieve useful results.

We have reported a differential-flow enthalpimetric system for urea using urease (61) and for glucose using the hexokinase-catalyzed phosphorylation (62). Both systems appear to be clinically useful. The urea system was capable of 40 analyses per hr (baseline resolution), was simple and economical to operate, and was essentially reagentless. The hexokinase–glucose system was complicated by the presence of ATP and Mg in the flow stream, but did function in the clinically useful range. Both methods showed good correlation with the commonly used methods. A detailed study of the contribution of heat leak and thermal spreading has allowed significant design improvements in the system (12). A diagram of this system is shown in Figure 8.11. The thermal system is now capable of at least 60 analyses per hr with base-line resolution of the peaks and a peak height of 10 m°C in response to a 0.120 ml sample of 2 mM urea.

The use of the enzyme thermistor for a sensitive assay of heavy metals via inhibition of urease has been reported (63). The column activity was first measured by a 30 sec pulse of urea after which a 30 sec pulse of sample containing mercury, copper, or silver was introduced. The decrease in enzyme activity was linearly related to the metal concentration over the 5–100 μmol/l range for mercury. Column activity was recovered by a wash with sodium iodide and EDTA. A thermal detection system for environmental cyanide was reported based on rhodanese and injectase (64).

An interesting approach has recently been reported in which a thermopile constitutes the reaction "vessel" for catalase bound directly to it (65). When 10 μl of solution containing H_2O_2 was placed on the thermopile–enzyme reactor, heat proportional to the H_2O_2 concentration is

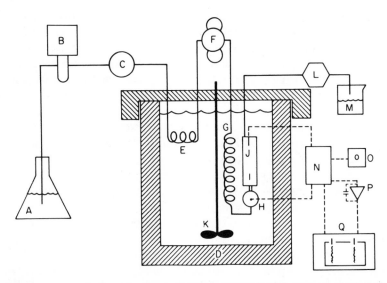

Fig. 8.11. Appartus designed for flow enthalpimetric detection of serum constitutents based on their specific degradation in an IMER. (A) reservoir; (B) pump; (C) pulse dampener–pressure gauge; (D) insulated water bath; (E) pre-equalibration coil; (F) sample injection valve; (G) equalibration coil; (H) reference thermistor; (I) adiabatic column (IMER); (J) measuring thermistor; (K) stirrer. [Reprinted with permission from R. S. Schifreen, D. A. Hanna, L. D. Bowers, and P. W. Carr, *Anal. Chem.* **49**, 1929 (1977). Copyright by the American Chemical Society.]

measured. Compared to other systems measuring the total heat evolved, the system is quite rapid, allowing a determination every 4 min. At present, however, the system has limited sensitivity requiring 0.012 M H_2O_2 and a very exothermic reaction for its only application.

8.3   APPLICATIONS OF ENZYME LAYERS

Although the major interest in enzyme layers or membranes has been their use in conjunction with an electrode, a number of interesting applications have appeared which explore other properties of enzyme reactions. In general, these methods are characterized by their simplicity and low cost. They are not readily amenable to high sample throughput. We have included in the following section a collection of techniques which have used immobilized enzymes in some form other than a reactor or electrode configuration.

8.3.1   Semisolid Surface Fluorescence

The classification of semisolid surface fluorescence techniques in a discussion of immobilized enzymes is rather arbitrary. A number of papers have appeared in which commercially available enzymes or enzyme kits have been dissolved in a buffer, placed on a silicon rubber pad, and lyophilyzed. The dried "immobilized" enzyme-containing pads are then used for an analysis by redissolving the enzyme and adding the sample. Hence, the silicon rubber surface provides an area for the reaction to take place much as a test tube provides for larger volumes of solution. This, of course, is within the bounds of our definition of an immobilized enzyme (see Chapter 4) and will be considered here for completeness.

In actuality, some benefits do accrue from placing the enzyme in the pads. The advantages include economy of reagent and sample, ease of use, and good storage stability of the pad. Two techniques have been used to prepare the reagent pads. The batch method involves placing the reagent onto a silicon rubber strip, drying the pad, and finally cutting the pad to the appropriate size. The other method requires application of solution to a pre-cut pad. The latter was found to be preferable. The pads were then lyophilyzed and stored in the refrigerator or in a dessicator. In general, 15 to 30 assays could be performed with the amount of reagent required for one solution-phase determination.

When required, the pads are brought to room temperature in a dessicator, reconstituted with 25–90 μl of water or buffer, and thoroughly mixed. An aliquot of sample, 3–25 μl in volume, is placed on the pad and mixed with the reagent. A blank solution may be applied to the second surface of the pad. The pad is then placed onto a specially designed holder and positioned in a fluorometer which is placed on its side. The rate of change of fluorescence is monitored and quantitation obtained by comparison to a standard curve.

Factors which affect the reproducibility of the method include drop shape of the reagent application, sample volume, pad shape, and lyophilyzation process (113). Other disadvantages include the necessity for dissolution of the reagent pad prior to the assay, relatively high background fluorescence in some assays, and a rather involved pad-preparation procedure. As can be seen from Table 8.4, Guilbault and his coworkers have applied the technique to a number of systems of clinical interest.

8.3.2 Enzyme Stirrers

Probably the most versatile application of immobilized enzymes is the immobilized-enzyme stirrer. The first such device was simply a glass- or Enzacryl®-immobilized urease preparation held onto a Teflon stirring bar with a nylon net (114). Used in conjunction with an ammonia air-gap electrode, a sensitive, simple end-point assay for urea was developed. A study of the stirrer revealed that from an initial response time and percent conversion of 2 sec and 70%, respectively, the performance declined to 9 min and 50% after 60 days (about 300 assays). A similar system with a specially designed stirrer body has recently been reported for the fluorometric determination of glucose (115) and the amperometric determination of cholesterol (116). In the former case, a study of the rate of conversion as a function of stirring rate indicated that an optimum stirring rate existed. This may pose a problem since obtaining a constant reproducible stirring rate is difficult with magnetic stirrers. The strong dependence of the reaction rate on stirring speed certainly aggravates this problem. Another problem was the production of air bubbles at the optimum stirring rate which caused noise in the fluorescence system. The glucose method, however, had a reported C.V. of 1.3% in the range of clinical interest. Thus, the immobilized-enzyme stirrer conveys the advantages of stability and economy of enzyme, is applicable to a wide variety of detection systems, and is adaptable to equipment readily available in most laboratories. Thus, almost any assay that is now performed with soluble enzymes could be implemented with this technique. It should be borne in mind that the stirrer requires washout between samples in the same way as the enzyme electrodes. The sample throughput rate is thus limited to less than 30 samples per hr. The immobilized-enzyme stirrer

Table 8.4 Applications of Enzyme Membranes and Thin Layers

Analyte	Enzyme (E.C. number)	Reactor type (and specifications)	Detection system	Comments	Reference
Cholinesterase	None	SSSF		Linear 25–1000 µg/ml, Stable at least 10 days	121
Acid or alkaline Phosphatase	None	SSSF	Fluorescence	4-Methylumbelliferone substrate. Reconstitution with different buffer allows quantitation of either enzyme. 1 µl serum required.	122
Acid or alkaline Phosphatase	None	SSSF	Fluorescence	Naphthol AS-B1 phosphate substrate Influence of drop volume and shape on background fluorescence studied.	113
Cholesterol	Cholesterol esterase Cholesterol oxidase	Enzyme stirrer	Amperometric		116
Creatine kinase	None	SSSF	Fluorescence (NADH)	Calbiochem "statzyme" CK reagent kit used. 10 µl reagent, 3 µl serum required. Linear range to over 500 I.U./l.	123
Creatine kinase	Luciferase	Thin layer of alkylamine CPG fixed to glass rod	Chemiluminescense	Determination of ATP $(10^{-8}–10^{-3}$ M). Measure CK after 20 sec from solution-containing rod.	124.
Ethanol	Luciferase	Thin layer of alkylamine CPG on glass rod	Chemiluminescence	Determine 0.0004% ethanol with ADH coupling to bacterial luciferase–oxidoreductase system.	125

Substance	Enzyme	Format	Detection	Comments	Ref.
Ethanol	Alcohol dehydrogenase	Enzyme stirrer	Fluorescence	Requires serum deproteinization. 50–300 mg/dl linear range. Rate measurement.	126
Glucose	Hexokinase Glucose-6-phosphate dehydrogenase	SSSF	Fluorescence	Glucose "Stat Pak" reagents. Good linear range. S.D. = 20 mg/100 ml (3% C.V.)	127
	Glucose oxidase Peroxidase	Enzyme stirrer	Fluorescence	Homovanillic acid fluorescence; measurement at 3 min intervals, pH 9.0, glycine buffer.	115
Glucose	Luciferase	Thin layer of alkylamine CPG fixed to glass rod	Chemiluminescence	Determination of ATP $(10^{-8}–10^{-3}\ M)$. Measure CK after 20 sec from solution-containing rod.	124
Glucose	Glucose oxidase	Membrane sandwich	Absorbance	Glucose oxidase trapped between auto analyzer dialysis membranes. Stable for 10 days. Nonlinear over 10–30 mg/dl range.	128
γ-Glutamyl transpeptidase	None	SSSF	Fluorescence		129
Glutarate–oxaloacetate transaminase	Malate dehydrogenase	SSSF		10 Dade tablets in 3 ml of H_2O. 30 μl of reagent, dry, 50 μl H_2O, 10 μl serum. Plot F/min versus activity. Linear 2.2–106 I.U./l. Stable 3–4 days.	130
Glutarate–pyruvate transaminase	Lactate dehydrogenase	SSSF	Fluorescence (NADH)		130

437

Table 8.4 (*Continued*)

Analyte	Enzyme (E.C. number)	Reactor type (and specifications)	Detection system	Comments	Reference
Hexokinase, Glucophospho-isomerase	Glucose-6-phosphate dehydrogenase	Membrane (polyacrylamide covalent)	Fluorescence (NADH)	Immobilized enzyme in electrophoresis film. Used for detection of enzymes after separation.	131
α-Hydroxybutyrate dehydrogenase	None	SSSF	Fluorescence (NADH)		130
Lactate dehydrogenase	None	SSSF	Fluorescence (HADH)	50 µl 1 mM NAD, 20 µl 1 mM lactate, 20 µl sample. Linear range 160–820 I.U./ml. 3% accuracy. Stable 30 days minimum.	132
NADH	Alcohol dehydrogenase (1.1.1.1)	Membrane	Single ion Monitoring (Mass spectrometric)	Monitor NADH by measuring volatile ethanol produced. Obtain 9×10^{10} cps/M. No calibration curve.	120
Nitrate	NADH-nitrate reductase	SSSF	Fluorescence	1–7.5 ppm detection limit.	133
Urea	Urease	Enzyme stirrer	Potentiometric	70% of theoretical pH observed. 2 min response time.	114
Urea	Urease	Membrane	Single ion monitoring (mass spectrometric)	CO_2 produced in urease reaction quantified after diffusion into mass spectrometer. Determination possible every 30 sec. Linear to 10^{-6}–10^{-3} M.	120
	Urease Glutamate dehydrogenase	SSSF	Fluorescence (NADH)		165
Uric acid	Uricase	Membrane	Fluorescence	No stability or data on auto analyzer curves given.	151

is not a trivial methodological modification of a batch-type reactor. In essence, the mass-transfer rate to the catalyst is greatly improved since the catalyst particles are not free to move with the fluid (see Chapter 6).

8.3.3 Reactor–Separators

Although immobilized enzymes have been used effectively and extensively as reagents, the potential use of immobilized enzymes in separations has been largely ignored. Barker has outlined a number of systems in which an enzyme can be immobilized onto a membrane which has some degree of separation capacity (117). The term "reactor–separator" has been used to describe this combination of an enzyme and a separation membrane. A number of techniques with gas-sensitive electrodes make use of the fact that only gaseous products of the enzyme reaction pass through the membrane and are detected. Enzymes have also been immobilized on cellulose acetate hollow-fiber reactor tubes with a resulting removal of product from the flow stream (118).

A quantitative study of a urease–anion-exchange membrane reactor–separator system was recently reported by Blaedel and Kissel (119). A mathematical model was presented for the enrichment of an acceptor solution based on the diffusion of substrate through an anion-exchange and immobilized-enzyme membrane. The final equation for the concentration of product in the recipient solution is rather complex, but under appropriate conditions, that is, relatively high enzyme concentration in the membrane, the rate of buildup of product in the acceptor solution is given by

$$P_2 = P_2^0 + \left(\frac{n\bar{D}_{S_1a}\delta_{S_1a}\bar{A}}{\bar{D}L_2} \right) S_1 t \tag{8.18}$$

where P_2^0 is the product concentration in the recipient solution at the initiation of steady state, n is the stoichiometric coefficient, $\bar{D}_{S_1a}$ is the diffusion coefficient of substrate in the anion exchange membrane, δ_{S_1a} is the partitioning factor between the donor solution and the anion-exchange membrane, $\bar{A}$ is the area of the membrane, L_2 is the volume of the recipient solution, t is time, and the remainder of the terms are given in Figure 8.12. The potential analytical utility of these devices arises from the theoretically linear relationship between the product buildup and substrate concentration. Using a urease–anion-membrane model system, Blaedel and Kissel found that a linear relationship did indeed exist between the ammonium ion enrichment of the recipient solution and the urea concentration of the donor solution over the range of 10^{-5} to 10^{-1} M urea. It is interesting to note that the urea concentration at the upper

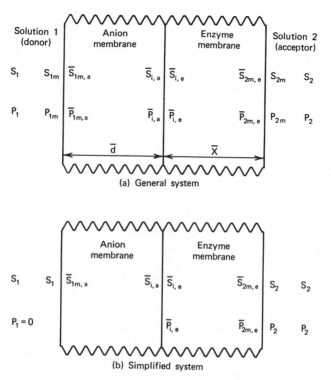

Fig. 8.12. Schematic of the reactor–separator membrane system at steady state. [Reproduced with permission from W. J. Blaedel and T. R. Kissel, *Anal. Chem.* **47**, 1602 (1975). Copyright by the American Chemical Society.]

end of this range exceeds the K_M of the enzyme, indicating that the reaction rate is controlled by diffusion. It was also observed that the efficiency of the reactor–separator was dependent on the anion-exchange membrane used. This would be expected due to differences in thickness, in partitioning ability, and due to mechanical difficulties with the membrane itself. This approach has a great deal of potential, however, since its use with a small recipient volume combined with a detector, or a tubular reactor–separator with detection downstream in the recipient solution might provide excellent methods for future clinical and industrial on-line determinations.

An interesting application of an enzyme membrane has been reported by Weaver and his coworkers (120). In this application, an enzyme layer in contact with solution was used to convert either urea to ammonia and CO_2 or alcohol to an aldehyde and NADH. The second site of the membrane was exposed to the vacuum at the inlet of a quadrapole mass spectrometer. The mass spectrometer was then used as a specific ion recorder to monitor the steady-state concentration of the product species. For the urea–urease system, monitoring the CO_2 allowed a measure of urea concentration every 30 sec. The measurement of NADH in the alcohol dehydrogenase system required 80 sec per sample because of the lower diffusion coefficient of the NADH through the membrane.

8.3.4 Cofactor Regeneration

An area which has tremendous implications in the area of immobilized enzymes is the recycling retention and regeneration of enzyme cofactors. For industrial production processes, this regeneration of the expensive cofactors is an economic necessity. An overview of industrial cofactor recycling has recently been presented (134). In many ways, the regeneration of cofactors for analytical purposes is a different problem from that faced by production engineers. As an example, the engineer may choose to cycle a cofactor numerous times on a single pass through an enzyme reactor. The analyst, on the other hand, requires production of an interference-free, measurable species proportional to the amount of analyte present. Thus, recycling on a single pass through the enzyme reactor would be untenable unless some other measurable species was generated. Some of the analytical aspects of recycling were discussed in Chapter 3.

In a discussion of recyclability of cofactors, it is convenient to classify the various cofactors according to their function in the enzyme reaction and to the chemical requirements for their regeneration. Cofactors may take part in an enzyme reaction in one of two ways: they may act to facilitate an "activated" enzyme which can then catalyze its reaction, or they may provide either chemical groups or reducing equivalents for the alteration of the substrate. Probably the best-known example of the former group is pyridoxyl phosphate. Pyridoxyl phosphate functions in transferase and decarboxylase reactions by forming a transient Schiff's base with the enzyme to facilitate the transfer of the chemical group. With each enzyme turnover, the pyridoxyl phosphate is returned to its initial state ready for another cycle. As such, it is required in molar concentration approaching that of the enzyme, that is, catalytic amounts. In contradistinction, cofactors such as ATP and NAD, which fall into the latter group above, are directly involved in the substrate change and are consumed in stoichiometric amounts. Due to the cost and the amount of cofactor required in a cofactor-requiring analytical system, the cofactor generally constitutes the major portion of the expense. A list of some common cofactors and their cost is shown in Table 8.5.

Table 8.5 Common Cofactors and Considerations of Their Analytical Recyclability

Cofactor	Amount required[a]	Recycling class	Molecular weight (daltons)
ADP	S	3	
ATP	S	3	507
AMP			
Biotin	C	1	244
CoA	S	3	768
B_{12}	C,S	1,3	1580
CTP	S	3	481
FAD ox	C,S	1,2	786
red		3	
FMN ox	C,S	1,2	492
red		3	
Folic acid	S	3	445
Glutathione, reduced	C,S	1,3	307
Glutathione, oxidized	C,S	1,3	612
Lipoic acid, reduced	S	2	206
Lipoic acid, oxidized	S	2	204
NADH	S	3	664
NAD	S	2	663
NADPH	S	3	744
NADP	S	2	743
Pyridoxy P	C	1	247
S-adenosyl methionine	S	3	435
Thiamine pyrophosphate	C	1	496
UTP	S	3	484

[a] S means stoichiometric; C means calalytic.

The second consideration, the chemical "effort" required for regeneration, is of interest only for chemically modified cofactors. Regeneration of the cofactor can be accomplished either enzymatically or chemically, but for an analytical system of either type, several important issues must be considered: the ease of measuring the recycling agent, the cost of the recycling system, and the compatibility of the chemical or enzyme system and its products with the reaction to be measured (e.g., pH, inhibition, separability of recycling from primary reaction species). The simplest regeneration systems are those which make use of oxygen as the regeneration substrate, either chemically or enzymatically. These systems easily meet all of the above requirements and are summarized in Chapter 3. For some species, such as ATP, no means of coupling to oxygen is available. In these cases, a careful assessment of the regeneration system using the above criterion is essential. In Table 8.5, the common cofactor systems are listed along with a number of important parameters of analytical interest.

There are two immediately apparent features of Table 8.5. First, those cofactors which are required in stoichiometric amounts are, in general, going to be the *major cost expenditure in the analytical system* and are thus primary candidates for recycling. Second, in every case listed, the reduced form of a cofactor is considerably more expensive than the oxidized form. From this consideration, it would be worthwhile to generate the reduced form of the cofactor whether it is to be recycled or not. Hornby and his coworkers have used this approach to generate NADH "on line" with an immobilized alcohol dehydrogenase tubular reactor. The NADH formed was then used in the determination of the serum enzyme glutamate oxaloacetate transaminase. The system, employing both immobilized malate dehydrogenase and alcohol dehydrogenase reactors, is shown in Figure 8.13. The use of generating reactors with inexpensive feedstock such as ethanol would appear to have promise in decreasing the cost of assays simply and efficiently.

8.4.1 Methods of Regeneration

Cofactor regeneration, or generation in the above example, can be achieved either enzymatically or chemically. Both types of regeneration

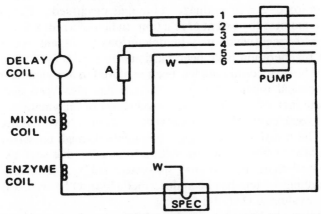

Fig. 8.13. Schematic diagram of continuous-flow analyzer in which NADH is produced by an IMER. A packed bed of immobilized alcohol dehydrogenase is inserted at position A and a solution of NAD and ethanol pumped through the column. The enzyme coil in the drawing is a tubular reactor of malate dehydrogenase required for the determination of glutamateoxaoacetate transaminase activity. [Reproduced from W. E. Hornby, J. Campbell, D. J. Inman, and D. L. Morris, in *Enzyme Engineering II* (Plenum Press, New York, 1974), courtesy of Plenum Press.]

have been primarily directed toward the pyridine nucleotide cofactors. Analytically, the most successful system has been the alcohol dehydrogenase (AD)/ethanol system. NAD has been cycled in excess of 800 times per minute because of the high turnover number of AD (135). In this system, the acetaldehyde formed was removed and quantified with a semicarbazide reagent. It should be noted that to exhibit oxidizing activity, the thermodynamically favored reaction, AD requires an aldehyde substrate. Other enzyme systems which have been used for NAD regeneration include lactate dehydrogenase, malate dehydrogenase, and isocitrate dehydrogenase. Two systems using co-immobilized enzyme reactors have been reported (136, 137).

Methods for regeneration of cofactors other than NAD have not been exploited extensively. The only species which have been successfully recycled are ATP, NADH, and NADPH. The latter two species require substrates other than oxygen (type 3 in Table 8.5) but are relatively easy to produce through, for example, the AD/ethanol system. ATP requires a phosphate donor for recycling. Two systems have been reported involving acetate kinase/acetyl phosphate (138, 139) and carbamyl phosphokinase/carbamyl phosphate (140). Examples of type 3 reactions are shown in Table 8.6.

As noted above, regeneration of NAD through the use of O_2 has a number of advantages. The enzymatic oxidation of NAD using diaphorase (NADH oxidase) has been evaluated by several groups (141, 142). The use of a redox dye such as thiazolyl blue, as discussed in Chapter 3, can be used to monitor the extent of reaction. In lieu of a redox dye, hydrogen peroxide is formed which may be deleterious to the enzyme. Chemical means of regeneration have been particularly remiss (141). Only the reduction of NAD by dithionite has been explored to any significant extent.

The chemical regeneration of cofactors has also been widely used. Despite the lack of specificity and production of enzymatically inactive species in some cases, the chemical regeneration of NAD has been widely studied. The transfer of reducing equivalents through phenazine methosulfate (PMS) or phenazine ethosulfate to a redox dye such as thiazolyl blue is a well-characterized analysis system (143). A relatively efficient NAD recycling system has been described (144). Other redox dyes have also been evaluated (141, 145).

Coughlin et al., have described an electrochemically based system for NAD regeneration in a continuous-flow hollow-fiber reactor (146). The system is shown in Figure 8.14. The regeneration itself is accomplished in a somewhat batchwise manner after the analysis is made spectrophotometrically. With increased interest in the electrochemical behavior of NAD, the future would appear bright for this type of analytical approach.

Table 8.6 Selected Enzymatic Regeneration Reactions for Stoichiometric Type 3 Cofactors

Cofactor	Reaction
ATP	$AMP + ATP \xrightarrow{\text{adenylate kinase}} 2\ ADP$
	$2ADP + 2\ \text{acetylphosphate} \xrightarrow{\text{acetate kinase}} 2\ ATP + 2\ \text{acetate}$
CoA	$CoA + \text{acetylphosphate} \underset{}{\overset{\text{phosphotransferase}}{\rightleftharpoons}} \text{acetyl-CoA} + \text{phosphate}$
FADH	$\text{D-alanine} + FAD \xrightarrow[\text{oxidase}]{\text{D-amino acid}} \text{pyruvate} + NH_3 + FADH$
FMNH	$\text{lactate} + O_2 + FMN \xrightarrow[\text{oxidase}]{\text{L-lactate}} \text{acetate} + CO_2 + H_2O + FMNH$
NADH	$NAD + \text{ethanol} \underset{\text{dehydrogenase}}{\overset{\text{alcohol}}{\rightleftharpoons}} \text{acetaldehyde} + H^+ + NADH$
NADP	$\text{glucose-6-phosphate} + NADP \xrightarrow[\text{dehydrogenase}]{\text{glucose-6-PO}_4} \text{6-phosphogluconate} + NADPH + H^+$
	$\text{isocitrate} + NADP \underset{\text{dehydrogenase}}{\overset{\text{isocitrate}}{\rightleftharpoons}} \text{2-oxoglutarate} + NADPH + H^+$
UTP	$UDP + ATP \underset{\text{diphosphate kinase}}{\overset{\text{nucleoside}}{\rightleftharpoons}} UTP + ADP$

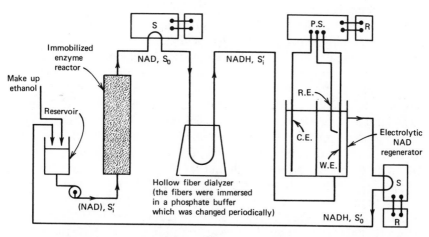

Fig. 8.14. Diagram of the immobilized-enzyme continuous-flow reactor incorporating continuous electrochemical regeneration of NAD. The feed solution was circulated at a constant flow rate with periodic adding of ethanol to the reservoir. The outside solution of the dialyzer and the analyte in the electrolytic cell were stirred magnetically (P)-pump; (S)-spectrophotometer; (R)-recorder; (P.S.)-potentiostat; (W.E.)-working electrode; (R.E.)-reference electrode; (C.E.)-counterelectrode. [Reproduced with permission of John Wiley & Sons, Inc., from R. W. Coughlin, M. Aizawa, B. Alexander, and M. Charles, *Biotechnol. Bioeng.* **17**, 515 (1975).]

Chemical regeneration of species other than NAD is almost nonexistent. The only exception is the recycling of NAD(P)H using dithionite (147).

8.4.2 Methods of Retention

The second aspect of cofactor recycling, the retention of the cofactor, can be accomplished in one of a number of ways: (1) use of a low-molecular-weight cut-off filter to retain the native species (141), (2) use of an ion-exchange membrane tube to selectively remove the charged cofactors, and (3) chemical coupling of the cofactor to an insoluble polymer (148), to a soluble polymer (149), to the same matrix as the enzyme (150), or to the cofactor-requiring enzyme itself. The use of an ultrafiltration membrane is unappealing since it would in all probability retain the reaction products as well. The use of an ion-exchange membrane is, as yet, unreported. It has been reported that regeneration of cofactors bound to an insoluble matrix is slow even using soluble enzymes (152). It would appear that if the cofactor is to be retained, either soluble polymer-bound cofactor or co-immobilized cofactor and enzyme in conjunction with a hollow-fiber dialyzer would have the greatest analytical utility. Several such systems have been proposed (153).

Although the rigors of industrial use require extremely high cofactor cycles for economic feasibility in production, such is not necessarily the case in analytical applications. For example, Baricos (134) points out that for post-reactor recovery and regeneration of cofactor with an efficiency of 99%, half of the cofactor will be lost after only 69 cycles. Although this is economically unacceptable for industrial reactors, it represents a considerable improvement for analytical purposes. Hence, external regeneration and reuse may not be untenable analytically. Consider also a column reactor containing both analytical and regenerating enzymes. If the analyte concentration is high and the K_M for the cofactor low, and if a species can be produced which "integrates" the number of molecules cycled, then analytical recycling in a single pass through a reactor without retention does not seem unreasonable. A reactor has been described in which 10,000 NAD turnovers were achieved in a single pass through the column (134).

8.4 CONCLUSION

In closing, it would seem worthwhile to compare immobilized-enzyme analyses to other types of analytical techniques. The applicability of the technique is very broad, ranging from ionic species such as nitrate to steroid molecules and proteins. The specificity of the method is dependent on the enzyme chosen for the application. It can range from almost totally specific, as is the case with clinical measurements involving urease, to almost nonspecific within a class such as the carbohydrates. In the latter case, the immobilized enzyme may be used after a preliminary separation. The general sensitivity, accuracy and precision of the method are dependent on the type of detection system used in the specific application. IMER's have been used for trace analyses with precision better than 2%. The analyses are in general nondestructive, removing only the species being quantified. Although the above considerations indicate that the technique is analytically quite useful, all of the above factors are true of soluble-enzyme systems as well as immobilized ones. The real power of the immobilized-enzyme based analysis is its low cost and its general ease of operation and maintenance. For some systems, the assay is essentially reagentless, requiring only a buffer solution. In other cases, additional reagents may be required. The major expense, the enzyme, is removed. In addition, because the enzyme is essentially a part of the instrument, it can be calibrated quite easily and the effects of differing soluble enzyme preparations avoided. Finally, equipment involving immobilized enzymes is generally quite simple and easy to operate and maintain. With the increasing availability of both commercially prepared instrumentation and immobilized-enzyme preparations, the usefulness of these materials should increase.

References

1. H. H. Weetal, *Anal. Chem.* **46,** 602A (1974).
2. A. H. Free, *Ann. Clin. Lab. Sci.* **7,** 479 (1977).
3. G. G. Guilbault, *Handbook of Enzymic Methods of Analysis* (Marcel Dekker, New York, 1976).
4. G. B. Broun, in *Enzyme Engineering,* Vol. 3, E. K. Pye and H. H. Weetall, Eds., Vol. 3 (Plenum Press, New York, 1978), p. 381.
5. L. D. Bowers and P. W. Carr, *Anal. Chem.* **48,** 544A (1976).
6. L. R. Snyder, *J. Chromatogr.* **125,** 287 (1976).
7. L. R. Snyder and J. H. Adler, *Anal. Chem.* **48,** 1017, 1022 (1976).
8. J. C. Giddings, *Dynamics of Chromatography,* Part 1 (Marcel Dekker, New York, 1965).
9. J. J. Kirkland, Ed., *Modern Practice of Liquid Chromatography* (Academic, New York, 1971).
10. M. J. E. Golay in *Gas Chromatography 1958, Proc. of the Second Symposium, Amsterdam, 1958* D. H. Desty, Ed. (Academic, New York, 1958), p. 36.

11. J. Knox in *High Pressure Liquid Chromatography in Clinical Chemistry,* P. F. Dixon, C. H. Gray, C. K. Lim, and M. S. Stoll, Eds. (Academic, New York, 1976).
12. R. S. Schifreen, A. D. Hanna, L. D. Bowers, and P. W. Carr, *Anal. Chem.* **49,** 1929 (1977).
13. B. C. Karger, M. Martin, and G. Guiochon, *Anal. Chem.* **46,** 1640 (1974).
14. R. E. Adams and P. W. Carr, *Anal. Chem.* **50,** 944 (1978).
15. J. F. Rustling, G. H. Luttrell, L. F. Cullen, and G. J. Papariello, *Anal. Chem.* **48,** 1211 (1976).
16. J. C. Sternberg in *Advances in Chromatography,* J. C. Giddings, and R. A. Keller, Eds., Vol. 2 (Marcel Dekker, New York, 1966), p. 206.
17. A. B. Littlewood, *Gas Chromatography,* 2nd ed. (Academic, New York. 1970).
18. H. W. Johnson and F. H. Strauss, *Anal. Chem.* **31,** 357 (1959).
19. I. Halasz, *Anal. Chem.* **36,** 1428 (1964).
20. S. R. Bakalyar and R. A. Henry, *J. Chromatogr.* **126,** 327 (1976).
21. H. Veening, *Crit. Rev. Anal. Chem.* **5,** 165 (1975).
22. L. N. Klatt, *Anal. Chem.* **48,** 1845 (1976).
23. P. T. Kissinger, *Anal. Chem.* **49,** 447A (1977).
24. L. R. Snyder, in *Gas Chromatography, 1970,* R. Stock, Ed. (The Institute of Petroleum, London, 1971).
25. E. Reisel and E. Katchalski, *J. Biol. Chem.* **239,** 1521 (1964).
26. G. P. Hicks and S. J. Updike, *Anal. Chem.* **38,** 726 (1966).
27. D. R. Senn, P. W. Carr, and L. N. Klatt, *Anal. Chem.* **48,** 954 (1976).
28. P. Cremonesi and R. Bovara, *Biotechnol. Bioeng.* **18,** 1487 (1976).
29. S. J. Updike and G. P. Hicks, *Science* **158,** 270 (1967).
30. M. K. Weibel, W. Dritschilo, H. J. Bright, and A. E. Humphrey, *Anal. Biochem.* **52,** 402 (1973).
31. H. J. Kunz and M. Stasny, *Clin. Chem.* **20,** 1018 (1974).
32. W. Dritschilo and M. K. Weibel, *Biochem. Med.* **9,** 32 (1974).
33. B. Watson and M. H. Keyes, *Anal. Letters* **9,** 217 (1976).
34. G. Johannson and L. Ögren, *Anal. Chim. Acta* **84,** 23 (1976).
35. E. K. Bauman, G. G. Guilbault, D. J. Kramer, and L. H. Goodson, *Anal. Chem.* **37,** 1378 (1965).
36. L. H. Goodson and W. B. Jacobs, in *Enzyme Engineering,* E. K. Pye and L. B. Wingard, Jr., Eds. Vol. 2 (Plenum Press, New York, 1974), p. 383.
37. L. H. Goodson and W. B. Jacobs, Midwest Research Institute Reprint No. MRI 1173.
38. D. T. Bostick and D. M. Hercules, *Anal. Chem.* **47,** 447 (1975).
39. D. C. Williams, G. F. Huff, and W. R. Seitz, *Clin. Chem.* **22,** 372 (1976).
40. J. P. Auses, S. L. Cook, and J. T. Maloy, *Anal. Chem.* **47,** 244 (1975).
41. W. E. Hornby, in *Summer Symposium on Enzymes in Analytical Chemistry Amherst, Ma,* June 14–17 (1977).
42. D. C. Williams, G. F. Huff and W. R. Seitz, Anal. Chem., **48,** 1003 (1976).
43. D. C. Williams and W. R. Seitz, *Anal. Chem.* **48,** 1478 (1976).
44. P. W. Carr, *Crit. Rev. Anal. Chem.* **5,** 519 (1972).

45. H. D. Brown, Ed., *Biochemical Microcalorimetry* (Academic, New York, 1969).
46. J. J. Jordan, in *Treatise on Analytical Chemistry*, I. M. Kolthoff and P. J. Elving, Eds., Part I, Vol. 8 (Interscience, New York, 1968), p. 5175.
47. A. Johanson, *Protides Biol. Fluids, Proc. Colloq.* **20,** 567 (1973).
48. A. Johansson, J. Lundberg, B. Mattiasson and K. Mosbach, *Biochim. Biophys. Acta* **304,** 217 (1973).
49. C. L. Cooney, J. C. Weaver, S. R. Tannenbaum, D. V. Faller, A. Shields and M. Jahnke, in *Enzyme Engineering,* E. K. Pye and L. B. Wingard, Jr., eds., Vol. 2 (Plenum Press, New York, 1974), p. 411.
50. J. C. Weaver, C. L. Cooney, S. P. Fulton, P. Schuler, and S. R. Tannenbaum, *Biochim. Biophys. Acta* **452,** 285 (1976).
51. K. Mosbach and B. Danielsson, *Biochim. Biophys. Acta* **364,** 140 (1974).
52. L. M. Canning, Jr., and P. W. Carr, *Anal. Letters* **8,** 359 (1975).
53. K. Mosbach, B. Danielsson, A. Bogerud and M. Scott, *Biochim. Biophys. Acta* **403,** 256 (1975).
54. B. Danielsson, K. Gadd, B. Mattiasson, and K. Mosbach, *Anal. Letters* **9** (11). 987 (1976).
55. H. L. Schmidt, G. Krisam, and G. Grenner, *Biochim. Biophys. Acta* **429,** 283 (1976).
56. G. Krisam and H. L. Schmidt, in *Applications of Calorimetry in the Life Sciences,* W. de Gruyter, Ed. (Verlag, Berlin, 1977).
57. G. Krisam, *Z. Anal. Chem.* **280,** 130 (1978).
58. B. Mattiasson, B. Danielsson and K. Mosbach, *Anal. Letters* **9,** 217 (1976).
59. B. Mattiasson, B. Danielsson, and K. Mosbach, *Anal. Letters* **9,** 867 (1976).
60. B. Danielsson, K. Gadd, B. Mattiasson, and K. Mosbach, *Clin. Chim. Acta* **81,** 163 (1977).
61. L. D. Bowers, L. M. Canning, K. M. Sayers, and P. W. Carr, *Clin. Chem.* **22,** 1314 (1976).
62. L. D. Bowers, P. W. Carr, and R. S. Schifreen, *Clin. Chem.* **22,** 1427 (1976).
63. B. Mattiasson, B. Danielsson, C. Humansson, and K. Mosbach, *FEBS Letters* **85,** 203 (1978).
64. B. Mattiasson, K. Mosbach, and A. Svensson, *Biotechnol. Bioeng.* **19,** 1643 (1977).
65. S. N. Pennington, NSF/RA 760032 Enzyme Technol. Grantees-Users Conference; PE 265 548 (1975), p. 124.
66. G. Johansson, K. Edstrom, and L. Ögren, *Anal. Chim. Acta* **85,** 55 (1976).
67. T. T. Ngo, *Can. J. Biochem.* **54,** 62 (1976).
68. T. T. Ngo, *Int. J. Biochem.* **6,** 633 (1975).
69. S. Ikeda, Y. Sumi, and S. Fukui, *FEBS Letters* **47,** 295 (1974).
70. M. S. Denton, W. D. Bostick, S. R. Dinsmore, and J. E. Mrochek, *Clin. Chem.* **24,** 1408 (1978).
71. L. P. León, M. Sansur, L. R. Snyder, and C. Horvath, *Clin. Chem.* **23,** 1556 (1977).
72. B. Mattiasson, K. Mosbach, and A. Svensson, *Biotechnol. Bioeng.* **19,** 1556 (1977).

73. W. E. Hornby, H. Filippusson, and A. MacDonald, FEBS *Letters* **9**, 8 (1970).
74. D. J. Inman and W. E. Hornby, *Biochem. J.* **129**, 255 (1972).
75. J. Campbell, W. E. Hornby, and D. L. Morris, *Biochim. Biophys. Acta* **384**, 307 (1975).
76. L. P. León, S. Narayanan, R. Dellenbach, and C. Horvath, *Clin. Chem.* **22**, 1017 (1976).
77. K. F. O'Driscoll, A. Kapoulas, A. M. Albisser, and R. Gander, in *Hydrogels for Medical and Related Applications*, J. D. Andrade, Ed., ACS Symposium Series, Number 31 (1976).
78. H. W. Levin and M. K. Weibel, *Food Eng.* **47**, 58 (1975).
79. Leeds and Northrup Co., New Product Information Sheet C2-5111-TP.
80. R. A. Messing, *Biotechnol. Bioeng.* **16**, 897 (1974).
81. M. Werner, R. J. Mohrbacher, C. J. Riendeau, E. Mruador, and S. Cambiaghi, *Clin. Chem.* **25**, 20 (1979).
82. M. D. Joseph, D. J. Kasprzak, and S. R. Crouch, *Clin. Chem.* **23**, 1033 (1977).
83. D. L. Morris, J. Campbell, and W. E. Hornby, *Biochem. J.* **147**, 593 (1975).
84. C. C. Garber, D. Feldbruegge, R. C. Miller, and R. N. Carey, *Clin. Chem.* **24**, 1186 (1978).
85. S. K. Dahodwala, M. K. Weibel, and A. E. Humphrey, *Biotechnol. Bioeng.* **18**, 1679 (1976).
86. W. D. Fleischmann and H. Scherz, *Microchim. Acta* (*Wien*) **2**, 443 (1976).
87. W. E. Hornby, J. Campbell, D. J. Inman and D. L. Morris, in *Enzyme Engineering*, E. K. Pye and W. B. Wingard, eds. Vol. 2 (Plenum Press, New York, 1974).
88. R. Toftgård, T. Anfällt, and A. Granéli, *Anal. Chim. Acta* **99**, 383 (1978).
89. P. V. Sundaram and W. Hinsch, *Clin. Chem.* **25**, 285 (1979).
90. S. E. Brolin, A. Ågren, B. Ekman, and I. Sjöholm, *Anal. Biochem.* **78**, 577 (1977).
91. D. J. Inman and W. E. Hornby, *Biochem. J.* **137**, 25 (1974).
92. Leeds and Northrup Co., New Product Information Sheet C2-5113-TP.
93. L. Ogren and G. Johansson, *Anal. Chim. Acta* **96**, 1 (1978).
94. W. Marconi, F. Bartoli, S. Gulinelli, and F. Morisi, *Process Biochem.* **9**, 22 (1974).
95. L. J. Forrester, D. M. Yourtee, and H. D. Brown, *Anal. Letters* **7**, 599 (1974).
96. B. F. Rocks, *Proc. Soc. Anal. Chem.* **10**, 164 (1973).
97. J. N. Miller, B. F. Rocks, and D. T. Burns, *Anal. Chim. Acta* **86**, 93 (1976).
98. H. H. Weetall and M. A. Jacobson, *Proc. IV IFS Ferment. Technol. Today*, 361 (1972), p. 361.
99. T. L. Newirth, M. A. Diegelman, E. K. Pye, and R. G. Fallen, *Biotechnol. Bioeng.* **15**, 1089 (1973).
100. J. W. Finley and A. C. Olsen, *Cereal Chem.* **52**, 500 (1974).
101. Leeds and Northrup Co., New Product Information Sheet, C2.5112-TP.
102. S. Ikeda and S. Fukui, *FEBS Letters* **41**, 216 (1974).

103. P. V. Sundaram and W. E. Hornby, *FEBS Letters* **10**, 325 (1971).
104. H. Filippusson, W. E. Hornby, and A. MacDonald, *FEBS Letters* **20**, 291 (1972).
105. D. R. James and B. Pring, *Clin. Chim. Acta* **62**, 435 (1975).
106. D. J. Hanson and N. S. Bretz, *Clin. Chem.* **23**, 477 (1977).
107. J. S. Coliss, and J. M. Knox, *Med. Lab. Sci.* **35**, 275 (1978).
108. P. V. Sundaram, M. P. Igloi, R. Wasserman, and W. Hinsch, *Clin. Chem.* **24**, 234 (1978).
109. S. Lovett, *Anal. Biochem.* **64**, 110 (1975).
110. W. Dritschilo and M. K. Weibel, *Biochem. Med.* **11**, 242 (1974).
111. L. P. Leon, J. B. Smith, L. R. Snyder, and C. Horvath, *Clin. Chem.* **24**, 1023 (1978).
112. P. V. Sundaram, M. P. Igloi, R. Wassermann, and W. Hinsch, *Clin. Chem.* **24**, 1813 (1978).
113. G. G. Guilbault and A. Vaughn, *Anal. Chim. Acta* **55**, 107 (1971).
114. G. G. Guilbault and W. Stokbro, *Anal. Chim. Acta* **76**, 237 (1975).
115. S. W. Kiang, J. W. Kuan, S. S. Kuan and G. G. Guilbault, *Clin. Chem.* **22**, 1378 (1976).
116. H. Huang, S. S. Kuan, and G. G. Guilbault, *Clin. Chem.* **23** 671 (1977).
117. S. A. Barker and R. F. Burns, *Chem. Ind.* (*London*) **16**, 801 (1973).
118. H. P. Gregor and P. W. Rauf, Report of the 2nd NSF Enzyme Grantees Conference, Purdue University, Lafayette, IN., 1974.
119. W. J. Blaedel and T. R. Kissel, *Anal. Chem.* **47**, 1602 (1975).
120. J. C. Weaver, M. K. Mason, J. A. Jarrell, and J. W. Peterson, *Biochim. Biophys. Acta* **438**, 296 (1976).
121. G. G. Guilbault and R. L. Zimmerman, *Anal. Letters* **3**, 133 (1970).
122. B. Reitz and G. G. Guilbault, *Clin. Chem.* **21**, 1991, (1975).
123. H. K. Lau and G. G. Guilbault, *Clin. Chem.* **19**, 1045 (1973).
124. Y. Lee, I. Jablonski, and M. DeLuca, *Anal. Biochem.* **80**, 496 (1977).
125. C. Haggerty, E. Jablonski, L. Stav, and M. DeLuca, *Anal. Biochem.* **86**, 162 (1978).
126. J. C. W. Kuan, S. S. Kuan, and G. G. Guilbault, *Anal. Chim. Acta* **100**, 229 (1978).
127. S. W. Kiang, J. W. Kuan, S. S. Kuan and G. G. Guilbault, *Clin. Chem.* **21**, 1799 (1975).
128. J. Campbell, A. S. Chawla, and T. M. S. Chang, *Anal. Biochem.* **83**, 330 (1977).
129. B. Reitz and G. G. Guilbault, *Clin. Chem.* **21**, 715 (1975).
130. B. Reitz and G. G. Guilbault, *Anal. Chim. Acta* **77**, 191 (1975).
131. R. A. P. Harrison, *Anal. Biochem.* **61**, 500 (1974).
132. R. L. Zimmerman and G. G. Guilbault, *Anal. Chim. Acta* **58**, 75 (1972).
133. C. H. Kiang, S. S. Kuan, and G. G. Guilbault, *Anal. Chem.* **50**, 1323 (1978).
134. W. H. Baricos, R. P. Chambers, and W. Cohen, *Anal. Letters* **9**, 257 (1976).
135. M. P. Schulman, N. K. Gupta, A. Omachi, G. Hoffmann, and W. E. Marshall, *Anal. Biochem.* **60**, 302 (1974).

136. P. Sure, B. Mattiasson, and K. Mosbach, *Proc. Natl. Acad. Sci.* USA, **70**, 2534 (1973).
137. J. Wykes, P. Dunhill, and M. D. Lilly, *Biotechnol. Bioeng.* **17**, 51 (1975).
138. C. R. Gardner, C. K. Colton, R. S. Langer, B. K. Hamilton, M. C. Archer, and G. M. Whitesides in *Enzyme Engineering,* E. K. Pye and L. B. Wingard, Jr., Eds. Vol. 2 (Plenum Press, New York, 1974), pp. 209–216.
139. G. M. Whitesides, A. Chumurny, P. Garrett, A. Karnotte, and C. K. Colton, in *Enzyme Engineering,* E. K. Pye and L. B. Wingard, Jr., Eds. Vol. 2 (Plenum Press, New York, 1974), pp. 217–222.
140. D. L. Marshall in *Enzyme Engineering,* E. K. Pye and L. B. Wingard, Jr., Eds., Vol. 2 (Plenum Press, N.Y., 1974), pp. 223–228.
141. R. P. Chambers, E. M. Walle, W. H. Baricos, and W. Cohen in *Enzyme Engineering,* E. K. Pye and H. H. Weetall, Eds., Vol. 3 (Plenum Press, New York, 1976).
142. D. D. Koskins, H. R. Whitely, and B. Mackler, *J. Biol. Chem.* **237**, 2647 (1962).
143. S. Pinder, J. B. Clard, and A. L. Greenbaum, *Meth. Enzy.* **18**, 20 (1971).
144. C. Bernofsky and M. Swan, *Anal. Biochem.* **53**, 452 (1973).
145. M. O. Mansson, B. Mattiasson, S. Gestrelius, and K. Mosbach, *Biotechnol. Bioeng.* **18**, 1145 (1976).
146. R. W. Coughlin, M. Aizawa, B. Alexander, and M. Charles, *Biotechnol. Bioeng.* **17**, 515 (1975).
147. J. B. Jones, D. W. Sneddon, W. Higgins, and A. L. Lewis, *JCS Chem. Commun.* **5**, 856 (1972).
148. P. O. Larsson and K. Mosbach, *Biotechnol. Bioeng.* **13**, 393 (1971).
149. M. Weibel, C. Fuller, J. Stadel, A. Buckman, T. Doyle, and H. Bright, in *Enzyme Engineering,* E. K. Pye and L. B. Wingard, Jr., Eds, Vol. 2 (Plenum Press, New York, 1974) pp. 203–208.
150. S. Gestrelius, M. Mansson, and K. Mosbach, *Eur. J. Biochem.* **57**, 529 (1975).
151. P. Kamoun and O. Douay, *Clin. Chem.* **24**, 2033 (1978).
152. P. O. Larsson and K. Mosbach, *FEBS Letters* **46**, 119 (1974).
153. M. K. Weibel, *Enzyme Engineering,* E. K. Pye and L. B. Wingard, Jr., Eds, Vol. 2 (Plenum Press, New York, 1974), pp. 385–392.
154. P. V. Sundaram and M. P. Igloi, *Clin. Chim. Acta* **94**, 295 (1979).
155. B. Watson, D. N. Stiffel, and F. E. Semersky, *Anal. Chim. Acta* **106**, 233 (1979).
156. E. Bisse and D. J. Vonderschmitt, *FEBS Letters* **81**, 326 (1977).
157. P. V. Sundaram, B. Blumenberg, and W. Hinsch, *Clin. Chem.* **25**, 1436 (1979).
158. I. Karube, K. Hara, I. Satoh, and S. Suzuki, *Anal. Chim. Acta* **106**, 243 (1979).
159. I. Satoh, I. Karube, S. Suzuki, and K. Aikawa, *Anal. Chim. Acta* **106**, 369 (1979).
160. W. Hinsch and P. V. Sundaram, *Clin. Chim. Acta* **104**, 87 (1980).

161. F. Gorus and E. Schram, *Arch. Int. Physiol. Biochem.* **85,** 981 (1977).
162. P. V. Sundaram, *J. Solid Phase Biochem.* **3,** 185 (1978).
163. P. V. Sundaram and S. Jayaraman, *Clin. Chim. Acta* **94,** 309 (1979).
164. P. V. Sundaram, *Biochem. J.* **179,** 445 (1979).
165. J. C. W. Kuan, H. K. Y. Lau, and G. G. Guilbault, *Chim.* **21,** 67 (1975).

The following references have appeared after completion of the manuscript but are included for completeness.

G. G. Guilbault, "Preparation and Analytical Uses of Immobilized Enzymes," *Acc. Chem. Res.* **12**(9), 344 (1979).

G. Johanson, K. Edström, and L. Ögren, "An Enzyme Reactor Electrode for Determination of Amino Acids," *Anal. Chim. Acta.* **85,** 55 (1976).

J. C. Kuan, S. S. Kuan, and G. G. Guilbault, "Determination of Plasma Glucose with Use of a Stirrer Containing Immobilized Glucose Dehydrogenase," *Clin. Chem.* **23,** 1058 (1977).

H. Jaegfeldt, A. Torstensson, and G. Johansson, "Electrochemical Oxidation of Reduced Necotinamide Adenine Dinucleotide Directly and after Reduction in an Enzyme Reactor," *Anal. Chim. Acta.* **97,** 221 (1978).

M. Bellal, J. Boudrant, and C. Cheftel, "Tubular Enzyme Reactor Using Glucoamylase Adsorbed onto an Anionic Resin," *Ann. Technol. Agric.* **27,** 469 (1978).

M. D. Legoy, D. D. Thomas, "Cofactor Regeneration in Artificial Enzyme Membranes: Potentialities for Analytical and Reactor Applications," *Enzyme Eng.* **3,** 93–99 (1978).

J. E. Prenosil, "Immobilized Glucose Oxidase-Catalase and Their Deactivation in a Differential-Bed Loop Reactor," *Biotechnol. Bioeng.* **21,** 89 (1979).

R. Chirillo, G. Caenaro, B. Pavan, and A. Pin, "The Use of Immobilized Enzyme Reactors in Continuous-Flow Analyzers for the Determination of Glucose, Urea, and Uric Acid," *Clin. Chem.* **25**(10) 1744–8 (1979).

S. Okuyama, N. Kokubun, S. Higashidate, D. Uemura, and Y. Hirata, "A New Analytical Method for Individual Bile Acids Using High Performance Liquid Chromatography and Immobilized 3.Alpha-Hydroxysteroid Dehydrogenase in Column Form," *Chem. Lett.* **12,** 1443–1446 (1979).

S. P. Fulton, C. L. Cooney, and J. C. Weaver, "Thermal Enzyme Probe with Differential Temperature Measurements in a Laminar Flow-Through Cell," *Anal. Chem.* **52,** 505 (1980).

R. Bais, N. Potezny, J. B. Edwards, A. M. Rofe, and R. A. J. Conyers, "Oxalate Determination by Immobilized Oxalate Oxidase in a Continuous Flow System," *Anal. Chem.* **52,** 508 (1980).

L. P. Léon, D. K. Chu, L. R. Snyder, and C. Horváth, "Continuous-Flow Analysis for Glucose in Serum, with Use of Hexokinase and Glucose-6-Phosphate Dehydrogenase Coimmobilized in Tubular Form," *Clin. Chem.* **26,** 123 (1980).

J. F. K. Huber, K. M. Jonker, and H. Poppe, "Optimal Design for Tubular and Packed Bed Homogeneous Flow Chemical Reactors for Column Liquid Chromatography," *Anal. Chem.* **52,** 2 (1980).

A. S. Attiyat and G. D. Christian, "Biamperometric Determination of Ethanol, Lactate, and Glycerol Using Immobilized Enzymes in Flow Streams," *Analyst* **105,** 154 (1980).

INDEX

455

L1196 (013) $\frac{614}{21}$